# HIDDEN UNITY IN NATURE'S LAWS

As physics has progressed through the ages it has succeeded in
explaining more and more diverse phenomena with fewer and fewer
underlying principles. This lucid and wide-ranging book explains how
this understanding has developed by periodically uncovering unexpected
"hidden unities" in nature. The author deftly steers the reader on
fascinating path that goes to the heart of physics – the search
discovery of elegant laws that unify and simplify our
the intricate universe in which we live.

Starting with the ancient Greeks, the auth
major concepts in physics right up to the pres
presentation is crisp and informative, and only
mathematics is used. Any reader with a backgro            ...atics or
physics will find this book provides fascinating in        ...to the
development of our fundamental understanding of the world, and the
apparent simplicity underlying it.

John C. Taylor is professor emeritus of mathematical physics at the
University of Cambridge. A pupil of the Nobel Prize–winner Abdus
Salam, Professor Taylor has had a long and distinguished career. In
particular, he was a discoverer of equations that play an important role
in the theory of the current "standard model" of particles and their
forces. In 1976, he published the first textbook on the subject, *Gauge
Theories of Weak Interactions*. He has taught theoretical physics at
Imperial College, London, and the Universities of Oxford and
Cambridge, and he has lectured around the world. In 1981 he was
elected a Fellow of the Royal Society.

# HIDDEN UNITY
# IN NATURE'S LAWS

## JOHN C. TAYLOR

*University of Cambridge*

CAMBRIDGE
UNIVERSITY PRESS

PUBLISHED BY THE PRESS SYNDICATE OF THE UNIVERSITY OF CAMBRIDGE
The Pitt Building, Trumpington Street, Cambridge, United Kingdom

CAMBRIDGE UNIVERSITY PRESS
The Edinburgh Building, Cambridge CB2 2RU, UK
40 West 20th Street, New York, NY 10011-4211, USA
477 Williamstown Road, Port Melbourne, VIC 3207, Australia
Ruiz de Alarcón 13, 28014 Madrid, Spain
Dock House, The Waterfront, Cape Town 8001, South Africa

http://www.cambridge.org

First published 2001
Reprinted 2002

Printed in the United States of America

*Typeface* Sabon 11/13 pt.    *System* LaTeX $2_\varepsilon$    [TB]

A catalog record for this book is available from the British Library.

*Library of Congress Cataloging in Publication Data*
Taylor, John C. (John Clayton), 1930–
Hidden unity in nature's laws / John C. Taylor.
p.   cm.
ISBN 0-521-65064-X – ISBN 0-521-65938-8 (pbk.)
1. Physics – History.   I. Title.
QC7.T39   2001
530'.09 – dc21                                    00-041458

ISBN 0 521 65064 X hardback
ISBN 0 521 65938 8 paperback

# CONTENTS

# PREFACE

I have tried to write a non-technical tour through the principles of physics. The theme running through this tour is that progress has often consisted in uncovering "hidden unities". Let me explain what I mean by this phrase, taking the example (from Chapter 3) of electricity and magnetism. The unity here is hidden, because at first sight there seemed to be no connection between the two. The invention of the electric battery at the beginning of the nineteenth century ushered in a new period of research that showed that electricity and magnetism are interconnected when they change with time. This did not mean that electricity and magnetism are the same thing. They are certainly different, but they are two aspects of a unified whole, "electromagnetism". In general, it makes no sense to talk about one without the other.

This pattern of unification is fairly typical. Every time such a unification is achieved, the number of "laws of nature" is reduced, so that nature looks not only more unified but also, in some sense, simpler. More and more apparently diverse phenomena are explained by fewer and fewer underlying principles. This is the message I have tried to get across.

This book has a second theme. Quite often, different branches of physics have seemed to contradict each other when taken together. The contradiction is then resolved in a new, consistent, wider theory, which includes the two branches. For example, Newton's theory of motion and of gravitation conflicted with electromagnetism, as it was understood in the nineteenth century. The resolution lay in Einstein's theories of relativity. There are several other instances of progress by resolution of contradictions in this book.

Much of modern physics is expressed in terms of mathematics. But I have tried to avoid writing equations in mathematical symbols. I have attempted to do this by translating the equations either into words or into pictures. Geometry seems to be playing a bigger and bigger role in modern physics, so pictures are quite appropriate. In any case, mathematical symbols can never be the whole story. You can write down as many elegant equations as you like, but somewhere there has to be a framework for connecting these symbols to real things in the world. To provide this, I do not think there is any substitute for ordinary language.

I have presented things from a partially historical point of view. It is sometimes said that the sciences are different from the arts in that contemporary science always supersedes earlier science, whereas no one would dream of saying that Pinter had superseded Chekhov or Stravinsky Mozart. There is some truth in this. It is possible to imagine somebody learning Einstein's theory of gravitation without having heard of Newton's, but I think such a person would be that much the poorer. It would be a bit like being dropped on the top of a mountain by helicopter, without the pleasure and effort of climbing it.

I have very briefly introduced some of the great physicists, hoping the reader may be intrigued by them and admire them as I do. But my "history" would irritate a real historian of science. I have mainly (but not entirely) concentrated on things that, from the contemporary perspective, have proved to be on the right track – no doubt a very unhistorical way to proceed. Also, I suspect that I have given a disproportionate number of references to British physicists.

For the main part, I have limited myself to theories that are comparatively well understood and accepted. This does not mean that they are certain or completely understood: I do not think anything in science is like that. But it is difficult enough to try to simply explain topics that one thinks one understands (sometimes finding in the process that one does not understand them so well), without burdening the reader with speculations that may be dead tomorrow. Nevertheless, in the later chapters, I have allowed myself to describe some subjects on which a lot of physicists are presently working, even though nothing really firm has been decided. I hope I have made clear what is established and what is speculative.

There is an extensive Glossary, which includes thumbnail biographies, as well as reminders of the meaning of technical terms. The Bibliography lists books that I have referred to or quoted from or enjoyed or otherwise recommend.

I want to thank people who have generously given their time to read some of my chapters and to point out errors or suggest improvements. These people include David Bailin, Ian Drummond, Gary Gibbons, Ron Horgan, Adrian Kent, Nick Manton, Peter Schofield, Ron Shaw, Mary Taylor, Richard Taylor, Neil Turok, Ruth Williams and Curtis Wilson. Of course, they are not responsible for the deficiencies that remain.

# MOTION ON EARTH AND IN THE HEAVENS

*How modern science began when people realized that the same laws of motion applied to the planets as to objects on Earth.*

## 1.1 Galileo's Telescope

In the summer of 1609, Galileo Galilei, professor of mathematics at the University of Padua, began constructing telescopes and using them to look at the Moon and stars. By January the next year he had seen that the Moon is not smooth, that there are far more stars than are visible to the naked eye, that the Milky Way is made of a myriad stars and that the planet Jupiter has faint "Jovian planets" (satellites) revolving about it. Galileo forthwith brought out a short book, *The Starry Messenger* (the Latin title was *Sidereus Nuncius*), to describe his discoveries, which quickly became famous. The English ambassador to the Venetian Republic reported (I quote from Nicolson's *Science and Imagination*):

> I send herewith unto his Majesty the strangeth piece of news . . . ;
> which is the annexed book of the Mathematical Professor at
> Padua, who by the help of an optical instrument (which both
> enlargeth and approximateth the object) invented first in Flanders,
> and bettered by himself, hath discovered four new planets rolling
> around the sphere of Jupiter, besides many other unknown fixed
> stars; likewise the true cause of the *Via Lactae*, so long searched;
> and, lastly, that the Moon is not spherical but endued with many
> prominences. . . . So as upon the whole subject he hath overthrown
> all former astronomy . . . and next all astrology. . . . And he runneth

a fortune to be either exceeding famous or exceeding ridiculous. By the next ship your Lordship shall receive from me one of these instruments, as it is bettered by this man.

Galileo's discoveries proved to be at least as important as they were perceived to be at the time. They are a convenient marker for the beginning of the scientific revolution in Europe. By 1687, Isaac Newton had published his *Mathematical Principles of Natural Philosophy and the System of the World* (often called the *Principia* from the first word of its Latin title), and the first phase of the revolution was complete. The laws of motion and of gravity were known, and they accounted for the movements of the planets as well as objects on Earth.

## 1.2   The Old Astronomy

Let us review what was known before the seventeenth century about motion and astronomy. I will try to describe what humankind has known for thousands of years, forgetting modern knowledge gained from telescopes, space travel and so on. I will also ignore exceptions and refinements. The basic facts are obvious, qualitatively at least, to anyone. On the Earth, these facts are simple. Solid objects (and liquids) that are free to do so fall down. Otherwise, an effort of some sort is needed to make something move. A stone, once thrown, moves through the air some distance and then falls to the ground. But also a heavy object in motion, like a drifting ship, requires effort to stop it quickly.

The facts about the motion of the stars take longer to tell. I shall describe things as they appear from the Earth, as they would have been perceived say 3,000 years ago.

Thousands of "fixed" stars are visible to the naked eye. These all rotate together through the night sky along parallel circles from east to west. It is as if there were some axis, called the *celestial axis*, about which they all turned. The Pole Star, being very near this axis, hardly moves at all. Stars near the axis appear to move in smaller circles; stars further away in larger ones. The stars that appear to move on the largest circle are said to lie near the *celestial equator* (see Figure 1.1). The time taken to complete one of these

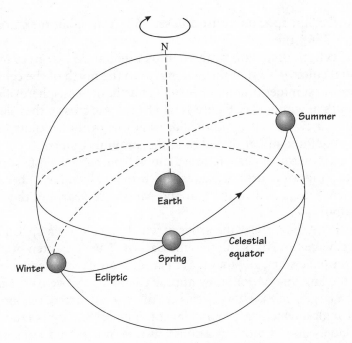

FIGURE 1.1 The "sphere of the fixed stars", which appears to rotate westward daily (as indicated by the arrow at the top). The Sun, relative to the stars, circuits eastward annually along the ecliptic.

apparent revolutions, 23 hours, 56 minutes, 4 seconds, is called a *sidereal day*.

The motions of the Sun, Moon and planets are more complicated. I shall describe their apparent motions *relative* to the fixed stars, because this is slower and somewhat simpler than the motion relative to Earth. The positions of the Moon and planets can easily be compared with those of the stars. The Sun is not usually visible at the same time as the stars, but we can work out what stars the Sun *would* be near, if only we could see them.

Relative to the stars, then, the Sun moves from west to east round a circle, called the *ecliptic*, taking $365\frac{1}{4}$ days to complete a circuit. Since

$$365\frac{1}{4} \times (24 \text{ hours}) = 366\frac{1}{4} \times (23 \text{ hours } 36 \text{ minutes } 4 \text{ seconds}),$$

this means that the Sun appears to circle the Earth in 24 hours. In a year the Sun appears to rise and set $365\frac{1}{4}$ times, but the stars rise and set $366\frac{1}{4}$ times.

The ecliptic (the path of the Sun) is tilted at $23\frac{1}{2}$ degrees to the celestial equator, so that the Sun moves to the north of the celestial equator in summer (the summer of the northern hemisphere) and to the south in winter. (See Figure 1.1.) The ecliptic crosses the celestial equator at two points, and the Sun is at one of these points at the spring equinox and at the other at the autumn equinox.

The Moon too appears to move round from east to west, near the ecliptic, and, of course, it waxes and wanes. The interval between two new moons (when the Sun and Moon are nearly in the same direction) is $27\frac{1}{3}$ days.

Lastly there are the planets, five of which were known up to 1781: Mercury, Venus, Mars, Jupiter and Saturn. They are often brighter than the fixed stars, and they move in much more complicated ways. Like the Sun and Moon, they appear to move relative to the fixed stars in large circles. These circles are tilted relative to the ecliptic by small angles, which vary from planet to planet. But, unlike the Sun, the planets do not move at a constant rate, nor even always in the same direction. Most of the time, they appear to move, like the Sun, west to east relative to the stars, but at rates that vary greatly from time to time and from planet to planet. Sometimes they appear to slow down and stop and go east to west temporarily. As examples, as seen from Earth, Venus completes a circuit relative to the stars in 485 days and Mars in 683 days. (This apparent motion comes about from a combination of the planet's true motion with the Earth's. The true periods of Venus and Mars are 225 and 687 days.)

What was made of all this before modern times? Ancient civilizations, like the Babylonian, the Chinese and the Mayan, had officials who kept very accurate records of the movements of the heavenly bodies. They noticed regularities from which, by extrapolating to the future, they were able to predict events like eclipses. One practical motive for their interest was to construct an accurate calendar. This is a complicated matter, because there are not a whole number of days in a year or in a month, nor a whole number of months in a year. Navigation was another application of astronomy. Astrology was yet another.

Yet these peoples did not try to *explain* their astronomical observations, except in terms of what we would call myth. The first people known to have looked for an explanation were from the Greek cities bordering the Aegean in the sixth and fifth centuries B.C. The problem of decoding the (Sir Thomas Browne quoted in Nicolson's book)

Strange cryptography of his [God's] starre Book of Heaven

occupied some peoples' minds for about 2,200 years before it was solved. It needs an effort of our imagination to appreciate how difficult the problem was.

Some things were understood quite early, for example, that the Earth is round, and that the Moon shines by the reflected light of the Sun, the waxing and waning being due to the fraction of the illuminated side of the Moon that is visible from the Earth. For example, the full Moon occurs when the Earth is nearly between the Moon and the Sun, so that the whole of the illuminated side of the Moon is facing the Earth. In the fifth century B.C., Anaxagoras (who was expelled from Periclean Athens for teaching that the Sun was a red-hot rock) understood the cause of eclipses. An eclipse of the Sun is seen from a place on Earth when the Moon comes between the Earth and Sun and casts its shadow at that place. (Because the Moon is small compared to the Sun, the region in shadow on the Earth is small.) The Moon's path is tilted with respect to the ecliptic (the Sun's path), so an eclipse does not happen every month. The two paths cross each other at two points called *nodes*. An eclipse of the Sun occurs only when the Sun and Moon happen to be both simultaneously in the direction of one of these nodes. An eclipse of the Moon occurs when the Moon comes into the Earth's shadow. This happens only when, simultaneously, the Moon is in the direction of one node and the Sun in the direction of the other.

## 1.3   Aristotle and Ptolemy: Models and Mathematics

I will now move on to the ideas of Aristotle in the fourth century B.C. He had amongst other things a full theory of motion and of astronomy, which was (with some amendments) enormously influential for some 2,000 years. The story of the Scientific Revolution in the seventeenth century is in some ways the story of the escape from the influence of Aristotle's physics.

Aristotle contrasted "natural" motion and "forced" motion. On Earth, the natural motion of heavy bodies (made of the elements earth and water) was towards the centre of the Earth (which was considered also to be the centre of the universe). In the heavens, the natural motion was motion in a circle at constant speed. On Earth, there were also forced departures from natural motion, caused by efforts like pushing, pulling and throwing. In the heavens, only the natural circular motion could occur, lasting eternally unchanged. Thus the heavens were perfect and the "sublunary" regions were not. Stones fall, but stars do not.

To explain the complicated motions of the heavenly bodies, Aristotle invoked a system of great invisible spheres, nested inside each other, and each with its centre at the Earth. The spheres were made of a fifth element ("quintessential") different from the four "elements" (earth, water, air and fire), which he supposed to make up everything sublunary. Each sphere was pivoted to the one just outside it at an axis, about which it spun at a constant rate. The axes were not all in the same direction. The fixed stars were attached to the outermost sphere. Next inside was a system of four spheres designed to get right the motion of Saturn, the planet attached to the innermost of these four spheres. Aristotle, careful to be consistent, then put three spheres inside just to cancel out Saturn's motion. Then more spheres gave successively the motion of Jupiter, Mars, the Sun, Venus, Mercury and the Moon. He ended up with a total of 55 spheres. With this wonderful machinery, Aristotle could get the observed motions roughly right.

This theory may seem far-fetched to us. We do not find it easy to visualize these great, transparent, unalterable spheres. However the ancients thought about this cosmology, by the middle ages people had begun to envisage the celestial spheres as solid things. One then had an example of what we may call a *mechanical model*. We shall meet several such in the course of this book. It is an explanation based upon imagining a system built like a machine or a mechanical toy. It does nearly all that such a machine would do, except that some properties are pushed to extremes. The fifth element is a bit different from anything we know on Earth: more transparent than glass, and no doubt perfectly rigid.

Aristotle's model of planetary motion did not fit all the observations, and, by the second century A.D., it had been superseded by a synthesis due to Ptolemy of Alexandria. The Earth was still fixed at the centre, and motion in circles was still assumed to be the right thing in the heavens. But, to get the motions right, Ptolemy (following Apollonius and Hipparchus) took the planets to revolve in small circles ("epicycles") whose centres were themselves rotating about the Earth in bigger circles. (It is easy to see how, for example, a planet could sometimes reverse the direction of its apparent motion when the motion in the small circle was taking it backwards with respect to the motion in the large one.) There were other complications. The centres of the larger circles were not quite at the position of the Earth, and the circles were not traversed at quite constant speed (as viewed from their centres, at any rate). With a sufficient number of such devices, Ptolemy was able to fit the observed motions very accurately. Even his system did not get everything right at the same time. For example, the Moon's epicycle would make the apparent *size* of the Moon vary much too much, because its distance from the Earth varied too much.

Ptolemy provided no mechanical mechanism for the motions. His was more of a *mathematical* (specifically, geometrical) theory than a mechanical model. This too is something we will meet again. When people despair of imagining a physical model, they fall back on mathematics, saying: "Well the mathematics fits the facts, and maybe it is not possible to do better. Maybe we are not capable of understanding more than that".

Before leaving the ancient world, we should note one more piece of knowledge that had been gained. This was some idea of size. In the third century B.C., Eratosthenes, librarian at Alexandria, had determined the radius of the Earth from a measurement of the direction of the Sun at Alexandria at noon on midsummer day. It was $7\frac{1}{2}$ degrees from being vertically overhead. On the Tropic, 500 miles south, the Sun would be overhead at the same time. From this it follows that the circumference of the Earth is

$$\frac{360}{7.5} \times (500 \text{ miles}) = 24,000 \text{ miles}.$$

Ptolemy later made an estimate of the distance of the Moon, using its different apparent positions (parallax) as viewed from different places on the Earth. The distance of the Sun could be inferred from the extent of the Sun's shadow at a solar eclipse and the extent of the Earth's shadow at a lunar eclipse, but the ancient estimates were badly out.

Aristotle and Ptolemy had these beliefs in common: that the Earth was at rest, that the motion of the heavenly bodies had to be constructed out of unchanging *circular motion* but that the motion of bodies on Earth was of a quite different nature. These beliefs dominated scientific thought, first in Arab lands from the eighth to the twelfth century, then in medieval Europe until the sixteenth century.

The ancient world became aware that the Moon had weight, like objects on Earth, and there had to be a reason why it did not fall out of the sky. For example, Plutarch wrote,

> Yet the Moon is saved from falling by its very motion and the rapidity of its revolution, just as missiles placed in slings are kept from falling by being whirled around in a circle.

People were certainly aware of the shortcomings of the Aristotelian and Ptolomaic views. There were some strange coincidences in Ptolemy's theory. The periods of revolution were about one year in the *large* circles of the inner planets (Mercury and Venus) and also about one year in the *small* circles of the outer planets. Aristarchus in the third century (quoted by Archimedes) had suggested that everything would be simpler if the Sun, not the Earth, was at rest.

As regards motion on Earth, Aristotle's doctrine had great difficulties with something as simple as the flight of an arrow. This was not a "natural" motion towards the centre of the Earth (except perhaps at the end of its flight), so what was the effort keeping it in motion after it had left the bow? Aristotle said that a circulation of the air followed it along and kept it going. It is not hard to think of objections to this idea. In the sixth century, the Christian Philoponus of Alexandria made a particularly effective critique of Aristotle's physics. (See Lloyd's book *Greek Science after Aristotle*.)

In the middle ages, several attempts were made to improve on Aristotle's account of motion. Nevertheless, in the thirteenth century

Thomas Aquinas argued that Aristotelian physics was compatible with Christian theology, and the two systems of thought got locked together. When Galileo published his dialogues in the 1630s, it was still the Aristotelian viewpoint he was combating (represented in the dialogues by one of the disputants, Simplicio).

## 1.4   Copernicus: Getting Behind Appearances

Nicolas Copernicus, born in 1473, was a Polish canon who worked at the University of Cracow and later in Italy. He developed a Sun-centred theory of the Solar System, in which the Earth was just another planet, circulating the Sun yearly between Venus and Mars. (Actually, the centre of the planetary motions was taken to be slightly displaced from the Sun.) He assumed that the planetary motions had to be built up out of circular motions, and so he had a system of epicycles and so on, not much less complicated than Ptolemy's. Copernicus also assumed that the "fixed" stars were indeed fixed, their apparent daily motion being due to the Earth's spinning on its axis. He nursed his ideas for some 40 years and published his complete theory (in *De Revolionibus Orbium Coelesium*) only in the year of his death, 1543. Copernicus dedicated his book to Pope Paul III, but a colleague, Andreas Osiander, added a cautious preface saying that the Sun-centred system was not to be taken as the literal physical truth, but only as a geometrical device for fitting the observations.

In a Sun-centred system, many things fall into place. The reason that planets sometimes appear to reverse their motion relative to the stars and move "backwards" (that is, east to west instead of west to east) is that the forward motion of the Earth can, at certain times, make a planet appear, by contrast, to go backwards. In Ptolemy's system, the order of planets from the Earth was to some extent arbitrary, but in Copernicus' system there is a natural order of planets from the Sun, with the periods of revolution increasing with distance: Mercury (88 days), Venus (225 days), Earth (1 year, i.e., 365 days), Mars (1.9 years), Jupiter (11.9 years) Saturn (29.5 years). The fact that Mercury and Venus never appear far from the Sun is explained because they really *are* nearer the Sun than the Earth and other planets.

But what were the drawbacks of the Copernican system, given that it seems (to us) so much more natural than Ptolemy's picture? There were two objections, each of which might be thought to be fatal. Since the Earth moves (roughly) in a circle of radius 150,000,000 kilometres, we ought to be seeing the fixed stars from a different standpoint at different times of the year, and this should be evident. This effect is called *annual parallax*. The only way to avoid it is to assume, as Copernicus did, that the stars are at distances very large compared to this 150,000,000 kilometres, so that the annual parallax was too small to be seen. This seems a rather weak excuse: the effect is there, but unfortunately it is too small for you to see it. But it turned out to be true. The nearest star has an annual parallax of only a few hundred thousandths of a degree (i.e., its apparent direction varies by this amount at different times of the year). This is much too small to see without a good telescope. As we shall see in later chapters, very large numbers *do* turn up in nature, and as a result some things are very nearly hidden.

In fact, people had already used a weaker version of this argument, also concerned with a parallax effect. The view of the stars ought to be slightly different at different places on the Earth. For this effect to be unobservable, one must assume that the stars are very far away compared to the Earth's radius (6,378 kilometres).

The second argument against the Copernican system is this. The rotation of the Earth about the Sun gives it a speed of about 100,000 kilometres per hour, and the daily spin of the Earth gives a point on the equator a speed of 1,670 kilometres per hour. Why do we not feel these speeds? Why is the atmosphere not left behind? Why is a projectile not "left behind"? It appears to us obvious that the Earth is at rest. Copernicus of course recognized the difficulties with his theory:

> Though these views of mine are difficult and counter to
> expectation and certainly to common sense . . .

Galileo was the first to understand fairly clearly the fallacy underlying the second objection to Copernicanism.

It happened that some natural events occurred in the latter half of the sixteenth century that challenged the Aristotelian view. In 1572 there appeared a supernova, that is, a "new" star, which rapidly became very bright (visible in the daytime) and then faded in a few months. There was another in 1604. The Chinese had recorded another in 1054 (whose remnant now is probably the Crab Nebula), but for some reason there was no record of this in the West. A comet was seen in 1577. The Danish astronomer Tycho Brahe, for example, demonstrated that both the supernova and the comet were farther from Earth than the Moon (because they exhibited no observable parallax), contradicting Aristotle's belief that the heavens above the Moon were unchanging.

## 1.5   Galileo

Galileo Galilei was born in Pisa in 1564 (the same year as Shakespeare). In 1592 he became professor of mathematics at the University of Padua (part of the Republic of Venice). An unsuccessful attempt was made to patent the telescope (two lenses used together to view distant objects) in the Netherlands in 1608. Galileo heard of this in the summer of 1609 and immediately began to make, and improve, telescopes for himself. By the autumn, he had one magnifying 20 times and began making astronomical observations. Like Newton and like Enrico Fermi in our own time, Galileo must have combined theoretical genius with a flare for experiment. He saw that the Moon was rough, just like the Earth. He saw that Jupiter had four satellites (which Galileo tactfully named "Medician stars"), so the Earth was not unique in having a Moon. He saw that Venus waxes and wanes just as the Moon does. Venus is "full" when it is on the opposite side of the Sun from the Earth, but on the Ptolemaic system it would never be "full" since it would stay between the Sun and the Earth.

As mentioned at the beginning of this chapter, Galileo immediately published his *Sidereal Messenger* to report what he had seen. He became mathematician in Florence to Cosimo de Medici, grand duke of Tuscany. Although some people were convinced of the truth of Copernicanism, the universities remained Aristotelian. Twenty-two years later, in 1632, Galileo published *Dialogue on the Great World Systems* to make the case for the Copernican system. This work was written in Italian, in the form of a dialogue among three characters, and designed to be widely understood. A papal decree of 1616 had declared Copernicanism to be "erroneous" (not as bad as being heretical); but the new pope, Urban VIII, gave Galileo leave to write about it. However that may be, Galileo was brought before the Inquisition in 1633, made to abjure his "errors and heresies", and he spent the remaining nine years of his life effectively under house arrest.

In the *Dialogue*, the Aristotelian character Simplicio is subjected to a Socratic type of cross-examination (which he bears with cheerfulness and resilience). Sometimes Aristotle is criticized for not having done experiments, but it is not always clear whether Galileo has done them either. Sometimes the argument is about "thought experiments", such as were used in the twentieth century by Einstein and Heisenberg, for example. Aristotle said that heavy bodies move towards the centre of the Earth. What would happen if there were a hole right through the Earth to the antipodes and you dropped a stone down it? Would it come to rest at the centre? Aristotle said that heavier objects fall faster than lighter ones. What would happen if you tied two cannon balls together to make an object of twice the weight? Would that fall faster than each separately? What happens if you release a stone from the top of a mast of a moving ship? Is it left behind so that it hits the deck behind the mast?

Here is another of Galileo's thought experiments:

> Shut yourself up with some friend in the largest room below decks of some large ship and there procure gnats, flies, and such other small winged creatures. Also get a great tub full of water and within it put certain fishes; let also a certain bottle be hung up, which drop by drop lets forth its water into another narrow-necked bottle placed underneath. Then, the ship lying still, observe how these small winged animals fly with like velocity towards all parts of the room; how the fishes swim indifferently to all sides; and how the distilling drops all fall into the bottle placed underneath. And casting anything towards your friend, you need not throw it with more force one way than another, provided the distances be equal; and jumping abroad, you will reach as far one way as another. Having observed all these particulars, though no man doubt that, so long as the vessel stands still, they ought to take place in this manner, make the ship move with what velocity

you please, so long as the motion is uniform and not fluctuating this way and that. You shall not be able to discern the least alteration in the forenamed effects, nor can you gather by any of them whether the ship moves or stands still.

By considerations like this, Galileo disposes of the argument that a moving Earth would leave things near it behind. Provided everything moves uniformly together, we notice nothing.

Galileo emphasized a new idealized state of motion (he is talking about a ball rolling along a sloping flat surface – we might think of a billiard ball on a table):

But take notice that I gave as an example a ball exactly round, and a plane exquisitely polished, so that all external and accidental impediments may be taken away. Also I would have you remove all obstruction caused by the air's resistance and any other causal obstacles, if any other there can be.

In other words, Galileo is imagining motion in the absence of friction or resistance. This is a state of affairs that can perhaps never be achieved *exactly*, but by taking careful precautions we can get nearer and nearer to such an ideal situation. Aristotle would have probably dismissed it as being hopelessly unrealistic, but in science half the battle seems to be to find the right simplified starting point, then perhaps build on it by adding complications (like friction in the present case) later.

Galileo asserted that, in this ideal frictionless situation, motion with constant speed (along a straight line in a fixed direction) persists unchanged without the application of any force or effort. No force is required to keep a billiard ball moving with constant speed. A force is needed to start it (applied by the billiard cue perhaps) or to change it (by impact on the edge of the table perhaps). So far as it is not negligible, the force of friction changes (reduces) this constant velocity.

Actually, Galileo did not get it quite right. Instead of motion in a straight line, he thought that the natural thing was motion in a great circle (that is, a circle whose centre is at the centre of the Earth) on the Earth's surface. For motion on a scale small compared to the size of the Earth, this is almost the same as motion in a straight line. Thus Galileo had not thrown off the Greek belief in the importance of circular motion. (It seems to have been René Descartes who first got it quite right.) However, Newton attributed the correct law (Newton's first law of motion) to Galileo.

Now apply Galileo's idea to the moving Earth. If the Earth, the atmosphere and all the things on the Earth are moving with the same constant velocity, they will all continue to do so. Everything will go on moving together, and no one on the Earth will notice anything. Thus there is nothing against the Copernican system on this account.

(The previous paragraph is a slight oversimplification. Take the velocity of, say, Singapore (near the equator) due to the spin of the Earth on its axis. This is a constant 1,670 kilometres per hour, but its *direction* is changing. In fact, the rate of change of direction is

$$\frac{360 \text{ degrees}}{24 \times 60 \text{ minutes}} = 2\tfrac{1}{2} \text{ degrees per minute.}$$

To produce this change of direction, a force directed towards the centre of the Earth would be needed, but this force is only a small percentage of the force due to the Earth's gravity, so it is not a very noticeable effect.)

Galileo formulated another law of supreme importance. In direct opposition to Aristotle, he said that, in the absence of resistance due to the air, all bodies would fall downwards under gravity in identical ways. As mentioned, he produced some thought experiments (cannon balls tied together) in support of this claim. But he is known to have done much real experimentation too.

This law of Galileo's was to wait three centuries before being explained by Einstein.

## 1.6 Kepler: Beyond Circles

After Copernicus and Galileo, one feature of Aristotle's physics remained. That was the belief in the naturalness of circular motion. In fact, Johann Kepler had already shown that the planetary motions could be better understood (without epicycles and so on) if the planets moved on *ellipses* not circles. Kepler (1571–1630) in 1600 became assistant to, then succeeded, the great Danish astronomer Tycho Brahe, as mathematician to the German emperor Rudolph II in Prague until 1612. (Brahe had previously been granted, by King Frederick II, the Danish island of Hven for his magnificent observatory complex Uraniborg.)

An ellipse is a curve got by slicing through a cone. In a perspective drawing, a circle is represented by an ellipse. (Steeper slices through a cone produce parabolas and hyperbolas, which, unlike ellipses, extend indefinitely rather than closing up.) After the geometry of points, lines and circles, the Greeks also studied ellipses. It is somewhat ironic that the most important Greek writer on ellipses, Apollonius, also introduced epicycles (see Section 1.3) into astronomy, and the epicycles were needed on the assumption that *circles* were the things out of which to build planetary motion. An ellipse has two important points inside it called *foci*. For all points on an ellipse, the sums of the distances to the two foci are the same.

Kepler proposed a modification of the Copernican system embodying three principles, which have come to be called *Kepler's three laws*. Kepler had at his disposal Tycho Brahe's and his own detailed observations. He believed that astronomy should be part of physics, and that the motions of the planets should somehow be caused by the influence (perhaps magnetic in origin) of the Sun. His theoretical arguments were erroneous, but nevertheless they inspired him in his struggle to understand planetary motions (especially that of Mars) consistently with the observational data.

Kepler's first law is that each planet moves in an ellipse with the Sun at one of its foci. These ellipses replace all the circles of Aristotle, Ptolemy and Copernicus.

The second law replaces the Ptolemaic idea that the circles should be traversed at constant rates (an assumption that had been qualified anyway). Kepler said instead that the line joining the planet to the Sun should sweep out *area* at a constant rate. What this means is illustrated by Figure 1.2. Thus the old assumption of traversing at a fixed *distance* in a given time is replaced by a fixed area in a given time. It is clear from the diagram that the second law implies that a planet moves faster when it is nearer the Sun and slower when it is farther from the Sun. This is for the same cause that a spinning ice-skater speeds up when she draws in her arms.

A circle can be thought of as a special case of an ellipse, and in that special case Kepler's laws reduce to the assumptions of the old

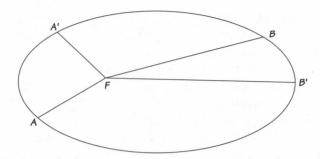

FIGURE 1.2   An example of Kepler's third law. The Sun is at the focus, $F$, of the elliptical orbit. The areas $FAA'$ and $FBB'$ are equal. The planet takes the same time from $A$ to $A'$ as from $B$ to $B'$.

astronomy. For example, the orbit of Venus deviates from being a circle by less than 1 percent, but other planets deviate more, up to 25 percent in the case of Pluto. It is often the case that a new scientific theory contains an old one within it as a special case. Looking back from the vantage point of the new theory, things are clear. But, locked within the old theory (as humankind had been for some 2,000 years in the present example), it requires someone of immense imagination to glimpse the new one.

Kepler's third law had no counterpart in the old astronomy. It connects the average distance of a planet from the Sun and the period of its revolution (its "year").

The third law states that, for any two planets, call them $P$ and $Q$, in the Solar System,

$$\frac{(\text{average distance of } P \text{ from the Sun})^3}{(\text{average distance of } Q \text{ from the Sun})^3} = \frac{(\text{period of } P)^2}{(\text{period of } Q)^2}.$$

As an example, the average distance of Pluto from the Sun is about 100 times that of Mercury, and Pluto's period is about 1,000 times Mercury's. These numbers agree with the law because $100^3 = 1,000^2$.

An equivalent way to state the Kepler's third law is:

$$\frac{(\text{average distance of a planet from the Sun})^3}{(\text{period of this planet})^2}$$

$$= \text{a fixed value for all planets of the Solar System.}$$

The value of this "fixed quantity" is actually

$$3.24 \times 10^{24} \ (\text{kilometres})^3 \ \text{per} \ (\text{year})^2,$$

as can be inferred from the the size of the Earth's orbit.

Kepler published his first two laws in 1609 and his third in 1619. Kepler and Galileo corresponded, and these two great and likeable men held each other in much esteem. Like Galileo, Kepler wrote in favour of Copernicanism (*Epitome astronomiae Copernicanae*). It is strange that Galileo's *Dialogue on the Great World Systems* (1632) makes no mention of Kepler's laws, or even of ellipses. The two men had very different scientific styles. Galileo was down-to-earth, and an exceptional communicator of science (writing often in Italian not Latin). Kepler (quoted in Baumgardt's book) had a more unworldly attitude:

> It may be that my book will have to wait for its reader for a hundred years. Has not God himself waited for six thousand years for someone to contemplate his work with understanding?

We should remember too that even the greatest of scientists get some things wrong. Galileo had a theory of the tides, with which he was very pleased. He thought it gave the most decisive argument for Copernicanism. It was wrong. Kepler thought that magnetism kept the planets moving in their orbits. He tried (like Pythagoras before him) to connect the planetary orbits with musical harmony. Also, earlier, his *Mysterium Cosmographicum* (1596) contained a beautiful explanation of the relative sizes of the planetary orbits. The Greeks had proved that there are exactly five regular solids (the "Platonic solids"). A regular solid has edges that all have the same length, faces that are all the same and corners that are all the same (with the same angles at them). Kepler, at that time thinking of the orbits as being on spheres, assumed that a regular solid was nested in between each neighbouring pair of planetary spheres, so that the faces touched the sphere inside and the corners lay on the sphere outside. The sequence went

> Mercury (octahedron), Venus (icosahedron), Earth (dodecahedron), Mars (tetrahedron), Jupiter (cube), Saturn.

This construction fitted the spacings between the planets moderately well. It was a theory that Plato and Pythagoras would have loved. Apart from anything else, it explained why there were six planets. Kepler must have been entranced by it. Of course, it was totally wrong. We know that there are more than six planets. Probably also the spacings between the planets owe a lot to accident (in the formation of the Solar System from the condensation of a cloud of gas and dust) and are not something we would expect to explain by a simple fundamental theory.

But this may not be quite right. Complicated causes can sometimes give simple results. Between the orbits of Mars and Jupiter there are a swarm of mini-planets, the asteroids. They have orbits in a spread of different sizes, and a corresponding distribution of orbital periods. But there are gaps in this distribution where, for example, the period is two-fifths or one-third of the period of Jupiter. How do these simple numbers get into such a complex dynamical system? Take an asteroid with the two-fifths period, for example. Suppose at some time it and Jupiter were at points on their orbits where they were as near as they could be. Then five orbits of the asteroid later and thus two orbits of Jupiter later they would be in just the same situation. At this position of closest approach, the gravitational force exerted by Jupiter on the asteroid (which is a small addition to the Sun's gravitational force on the asteroid) is at its greatest. It is likely that such a regular series of gravitational perturbations of the same kind would have been enough to throw the asteroid out of this particular orbit. This effect, of achieving a big result by a timed series of small impulses, is called *resonance*. It is like getting someone swinging on a garden swing by giving a series of little pushes each timed to occur at the same moment in the swing cycle.

There is another example in the Solar System in which simple ratios may be significant. The orbit of Pluto is quite eccentric and, although lying mainly outside that of Neptune, sometimes crosses inside. There are other small "Plutinos" in similar orbits. How have they avoided being ejected by gravitational tugs from (the much heavier) Neptune? Pluto and many of the Plutinos have periods close to three-halves that of Neptune. Consequently, it is possible that, everytime they cross Neptune's orbit, Neptune is at another part of its orbit.

Kepler, like Galileo again, suffered from the times he lived in. For example, between 1615 and 1621, Kepler's mother was charged with, and imprisoned for, witchcraft.

Let us return to the state of knowledge left by Galileo and Kepler. The question remained, What causes the planets to stay and move in their orbits? René Descartes (1596–1650), after a period as a professional soldier, spent 20 years in Holland and the remaining 4 years of his life in Stockholm, called there by Queen Christina. He stipulated that a body would continue with constant speed in a straight line if no force acted upon it. Therefore, a force was required to keep the planets in their curved orbits. Descartes had a mechanical explanation of this (in his *Principia Philosophiae*):

> Let us assume that the material of the heaven where the planets are circulates ceaselessly, like a whirlpool with the Sun at its centre, and that the parts which are near the Sun move more quickly than those which are a certain distance from it, and that all the planets (among whose number we include from now on the Earth) always remain suspended between the same parts of this heavenly matter; for only thus, and without using any other tools, shall we find a simple explanation of all things we notice about them.

This explanation was rather persuasive, especially immersed as it was in Descartes's complete system of philosophy. It was a mechanical explanation, like Aristotle's heavenly spheres. Like all mechanical explanations in science, it pushed the problem one stage back – to the question of what gave the "material of the heaven" *its* properties.

## 1.7 Newton

Isaac Newton was born in 1642, the year of Galileo's death. He attended Trinity College, Cambridge. The plague of 1664 caused him to return to his home in Lincolnshire. In the next two years, he began to develop his ideas about motion and the Solar System. Perhaps because of his remarkably suspicious, cautious and perfectionist character, Newton wrote almost nothing of his work for some 20 years, when he was coaxed by the second Astronomer Royal, Edmund Halley. The result was *Mathematical Principles of*

*Natural Philosophy* (1687) – a title that makes a large, but fully justified claim.

Newton's *Principia* is a remarkable work. It is written in an austere, magisterial style, giving the reader little help and admitting no human weakness. It includes three books. Book 1, after a few definitions, begins by stating three laws. In Newton's words, these are (*Newton's laws of motion*):

(i) Every body continues in its state of rest, or of uniform motion in a right line, unless it is compelled to change that state by forces impressed upon it.

(ii) The change of motion is proportional to the motive force impressed; and is made in the direction of the right line in which that force is impressed.

(iii) To every action there is always opposed an equal reaction: or, the mutual actions of two bodies upon each other are always equal, and directed to contrary parts.

Law i is Descartes's law of inertia.

In law ii the "motion" (which is now called *momentum*) is defined earlier in the *Principia* to be the "quantity of matter" (which we would call the *mass*) times the velocity. Mass ("quantity of matter") was not really well defined by Newton. For many purposes it is sufficient to say that mass is additive: that is, if two objects are put together to make a new one, the mass of the composite object is got by adding together the masses of the two original ones. In any case, law ii does not tell us anything unless we have some other method of knowing what the "motive force" is. For future reference, please note that the mass that enters in to the second law is sometimes called the *inertial mass*. This is to distinguish it from mass appearing in another context, which we shall meet shortly.

Law iii says, as an example, that if the Sun exerts a force on Jupiter, then Jupiter exerts an exactly opposite one on the Sun. Or, if two billiard balls collide, the momentary force of the first on the second is just the opposite of the force of the second on the first.

Assuming the truth of these three laws, Book 1 flows along (a bit like Euclid) with a series of mathematical proofs, giving the motions that would follow from various assumed forces. Newton shows immense mathematical power, sometimes using traditional

Euclidean geometry with great virtuosity, but also when it suits him using wholly new mathematical methods of his own invention.

Book 2 is a rather more miscellaneous collection of results. It treats of bodies moving when there is friction (and includes, for example, results of Newton's experiments on the oscillations of a pendulum damped down by the resistance of the air). Newton founds the science of the motion of fluids (liquids or gases). Assuming that sound consists of vibrations, with fluctuating pressure and density, he calculates what the speed of sound in air should be. In fact, he did not get quite the right answer because he did not realize that, in a sound wave, the pressure fluctuations involve changes of temperature as well as density.

Finally in Book 2, Newton uses his new science of fluid flow to work out the speed of a fluid in a whirlpool. Then comes the sting in the tail. He argues that, if Descartes's whirlpool explanation of planetary motion were true, the periods (times of revolution) of the planets should increase as the square of the sizes of their orbits. This contradicts Kepler's third law:

> Let philosophers then see how that phenomenon of the $\frac{3}{2}$th power [i.e., Kepler's third law] can be accounted for by vortices [i.e., whirlpools].

(In fact, Newton's argument is flawed.)

Descartes was to Newton as Aristotle had been to Galileo: the author of a hugely influential system that the younger man believed to be, and proved to be, wrong.

Newton's *The System of the World* begins with a thought experiment to show that the same gravity that causes a stone to fall on the Earth could make celestial bodies orbit round one another (say the Moon about the Earth). Consider an object (says Newton) projected from the top of a high mountain. What would happen if the speed of projection were increased? The greater the speed, the farther it would go before falling to Earth. At high enough speed it would end up by orbiting the Earth (neglecting the resistance of the air). There is thus a continuously varying range of situations, from dropping say a mile away to orbiting like the Moon. Today, this is no longer a thought experiment: we are familiar with artificial satellites being launched into orbit.

From this, and many other arguments, Newton goes on to build up to his law of universal gravitation, which I will now state in its modern form:

Every pair of particles of matter attract each other with a gravitational force, which is directed along the line joining them, and which has a strength given by

$$\frac{(\text{Gravitational constant}) \times (\text{mass of one}) \times (\text{mass of other})}{(\text{distance})^2}.$$

Nearly everything about this law is important; so I will go through its features one by one.

To get the force, one must divide by $(\text{distance})^2$. Hence this is called an *inverse-square law*. For example, if the distance is doubled, the force is divided by four. It is reasonable that the gravitational force should decrease with distance, but that it should decrease in exactly this way is not obvious. Newton showed that the inverse square law is necessary to explain Kepler's third law. I discuss this, and several other special properties of the law, in Appendix A.

If the particles mentioned in the law are sufficiently small, there is essentially no ambiguity about defining the distance between them. But for astronomy, we want to apply the law to large objects, like the Sun. For the solar system, it is still true that the distances between bodies is large compared to the size of the Sun, so any uncertainty about the meaning of distance is fairly unimportant. Most people would have been satisfied with this approximation, but not Newton. In the *Principia* he proves that, if you have two spheres, where the matter has spherical symmetry (i.e., is the same in all directions from the centre), and apply the inverse-square law to all the particles of matter in each sphere, then the total gravitational force goes down as the square of the distance between the *centres* (and is directed along the line joining the centres). So it is as if all the mass of each sphere were concentrated at its centre. This simple and convenient result is a special property of the inverse-square law (see Appendix A): it does not hold if the force depends on the distance in any other way.

Other people had guessed at the inverse-square law. In 1680, Robert Hooke had stated the law in a letter to Newton, and by 1684

Christopher Wren and Edmund Halley had also come to believe the law (see 'Espinasse's book). Hooke thought that Newton did not sufficiently acknowledge his priority, and this was a cause of a sad quarrel between the two men.

The next point to note about the law of gravitation is that *mass* appears in it – the same mass that comes into Newton's third law, relating force to rate of change of velocity. The masses in the law of gravitation are sometimes called the *gravitational masses* to distinguish them from the *inertial mass*, that occurs in the second law of motion. Newton asserts that these are the same thing. We can express this by the slogan

gravitational mass = inertial mass.

This rule is part of Newton's law of gravitation. As a consequence, if a particle is moving under a gravitational force, the mass of that particle does not affect its motion at all. It cancels out from both sides of the equation. If you double its mass, you double the force, but this produces just the same rate of change of velocity. So, as Galileo had said, all bodies fall in the same way under gravity (neglecting air resistance). Newton himself performed a similar test by comparing the rate of swing of pendulums with bobs made of "gold, silver, lead, glass, sand, common salt, wood, water and wheat". He found no difference in the rates, as there would have been if the gravitational masses of the weights were not all equal to their inertial masses. (He claimed an accuracy of one part in a thousand in these experiments.)

Newton did another test. That was to verify that the same law of gravitation, with the same "constant", gives the rate of swing of a pendulum on Earth *and* the rate of revolution of the Moon (as he knew, reasonably accurately, that the distance of the Moon was about 60 times the radius of the Earth). Again this tests that the gravitational and inertial masses are equal for the Moon as well as the pendulum.

Nowadays, we know of another simple demonstration that the mass of an object is irrelevant to its motion under gravity. An artificial satellite and an astronaut inside it orbit the Earth in identical orbits, so the astronaut can, with sufficient care, "float" in the middle of the spacecraft without hanging on.

The exact cancellation of masses has now been confirmed by experiment to very high accuracy. From the point of view of Newton's theory, it is an unexplained "accident". It was explained only by Einstein in 1915.

The *constant* in Newton's law of gravitation means a number that is fixed in all applications of the law. It is an example of a "constant of nature". It is universal, in contrast, say, to the gravitational force at the surface of the Earth, which is special to the Earth, depending on its mass and size. The numerical value of the gravitational constant depends on the units you use, whether you measure distance in metres or feet, and so on. In the metric system, its value (as now known) is

$$6.6726 \times 10^{-11} \ (\text{metre})^3 \text{ per kilogram per } (\text{second})^2.$$

This number is "small", about a 67-trillionth (where I use an American trillion equal to a million million, or $10^{12}$). I have put "small" in quotation marks because of its dependence on the choice of units. If, for example, one uses centimetres instead of metres, then the number is 67-millionth. However, there is a real sense in which the gravitational force is very "weak". For example, the magnetic force between two ordinary magnets is much, much stronger than the gravitational force (which we do not normally notice and would be very hard to measure). The reason that gravitation is nevertheless so important in the universe is that the gravitational forces due to all the particles in a big body all *add* up and never subtract, thus giving a substantial amount for planets and so on. Gravitation is *always* attractive. This is unlike, for example, electric and magnetic forces, which sometimes attract but sometimes repel.

To return to Newton's achievements in the *Principia*. Given his three laws of motion and his law of gravity, he proved (deploying his unequalled mathematical virtuosity) that all three of Kepler's laws follow. More precisely, I should say that Newton took a pair of bodies, like the Earth and the Moon, or the Sun and the Earth, and made the approximation of neglecting the other objects in the Solar System. This can be shown (and was shown by Newton) to be a good approximation, because the Moon is much nearer the Earth than the Sun, and because the Moon is much lighter than the Earth. Then he proved that, say, the Earth and the Sun each move in an ellipse about a point between them that is nearer the Sun than the Earth in the same proportion that the Earth is lighter than the Sun. (In fact, this point about which they both orbit is inside the Sun, only a few hundred kilometres from its centre; so it is a good approximation to say that "the Earth orbits about the Sun").

Kepler's second law, about sweeping out equal areas in equal times, depends only upon the *direction* of the gravitational force being along the line joining one body to the other. The other two laws depend upon the "inverse-square" distance dependence of the force. Kepler's third law, relating the periods of different planets, tests the universality of the gravitation, and the equality of gravitational mass and inertial mass for each planet.

Newton went on to calculate corrections to this two-body approximation. For example, he studied the perturbations in the motion of the Moon about the Earth, taking account of the Sun's gravitational force. Newton was never fully satisfied with what he did, and indeed this problem has exercised astronomers until the present day. It is still not known whether the Solar System is stable over timescales of hundreds or thousands of millions of years. Might, for example, a succession of "small" effects eventually mount up and cause the ejection of a planet right out of the Solar System? It is not known for certain. (See Peterson's *Newton's Clock*.)

Newton gave a roughly correct explanation of the tides. (The observed correlation between tides and the phases of the Moon had long been puzzling. Galileo had given an incorrect explanation of the tides.) The gravitational force of the Moon acts on the oceans as well as on the solid Earth. But it is slightly stronger on the side nearer the Moon and slightly less strong on the side away from the Moon. The result is to make the oceans bulge a little towards the Moon (i.e., out from the centre of the Earth) on the near side and to bulge a little away from the Moon (i.e., also out) on the far side. Since the Earth spins on its axis daily, these two bulges seem to move round the Earth daily (actually, a little slower than that, because of the monthly orbit of the Moon). The gravitational force of the Sun has the same sort of effect. The size of the effect goes up with the mass of the body producing it, but down as the *cube* of its distance. This (although I will not prove it here) is because it depends not only on the gravitational force, but also on how much this force varies from one side of the Earth to the other. It turns out that the tidal effect of the Sun is about one-third that of the Moon. Of course, in detail the tides depend on the local shapes of the seas.

There is an interesting history connected with Newton's work on the shape of the Earth, and the precession of the equinoxes. Because of the spin of the Earth on its axis, one would expect centrifugal force to make it bulge a little towards the equator, that is, to be *oblate*. Newton assumed that the Earth as a whole could be treated as a fluid. He pointed out that to assume otherwise would lead to a contradiction. Suppose the Earth were solid enough to remain an exact sphere, in spite of its rotation. Then

> if our Earth was not higher about the equator than at the poles, the seas would subside about the poles, and, rising towards the equator, would lay all things there under water.

The centrifugal acceleration at the equator can be worked out from the length of the day and the radius of the Earth. It comes out to be about $\frac{1}{289}$ times the acceleration due to gravity. This causes the Earth to bulge a little. But this bulge, in turn, means that the force of gravity is a little less at the equator than at the pole. Newton showed (on certain simplifying assumptions) that the ratio of the total acceleration (due to gravity and rotation) at the equator to that at the pole should be

$$1 - \frac{5}{4} \times \frac{1}{289} = 1 - \frac{1}{230}.$$

He showed also that this same number is the ratio of the radius to the pole to the radius to the equator, that is, the degree of oblateness. (Newton oversimplified: the correct fraction is $\frac{1}{297}$ not $\frac{1}{230}$.)

Newton's argument in *Principia*, especially for the factor $\frac{5}{4}$, is very hard to follow. According to Chandresekhar (who has had the energy and skill to work through all of the *Principia*), the obscurity is due to Newton's withholding much that he knew.

(Newton also treats the case of Jupiter. This is larger, is less dense and rotates faster. Each of these factors increases the degree of oblateness relative to the Earth. The result is an effect of the order of 10 percent, as compared to a fraction of 1 percent for the Earth.

Newton states that an oblateness of roughly this amount had been observed.)

Returning to the Earth, the reduced effective gravity at the equator means that pendulum clocks should run slower there, actually by a factor (according to Newton's numbers)

$$1 - \frac{1}{460}$$

compared to the poles. The slowing of the pendulum at the equator had been observed in Newton's time, but the shape of the Earth was more controversial. In France, measurements made by Cassini seemed to show that the Earth was *prolate* (elongated at the poles). But in the 1730s, Maupertuis (1698–1759) became convinced of the strength of Newton's reasoning. At his instigation, expeditions were arranged to Lapland and to Peru to compare the lengths of a degree of longitude near the arctic circle and near the equator. The Lapland expedition departed in 1736; it included Clairaut and Celsius as well as Maupertuis himself. The results showed that the Earth was oblate and, after more controversy, contributed a little to the acceptance of Newtonianism in France. Voltaire wrote

> Vous avez confirmé dans des lieu pleins d'ennui
> Ce que Newton connut sans sortir de chez lui.

The axis about which the Earth spins has a direction that is not quite fixed (see Figure 1.3). At present, the axis points almost to the Pole Star, but it did not always and will not always. The axis keeps (almost) the same angle to the ecliptic (the plane of the Earth's orbit), but it "precesses" round in a little circle, taking 26,000 years to get back again. Since the seasons are determined by the direction of the Earth's axis with respect to the Sun (summer in the north when the North Pole is towards the Sun), the precession of the axis causes a precession of the equinoxes. So, for example, the Sun at the spring equinox occupies a slightly different position in the zodiac each year, moving from one sign of the zodiac to the next in just over 2,000 years.

This precession would be inexplicable if the Earth were a perfect sphere, but Newton realized that it was possible with an oblate

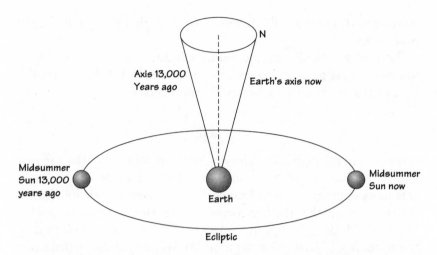

**FIGURE 1.3** Precession of the equinoxes. The Sun's apparent path is shown by the large circle. The Earth's axis precesses round the cone.

shape. The action of the Sun's gravitational force on the equatorial bulge causes it to precess, just as a tilted, spinning toy gyroscope precesses.

In the eighteenth century, French and Swiss mathematicians did most to perfect and generalize Newton's theory. Voltaire and his mistress Du Chatelet had been among those who popularized Newtonianism (which indeed became a plank of the Enlightenment). The calculation of the small disturbance of the Moon's orbit, caused by the Sun, had made Newton's "head ache" and was done more completely only in the mid-eighteenth century, by among others Alexis Clairaut (who had also been on the expedition to Lapland). It was Clairaut also who correctly predicted the return of Halley's comet in 1759.

In spite of the overwhelming success of Newtonianism later in the eighteenth century, his law of gravitation had at first been badly received on the Continent. This was for a very good reason. Newton offered no *explanation* of how gravity worked. How does the Sun somehow reach out across 100,000,000 kilometres and exert a pull on the Earth? How much better Descartes's theory of whirlpools, which *did* offer a mechanical explanation. Unfortunately, Descartes's theory (as Newton had tried to show) did not work, and Newton's worked superbly well.

Newton was well aware of this response to his work. His reaction (in the *Principia* at any rate) was sternly to refuse to speculate, and to confine himself to what he could prove either mathematically or experimentally. Whatever the final truth about these things, there can be no doubt that, at the time, Newton's strategy was the right one to yield progress, and Descartes's was not.

There is a sense in which Einstein's 1915 theory of gravity did finally provide an explanation of gravity, though as we shall see in Chapter 7, it is an explanation that lies somewhere in between a mechanical one, like Descartes's attempt, and a bare mathematical law, like Newton's.

After the work of Galileo, Kepler and Newton, no one could seriously doubt that the Earth orbited the Sun. But the first *direct* evidence for this was not found until 1728. James Bradley, the third astronomer royal, tried to measure the annual parallax (the apparent change in position as the Earth moves annually) of a star, $\gamma$ Draconis. This star lies near the axis of the Earth's orbit. Bradley sought to measure change in the apparent latitude of this star. The greatest difference should have been observed between mid-summer and mid-winter. In fact, Bradley observed a difference between spring and autumn. The amount was about $\frac{1}{170}$ of a degree. Bradley realized the cause of this unexpected result. It is due to the fact that the speed of light, although very large, is finite. The speed of the Earth in its orbit affects the apparent direction of the light from the star. This may be appreciated from the following analogy. Suppose, on a windless day, rain is falling vertically downwards. To someone riding a bicycle, the rain will seem to be falling in a direction sloping towards her. The rain is the analogue of the light from the star, and the motion of the bicycle is the analogue of the motion of the Earth. Thus Bradley found evidence for the Earth's motion, but a different sort of evidence from that he had sought.

The amount of Bradley's effect is determined just by the the speed of the Earth divided by the speed of light. (It has nothing to do with the distance of the star, as the parallax would have.) The speed of light had been estimated in 1676 by the Danish astronomer Olaf Rømer. He had observed that the motion of Jupiter's satellites seemed to run later when Jupiter was farther from Earth than when it was nearer. The time lag was about 17 minutes. This delay is due to the extra time taken by light to cross the Earth's orbit. Bradley's observation was consistent with Rømer's. In fact they both really measure the same thing: Bradley's angle is given by

$$\frac{17 \text{ minutes}}{365 \times 24 \times 60 \text{ minutes}} \times 180 \text{ degrees} = \frac{1}{170} \text{ degree.}$$

The star $\gamma$ Draconis does have an annual parallax, but this turns out to be only about 1 percent of the angle observed by Bradley. Such stellar annual parallaxes were not detected until the nineteenth century.

## 1.8   Conclusion

Newton's *Principia* of 1687, bringing to a conclusion more than two millennia of endeavours, gave a unified theory of all motion and of gravitational attraction. The whole theory is derived from just the three laws of motion and the inverse-square law of gravity. The essence of the laws of motion is that motion (strictly momentum) persists unless it is changed by a force. In the case of gravity, the force is worked out from Newton's inverse-square law of gravity. Newton achieved his success at the cost of renouncing understanding of a mechanism whereby the gravitational force is transmitted across space. Nevertheless, Newtonianism remained the model for most physical theories until well into the nineteenth century.

An odd and unexplained feature of Newton's theory is the equality of gravitational and inertial mass. This would, more than 200 years later, serve as a vital clue to Einstein.

The first phenomena discovered to be governed by a non-Newtonian type of mechanism were those of electricity and magnetism, in the middle of the nineteenth century. (See Chapter 3.) Only in 1915, at the hands of Einstein, was an "explanation" of gravity discovered (see Chapter 7). But first I will explain how the phenomenon of heat was brought within the scope of the Newtonian theory of motion. This was possible only after the atomic nature of matter had been reasonably well established. Another essential new ingredient was the understanding of the role of chance and randomness.

# 2

# ENERGY, HEAT
# AND CHANCE

*How heat is explained as random motion.*

## 2.1  Introduction

Heat is just random motion; more accurately, heat is any random process that involves energy. This short sentence needs a lot of explanation to make sense of it. It certainly requires a definition of *energy*. More difficult, and even controversial, is the word *random*.

The general idea that heat is a form of motion goes back at least to Galileo. It was a widespread notion in the seventeenth century, favoured by, for example, Robert Boyle (1627–1691). It fitted in with the Newtonian world view, and Newton himself believed (roughly correctly) that heat caused substances to expand because their atoms vibrated more.

There was an opposing theory that heat was a sort of "subtle fluid", often called *caloric* (the word was coined by the great French chemist Antoine-Laurent Lavoisier). This went along with the idea that there were also electric and magnetic "fluids". The caloric theory also suggested that, if the fluid could not be created or destroyed, then there should be, in some sense, "conservation of heat". This idea opened the way to a quantitative theory, with heat being something that could be measured. It was not obvious how to make the motion theory of heat quantitative in the same way, and throughout the eighteenth century the caloric theory seemed to many people to be more fruitful.

Not until the second half of the nineteenth century was the motion theory of heat fully understood and established. Several things were needed before this was possible:-

- Reliable thermometers and a definition of *temperature*
- A definition of *energy* and an understanding of the conservation of energy
- Some notion of the atomic nature of matter
- A quantitative measure of randomness, contained in the idea of *entropy*

I will explain all of these things in the remainder of this chapter.

## 2.2   Temperature and Thermometers

Everyday usage distinguishes fairly clearly between "heat" and "temperature". We would all be willing to accept that a red-hot needle had a higher temperature than a kettle of boiling water, but that the water contained more heat than the needle. One would not succeed in bringing a kettle of cold water to the boil by putting a red-hot needle into it: the needle would not contain enough "heat". We might make a distinction between "quantity of heat" on the one hand and "intensity of heat" on the other, the latter being measured by the temperature.

To define temperature at all requires a reasonably homogeneous and steady state of the substance concerned. To talk about the temperature of a coal fire, for example, would be a bit vague. It would probably vary from place to place and from moment to moment. But a thermos flask of coffee, stirred up and then left alone, has a fairly well-defined temperature.

Another property we expect of temperature is this. If we take two cups of water each at the same temperature and carefully mix them, the resulting water will also be at that common temperature. But if we mix hot and cold water the result will be at a third, intermediate temperature.

To be more quantitative about temperature, we need a thermometer. Expansion thermometers make use of the fact that most things expand when heated: the temperature is defined from the volume of the thermometric substance (e.g., alcohol or mercury). Galileo was one of the first to make a thermometer, using the expansion of air to push along a column of water. A thermometer has to be graduated in some way. For example, one may take two easily reproducible situations like boiling and freezing water and define these to be, say, 100 degrees and 0 degree, and then define other temperatures to be proportional to the volume of the mercury. For example, when the volume of the mercury is exactly the average of its volumes at 0 and at 100, the temperature is defined to be 50.

The trouble with this definition of temperature is its arbitrariness. There is a trivial arbitrariness: one might use 212 and 32 degrees, for example, instead of 100 and 0. But this is no more significant than changing from metres to feet. What is serious is that different substances might give different temperatures: 50 degrees by an alcohol thermometer might be different from 50 by a mercury one. But it turns out that thermometers are tolerably consistent in practice, and I will take them as providing a rough, provisional definition of temperature. Note too that thermometers like these only tell us temperature differences. The choice of melting ice to define 0 Celsius is arbitrary, if perhaps convenient. No "absolute" zero of temperature is defined at this stage, in the sense that we know what we mean by zero distance or zero speed.

With just these notions, one can do some quantitative things and ask some quantitative questions (both experimental and theoretical). Defining a unit of heat as, say, the heat required to raise the temperature of 1 gram of water by 1 degree Celsius, how much heat is needed to convert one gram of water into steam? (The answer is 537 units.) One end of a copper rod is kept in boiling water and the other end in melting ice. How does the temperature vary along the rod? How much heat flows along the rod per second? By the end of the eighteenth century questions like these could be posed and answered. Lavoisier and Laplace wrote together their *Mémoire sur la Chaleur* in 1783, giving a theory of heat founded on just these simple ideas.

## 2.3   Energy and Its Conservation

To put the motion theory of heat on a quantitative basis, the first requirement was a definition of *energy*. There are a cluster of words, *momentum, force, energy, work, power,* that are used in every-day language but have also been given a precise and specialized meaning in science. Sometimes, too, this scientific usage has in turn influenced common usage. For example, we are all aware that something that we call "energy" can be provided by electricity, oil or gas; that we cannot get it from nothing and should not waste it; that we can, for example, convert the energy in oil into electrical energy. All these usages are consistent with the scientific definition, and I think they would have meant little 300 years ago.

We have already defined *momentum* and *force* in Chapter 1. In the motion of, for example, the Solar System the positions and velocities of the planets are changing all the time. It is rather natural to ask whether there are nevertheless any unchanging quantities. Such a thing would be called a *constant of the motion* or a *conserved* quantity. One such was known to Newton. It is the total momentum, the sum of the momenta of all the planets and the Sun. (To be careful, one should remember that momentum is a vector quantity, as defined in Appendix B, and the "sum" of vectors is defined by the triangle rule in Figure B1.) That the total momentum is constant follows from Newton's second and third laws (see Section 1.7). To start with, consider just two bodies, say the Moon and the Earth. The force of the Earth's gravity on the Moon produces a rate of change of the Moon's momentum (by the second law). By the third law, the Moon's gravity exerts an equal and opposite force on the Earth, and so produces an equal and opposite rate of change of the Earth's momentum. Therefore, in the sum of both the momenta, these two rates of change cancel out. Considering in the same way the forces among all the pairs of bodies in the system, we see that the rate of change of the sum of all the momenta is zero. In other words, the total momentum is constant. (This neglects the gravitation of any star outside the Solar System. If such a force were not negligible, that star's momentum would have to be included.)

Another example of the constancy of total momentum is as follows. Let two billiard balls collide. Then the total momentum after the collision is the same as that before. Again this follows from Newton's third law. For the brief instant for which the balls are in contact, they exert equal and opposite forces on one another. (This neglects any frictional force that the billiard table might exert on the balls, or the balls on each other.)

In this example, conservation of momentum is not enough to fix motion after the collision uniquely. Suppose, for example, two similar balls collide head on, each with the same speed (but opposite direction). Then the total momentum is zero before the collision, and so is also zero afterwards. This implies that the speeds are equal afterwards (and the directions opposite), but it does not determine how big the speeds are. In fact, to a good approximation, the speed of each ball is the same afterwards as before, but its direction of motion is reversed by the bounce.

To explain things like this, Leibniz proposed that there was another conserved quantity, which he called *vis viva* ("living force" – but this use of the word *force* does *not* accord with modern scientific conventions). We now call this *kinetic energy* (meaning the energy of motion), and this is the term I shall use (the word *energy* was used for *vis viva* in 1807 by Thomas Young, whom we shall meet in Chapter 4). For one particle, the kinetic energy is defined to be

$$\frac{1}{2} \times \text{mass} \times (\text{speed})^2.$$

(Actually, Leibniz did not include the factor $\frac{1}{2}$, but the modern definition is better.) For a system of particles, the kinetic energy is defined to be a sum of terms like this, one for each particle.

Consider the kinetic energy in the billiard ball collision just mentioned. Before the collision the kinetic energies of the two balls are the same (because we have assumed that their masses and speeds are the same). Also, after the collision the kinetic energies of the two balls are the same as each other. If we now *assume* that the total kinetic energy is conserved, that is, the same after the collision as it was before, it follows that the speeds are the same afterwards as before; so in this example the motion after the collision would be completely fixed.

We may get some feeling for the difference between momentum and energy from the example of hammering a nail into a loose wooden post. The problem is to make the nail go in and not just knock the post backwards in a useless manner. Suppose you have a choice between a heavy hammer (comparable in weight to the post) and a light one, say a sixteenth of the weight. Suppose you can put the same amount of kinetic energy into the hammer head, whichever one you choose. Suppose all the momentum of the hammer head is transferred to the post (it is that loose). Then the heavy hammer will impart much of its kinetic energy to kinetic energy of the post, leaving not so much energy to drive in the nail. The light hammer, on the other hand, will give the post a quarter of the momentum, and a sixteenth of the kinetic energy, leaving most of the hammer's energy available to do work driving the nail into the wood (where some of that energy will eventually be converted into heat).

For some time there was controversy between supporters of momentum and supporters of *vis viva*, but it was soon realized that this was a silly argument, because both quantities are associated with useful conservation laws, as we shall see.

But *is* kinetic energy conserved? Suppose the collision were between two balls of wool instead of between two billiard balls. It is pretty clear that the speeds afterwards would be a lot less than they were before. Such a collision, in which kinetic energy goes down, is called *inelastic*.

And take a body falling near the surface of the Earth. If it is dropped from rest, it starts with no kinetic energy, but as it falls its speed increases and so does its kinetic energy. In this case, a conservation law can be recovered by making some definitions, which may appear at first sight rather like cheating. Define the *potential energy* to be (in this case)

potential energy = mass × (acceleration due to gravity) × height.

(The acceleration due to gravity at the Earth's surface is a fixed number, which is about 10 metres per second per second.) Define the *total energy*, or just the *energy*, by

energy = (kinetic energy) + (potential energy).

Then it is a simple deduction (which I will not give) from Newton's second law of motion that the energy remains constant as the body falls: the speed and so the kinetic energy get more but the height and so the potential energy get less, and these two changes cancel out.

So we have saved a conservation law by defining energy in such a way that it turns out to be constant in this example. Is this as empty a thing to do as it sounds? In just the present example, it probably is. But take something a little more complicated. Take a vehicle rolling along a switch-back, and suppose that it is so smoothly made and well oiled that friction can be neglected. We can retain the previous definitions of kinetic energy, potential energy and energy, and it turns out that energy is still conserved. Now this does tell us something. For example, wherever the switch-back track is at a certain height, the speed of the vehicle must be the same.

Now take a more complicated example: the Earth moving around the Sun. (To prevent complications, I will leave out the other planets and ignore the fact that the Sun actually moves a little.) Then, in the elliptic orbit (see Section 1.6), the kinetic energy certainly is not constant. The Earth moves faster when it is nearer the Sun and slower when it is farther away (see Figure 1.2). In this case, is there a potential energy that will fix up conservation of energy? There is. It is given by

$$-\left[\frac{\text{gravitational constant} \times \text{Sun's mass} \times \text{Earth's mass}}{\text{distance of Earth from Sun}}\right].$$

(Compare this with Newton's law of gravitation in Section 1.7.) We see that the potential energy increases (i.e., becomes less negative) farther away from the Sun.

Once again, conservation of energy tells us some things. For example, it tells us that whenever the Earth is a certain distance from the Sun its speed is the same.

Given *any* forces acting on a particle, can one always define a potential energy as we have in the preceding examples (the switch-back and the Earth moving round the Sun)? We cannot. Suppose there was a force which varied with position as in Figure 2.1. The particle might move under this force in some such orbit as shown in the figure. It is clear that the force is constantly accelerating the particle; so, when it gets back to a position from which it started, it must have more kinetic energy. If there were a potential energy (something depending just on the particles' position, as in the preceding examples), this would have to be the same when the particle got back to the same position. Consequently, the sum of the kinetic and potential energies could not be the same. Thus the assumption that there exists a potential energy such that total energy is

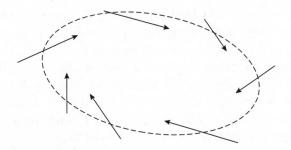

FIGURE 2.1   An example of a force that varies with position in such a way that a potential energy cannot be defined. The arrows indicate the directions of the force at different places, and the broken line shows a possible orbit under this force.

conserved prohibits the existence of a force like the one in Figure 2.1.

It is believed to be a law of nature that a potential energy can always be found so that total energy stays constant, for any *closed system*. Here, a "closed system" is one that does not have any significant interactions with anything outside itself. The reason for the last qualifying phrase about a closed system will be explained in the next paragraph. This law of nature (together with the idea of potential energy) was appreciated in about the middle of the nineteenth century, at first particularly by the German polymath Hermann von Helmholtz (1821–94), and by the Scotsmen William J.M. Rankine (1820–72) and William Thomson (later Lord Kelvin) (1824–1907). People had to sort out a clear definition of "energy" (and distinguish it from "force") as well as appreciate that it was conserved.

But what about apparent exceptions, the colliding balls of wool, mentioned earlier, that seem to lose energy in the collisions? A switch-back with friction, on which the vehicle would have less speed when it got back to the same height? An artificial satellite, equipped with a rocket motor, in orbit round the Earth? In each of these cases, we would say that we have tried to apply conservation of energy to a system that is not *closed*. We considered only the motion of the ball of wool as a whole. In fact, each little bit of wool can move relative to the others. What happens when two balls hit each other is presumably (at first, at any rate) that motion occurs *within* each ball – the balls squash, or whatever. This motion would have to be included in order to recover energy conservation. In the case of the switch-back with friction, the frictional forces may set the vehicle vibrating, squeaking, heating up and so forth. The motion of the vehicle as a whole does not make a closed system. Other possible motions need to be included to get conservation of energy. The artificial satellite might return to its starting position with more speed than it started with, if the rocket motor were operated appropriately. The motion of the satellite as a whole is not a closed system. The chemical energy in the rocket fuel would have to be included to get energy conservation.

Once again, this may sound like cheating. Whenever energy is not conserved, invent a new place where it can come from or go to. But there is no doubt that conservation of energy is a valid and useful concept. For one thing, there are many systems that are, to a good approximation, *closed*. For another, whenever we find energy to be apparently not conserved, the principle tells us that we are not dealing with a closed system, and that if we look hard enough we will find other changes that restore energy conservation. The examples just given have mentioned "heat" and "chemical energy". Clearly, we may have to look around quite widely to restore the balance of energy conservation.

It is perhaps no accident that the notion of energy came to the forefront after the Industrial Revolution. The concepts of *work* and *power* were used by engineers. If an elephant pushes at a lorry that has its brakes on, the elephant certainly exerts a force on the lorry, but the lorry may not move. If the brake is taken off, the lorry moves and so gets kinetic energy. This energy is equal to the force exerted multiplied by the distance moved in the direction of the force. This is illustrated more precisely in Figure 2.2. This latter quantity is called the *work* done (by the elephant, in this example). *Power* is defined as the rate of doing work. Clearly, the engineers of the Industrial Revolution would be interested in the power that could be obtained from a horse or a water-wheel or a steam engine.

If the situation depicted in Figure 2.1 could exist, the work done on the particle in the closed curve would not be zero. Conversely, the condition for a potential energy to exist is that the work done

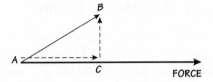

FIGURE 2.2 Work done by a force. The force is represented by the large arrow. In some (small) time, the object moves from $A$ to $B$. The displacement $AB$ is made up of two components, $AC$ in the direction of the force and $CB$ at right angles to the force. The work is given by multiplying the force by the length $AC$. Note that there is no reason for $AB$ to be in the direction of the force, unless the object happens to start from rest.

in any *closed* curve (getting back to its starting point) should be zero.

The conservation of energy is closely connected with the impossibility of perpetual motion. I will not try to define precisely what "perpetual motion" should mean, but the general idea is that you cannot get something for nothing, and if the energy account has to be strictly balanced this is indeed impossible. Here is an indication of how belief in the impossibility of perpetual motion hardened in the eighteenth century. In 1775 the Académie Royale des Sciences in Paris resolved (quoting from Elkana's book *The Discovery of the Conservation of Energy*)

> This year the Academy has passed the resolution not to examine
> any solution of problems on the following subjects: The
> duplication of the cube, the quadrature of the circle, the trisection
> of the angle, or any machine announced as showing perpetual
> motion.

(The first three items here are mathematical problems that are known to be impossible.) There is nothing so convincing of impossibility as repeatedly witnessing failed attempts.

The conservation of energy does more than rule out perpetual motion – the production of energy from nothing. It also works the other way, saying that energy is never lost either (although, as we shall shortly see, it may become inaccessible).

## 2.4   Heat as Energy

The caloric theory of heat had suggested something that could be measured: the quantity of caloric. The drawback of the motion theory of heat at first was that no such quantity obviously presented itself. But when the importance of energy was understood, one could say that, if heat were some sort of motion, then it had energy, and the "quantity of heat" could be reinterpreted as the amount of energy in the heat form of motion.

It was always clear that mechanical effects can in some circumstances produce heat. The bit of a drill heats up. A bicycle pump in use gets warm. Fire can be produced by rubbing wood against wood. These phenomena seem naturally to support the motion theory of heat. But the caloric theory could also, with some difficulty, be made to explain them: perhaps the friction somehow dislodges "trapped" caloric. After all, static electricity can be generated by rubbing (for instance, a plastic comb with a dry cloth), but no one believes that electricity is a form of motion.

Benjamin Thompson was born in New England. He was knighted by George III. He took up government service in Bavaria and was made a count of the Holy Roman Empire, taking the name *Rumford*. In 1798, he witnessed the boring of cannons in the arsenal at Munich and was impressed by the quantity of heat generated by the friction. He observed that the heat could boil water, and indeed an almost inexhaustible supply of heat seemed to be available. He was convinced that heat was a type of motion. But he did not convince his contemporaries, partly because the theoretical framework in terms of energy was not available to explain the observations. Parenthetically, in Rumford's paper to the Royal Society, he remarks (quoting from Elkana's book again),

> For fear I should be suspected of prodigality in the prosecution of my philosophical researches, I think it necessary to confirm to the Society that the cannon I made use of in this experiment was not sacrificed to it.

Count Rumford was much concerned with the "application of science to the common purposes of life". By his efforts the Royal Institution was founded in 1799.

One of those who did the most to establish quantitatively the connection between heat energy and mechanical energy was the Manchester scientist James Prescott Joule (1818–1889). One of his experiments measured the heating of water by a rotating paddle wheel immersed in it. He found the relationship between a unit of heat (say the *calorie*, the heat required to raise the temperature of one gram of water by 1 degree Celsius) and a mechanical unit of energy (say the *joule*, the kinetic energy of a two-kilogram mass moving at a speed of one metre per second). The modern value is 4.252 joules per calorie (as may be seen on any packet of breakfast cereal, which gives the energy per serving – chemical energy in this case – in both calories and joules). The important point is that same relationship (between heat and mechanical energy) is obtained whatever type of experiment is done.

Joule even had a simple image of the mechanism by which his paddle wheel heated up the water. He imagined the wheel hitting the atoms of water and so speeding up their motion – the motion that constituted the heat.

A century and a half after all this, we are all familiar with the commodity "energy" and the fact that it does not come free. We even know from our gas (chemical energy) and electricity (electrical energy) bills some of the units in which it is measured (1 kilowatt hour = 3,600,000 joules).

## 2.5   Atoms and Molecules

It certainly adds plausibility to the motion theory of heat if one has a concrete model for the nature of the heat motion. The simplest thing to understand is a gas (for simplicity, say a pure gas like oxygen). The gas consists of small particles, molecules, which are on average a long distance apart compared with their size (by a factor of order 100). The heat consists just in the kinetic energy of the random motion of the molecules. Most of the time, any given molecule is moving freely in a straight line, but every so often it hits another and bounces off in a new direction. The molecules are a bit like a cloud of gnats, with the scale reduced by a factor of about 1 million.

A molecule will sometimes move faster and sometimes slower, but it has an average speed. (I am not being precise about the way this average is defined.) The heat energy of the gas increases as the square of that average speed (since kinetic energy is defined with speed squared). If the gas is in a container, the molecules keep bouncing off it. The wall of the container has to exert a force on any molecule hitting it in order to make it bounce back. By Newton's third law, the molecule exerts an equal and opposite force on the wall of the container. The effect of all the molecules' bouncing is then an outward force on each unit area (say each square millimetre) of the wall of the container. This is nothing but the *pressure* of the gas.

We can make some simple deductions about the magnitude of the pressure. It goes up with the number of molecules per second that hit the unit area. This number in turn goes up with the the number of molecules in unit volume *and* with the average speed of the molecules (the faster they go the more frequently they hit the wall). For each collision, the velocity is sharply changed in direction, and the force necessary to do this also increases with the speed. Putting these two factors together, we see that the pressure increases with the number of molecules per unit volume and also with the *square* of the average speed, that is, with the average kinetic energy of a molecule. We can write this as an equation:

pressure = (constant)

$$\times \left[ \frac{\text{(average kinetic energy)} \times \text{(number of molecules)}}{\text{volume of gas}} \right].$$

Now let us suppose that we use a gas thermometer to define temperature. This means that we hold the pressure of the gas fixed and define the temperature to be (with some suitable scale) the volume the gas occupies. It follows from the equation that, with this definition of temperature,

average kinetic energy of a molecule

= (a constant) × temperature (T).

This equation gives us a physical interpretation of temperature in terms of molecular motion. Defining temperature this way provides what we did not have before, a meaning to *absolute zero*. It is the temperature at which the molecules are at rest and the pressure zero. (This statement ignores the effects of quantum theory.) We shall encounter in the next section another way to define this natural temperature scale.

The "constant" in the last equation is a multiple (for the simplest case, like a gas of helium atoms, the multiplying factor is $\frac{3}{2}$) of a number called *Boltzmann's constant*, after the Austrian Ludwig Boltzmann (1844–1906), who did the most to give a general theory of heat as random motion. His ideas met with strong criticism, and, perhaps partly as a consequence, he died by suicide. The constant is conventionally called $k$, and its value is now known to be

$$k = 1.380 \times 10^{-23} \text{ joule per kelvin}$$

(where the kelvin is a unit of temperature to be defined shortly). This equation nicely links together three of the great names in the theory of heat. Joule, from Manchester (see Section 2.4), gives his name to the unit of energy. Kelvin, from Scotland, gives his name to a unit of temperature. And the constant is named after Boltzmann, from Vienna. The tiny value of $k$ just reflects the fact that molecules are very light, so each one singly has very little kinetic energy.

The actual value of $k$ does not really tell us anything fundamental about nature. This is because any scale of temperature is to some extent arbitrary. The choice of 100 degrees between the freezing and boiling points of water, for example, is an arbitrary choice.

I stress that these considerations apply to an *ideal* gas in which the molecules are very small compared to the distances between them and in which the only energy is the kinetic energy of the motion of the molecules. A real gas approximates quite well to this ideal as long as it is at ordinary pressures and temperatures. But pressure and cooling can turn any gas into a liquid, and by the time this state is reached it is certainly far from being ideal. The molecules are close together, and the forces between them are important.

But what, in the middle of the nineteenth century, was the evidence for the existence of atoms and molecules? Speculation about atoms goes back to the earliest Greek philosophers. Those who believed in atoms tended to assume that between them there was just nothing, a "void" or "vacuum". But there have long been people who reject the notion of the void and prefer to think of some sort of continuous fluid pervading everything. Aristotle was one such, partly because he thought that in a vacuum things would move infinitely fast. The atomic theory was espoused by Epicurus (341–270 B.C.) and expounded in Lucretius' poem *On the Nature of Things* (first century B.C.). Stoic philosophy, on the other hand, considered matter to be continuous not atomic.

By the seventeenth century many people, like Robert Boyle, believed in atoms. Newton's success led naturally to the idea that everything could be explained by the motion of atoms, if only one knew the forces acting between them. But all this remained speculative until some quantitative evidence could be found for atoms.

One of the first such pieces of evidence came from chemistry. The Manchester scientist John Dalton (1766–1844) reasoned as follows (to put the argument in a modern form). When hydrogen reacts with chlorine to form hydrochloric acid, 1 gram of hydrogen always combines with 35.5 grams of chlorine to form 36.5 grams of hydrochloric acid (or, of course, the reaction can occur with any multiples of these weights). This can be understood if the reaction takes place between individual atoms. In this example, one atom of hydrogen combines with one of chlorine to form one molecule of hydrochloric acid (a molecule is a combination of atoms that forms the smallest unit of a chemical compound). Also one atom of chlorine weighs 35.5 times as much as one atom of hydrogen. (Actually, we now know that chlorine contains two different sorts of atom – isotopes – one weighing 35 and one 37 times as much as a hydrogen atom.) These numbers, 1, 35, 37, are called *atomic weights*. By comparing a lot of chemical reactions, the relative atomic weights of atoms of all the chemical elements can be found (of course, it is just a matter of convention to give hydrogen the atomic weight 1).

In 1811, the Italian scientist Amedio Avogadro suggested that any gas, at a given temperature, contains the same number of molecules in a given volume (strictly, this applies to an "ideal" gas, in the sense explained previously: for any real gas it is an approximation). Thus, also using Dalton's atomic weights, 70 grams of chlorine gas would contain as many molecules as 2 grams of hydrogen gas (this is because the molecules of hydrogen gas and chlorine gas actually contain two atoms each).

One thing remained before there was a really quantitative idea about atoms. That was to know the actual number of molecules in, say, 2 grams of hydrogen gas. This number is called *Avogadro's number* (and is conventionally denoted by the letter *N*). If it is known, then the weights of individual atoms can be immediately deduced. It is easy to find the value of $N \times k$, where $k$ is Boltzmann's constant, but to find each of these quantities separately is harder. We now know that $N$ is approximately

$$N = 6 \times 10^{23}.$$

By the end of the nineteenth century there were many indirect arguments for the existence of atoms and molecules, and estimates of the value of Avogadro's number and of the weight and sizes of molecules. But it was still possible for some people to be sceptical and to criticize the molecular theory as being unfounded speculation. One such person was the influential Viennese scientist and philosopher Ernst Mach (1838–1916). He believed that scientists should restrict themselves to what can actually be observed (though he did admit that talk of atoms could be a useful tool, a little reminiscent, perhaps, of the way some sixteenth-century thinkers could admit Copernicanism as a mathematical tool but not the actual physical truth). The reactions of scientists to Mach's philosophy of science fluctuated through the twentieth century.

If there was any one year when the reality of molecules and their random heat motion was most decisively proved, it was 1905. This was the year when Albert Einstein, born in Ulm in 1879 and a Swiss citizen in 1901 (unfit for military service because of flat feet and varicose veins), working in the patent office in Bern since 1902, completed his doctoral thesis for the University of Zürich. In this year also, he proposed the quantum nature of light and invented the theory of relativity. These are not what we are concerned with now. In 1905, Einstein gave two ways of determining Avogadro's number. I will describe one of these, which makes random heat motion real in a particularly simple fashion.

Very fine particles can be put in a "suspension" in a liquid like water. This means that the particles fall to the bottom only very slowly.

(Some medicines are in the form of suspensions, and the bottle is to be shaken to restore the particles to suspension.) Illuminated with a bright light and viewed through a microscope, the particles, if small enough, can be seen to be in continual zig-zag motion. (The particles have to be less than about a thousandth of a millimetre in size for the motion to be observable.) Fine pollen grains can be suitably small, and the effect was especially studied by the British botanist R. Brown (1773–1858) and is now called *Brownian motion*. It was natural to speculate that the observed motion of the suspended grains might somehow be due to buffeting by the molecules of the water, as a result of their random heat motion, but the speed of the grains is too small to be observed directly. It is only a cumulative effect of these random speeds that is seen. Einstein showed how this happens.

Assume that the suspended particles can be treated as a gas (except that they each weigh some $10^{16}$ as much as a molecule), so that they have each the same average kinetic energy as a gas molecule. Taking a mass of $10^{-7}$ (one ten-millionth) of a gram, the average speed comes out to be in the region of millimetres per hour. How can such a small thing be observable? The answer lies in the way repeated small random motions mount up with time. Let us first simplify this random heat motion and treat it as a succession of little steps. Let us say, for example, that each second a particle moves $10^{-4}$ millimetre in a random direction. What is the cumulative effect of these little displacements? At first, one might be tempted to argue that the particle will not move much, because the motions are as likely to be in one direction as another and so will soon cancel out.

This would be wrong. In fact there is a relation

$$(\text{distance moved})^2 = (\text{a constant}) \times (\text{time elapsed}).$$

A graph of this relation is shown in Figure 2.3. The distance moved with time continually increases, but does so at a slower and slower rate (the graph gets less and less steep). An explanation of this graph is given in Appendix C.

The "constant" in Einstein's equation depends upon the energy of an individual molecule, and so careful observations of Brownian motion, combined with Einstein's theory of it, gave the first reliable

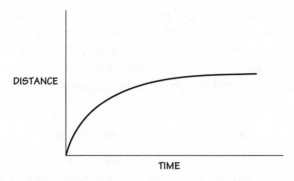

DISTANCE

TIME

FIGURE 2.3   A graph of the relation between distance and time in Brownian motion. The curve is in fact part of a parabola.

values for Avogadro's number, $N$, and hence for Boltzmann's constant, $k$.

There is a slight generalization of the foregoing, which we shall need in later chapters. I mentioned that the the average kinetic energy of a gas molecule was $\frac{3}{2}kT$, where $k$ was Boltzmann's constant. The factor 3 here arises because the molecule moves in three-dimensional space and so has three components to its velocity. There is a generalization of this rule, which is valid subject to some conditions that I will not go into. This generalization (it is called the *equipartition theorem*) states that

$$\text{average thermal energy} = (\text{number of degrees of freedom}) \times \frac{1}{2}kT.$$

I must explain what is meant by *number of degrees of freedom*. It is the number of pieces of data one would have to specify in order to define the motion at some initial time. For a particle moving in space, this number is 6, the 3 numbers needed to fix position and the 3 numbers needed to fix the velocity (the former three are not "relevant" for freely moving particles, hence the 3 here not 6). But if we were talking about a molecule made of, say, two atoms (like a hydrogen molecule), we would need some more quantities (angles) to fix the orientation of the molecule, and the number of degrees of freedom would be bigger.

All this chapter is about the energy of mechanical motion. But we shall see in subsequent chapters that energy is associated with other physical effects. There are electrical energy, magnetic energy, energy in light and so on. In these cases, too, there may be random thermal energy, and the concepts of temperature and entropy are relevant. The first evidence that classical physics was incomplete arose from an attempt to apply the last equation to light.

## 2.6   Steam Engines and Entropy

Heat is a form of energy, but that is not the whole story. In order to understand the way heat works, another concept is needed, that of *entropy*. Entropy is one of those scientific technical terms, like *relativity*, the *uncertainty principle*, *evolution*, more recently *chaos*, that have from time to time captured the literary imagination. Entropy is often thought to mean something like "disorder", and as we shall see this is roughly right.

The story of entropy, surprisingly for such a subtle concept, begins with the down-to-earth subject of steam engines. The steam engine had been invented by the Scottish engineer James Watt (1736–1819) and improved particularly by Cornish mining engineers. At the end of the Napoleonic wars, there was interest in France in the superior British steam engines. How did one get the maximum "motive power" (what we would call mechanical *work*) out of a given amount of coal or wood? That is, how did one maximize the *efficiency* of the engine? There is plenty of heat about, in the sea, for example. Can it all be used to produce mechanical work?

The decisive step in answering these questions was made by Sadi Carnot, born in Paris in 1796, the son of Lasare, who held high office during the revolution and under Napoleon and was an intellectual who himself worked on a general theory of the efficiency of machines. Sadi was certainly impressed by British engines. He wrote:

> If you were to deprive England of her steam engines, you would deprive her of both coal and iron; you would cut off the source of all her wealth, totally destroy her means of prosperity, and reduce this nation of huge power to insignificance.

Carnot asked himself the question,

Can we set a limit to the improvement of the heat engine, a limit which, by the very nature of things, cannot in any way be surpassed?

To answer this, he imagined an idealized heat engine (a heat engine might be a steam engine, or it might use another working medium like hot air). Any real engine would have complications like friction or heat loss that could only reduce its efficiency. The essential points about Carnot's ideal engine are the following.

There is a source of heat (the furnace, for example), which I will call $H$, and it is maintained (for example, by burning coal or oil) at a fixed temperature $t_H$. There is a colder region $C$, which is maintained at another fixed temperature $t_C$, less than $t_H$. This might be the "condenser" in a steam engine, where the steam is turned back into water, or it might just be the surrounding air. Carnot realized that the essence of a heat engine is that it produces mechanical work by allowing heat to "drop" from $H$ to $C$. There is a partial analogy with a waterwheel, which produces work by allowing water to drop from the upper mill stream to the lower one, and this analogy may indeed have influenced Carnot. (As a historical aside, it is interesting that Carnot believed – initially at least – in the false caloric theory of heat; but this did not stop him from getting things nearly right.)

One way a heat engine might work is in a series of *cycles*, so that at the end of each cycle the state of the engine is exactly the same as at the beginning. The overall change during one cycle is that some amount of heat has been taken in from $H$, some other amount of heat has been given out to $C$, and an amount of mechanical work has been done by the machine (for example, in pumping water out of a Cornish tin mine). Just by conservation of energy, we know that

$$\text{heat in} = (\text{heat out}) + (\text{work done}).$$

(In writing this equation, I have used the same units for heat as for work: if necessary, we must first convert calories into joules.) The "heat in" is determined by the amount of fuel that is used. It is natural to define the efficiency to be (making use of the conservation of energy)

$$\text{efficiency} = \left[ \frac{\text{work done}}{\text{heat in}} \right]$$

$$= \frac{\text{heat in} - \text{heat out}}{\text{heat in}}$$

$$= 1 - \frac{\text{heat out}}{\text{heat in}}.$$

The engine might consist of a cylinder of air with a piston that moves back and forth. The cylinder would some of the time be in contact with $H$, taking in heat, with the air expanding and pushing out the cylinder. Some of the time the heat supply would be cut off, but the air would still be expanding and the temperature dropping. Some of the time the cylinder would be in contact with $C$, the air would be compressed, and heat would be given out. And so on. All this is indicated schematically in Figure 2.4.

But these details do not matter. What is important is that the working of this ideal engine should be *reversible*. That is to say, the cycle could just as well be run backwards, with mechanical work being done *on* the engine so that it takes heat in from C and sends heat out to H. With this mode of running, it is a refrigerator if it is thought of as abstracting heat from C, or a heat pump if it is thought of as supplying heat to $H$. In the analogous case of a waterwheel, we can imagine its being used, by the application of mechanical work, to pump water up from the lower-level stream to the higher.

Now comes the crux of Carnot's argument. For a given $H$ and $C$, imagine two engines working between these reservoirs. Suppose these engines could have different efficiencies. Run the more efficient one as a heat engine, but the less efficient one in the refrigerator/heat pump mode. Let us suppose that we arrange that the heat given out to $H$ by the heat pump is used as heat in by the heat engine. Then, because of the assumed greater efficiency, *more* work is produced by the heat engine than is used up by the refrigerator. The combined effect, after one cycle of each engine, is that net mechanical work has been made available, and that net heat has been abstracted from C. If this were possible, we could obtain almost

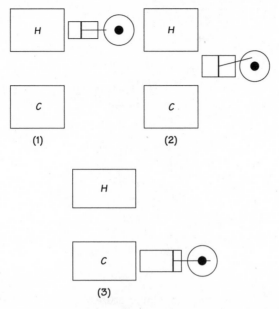

FIGURE 2.4   A schematic illustration of an engine working in a Carnot cycle. The source of heat is *H* and the "source of cold" is *C*. The engine is illustrated as a cylinder with a piston in it. Three stages in a cycle are shown. In (1), heat is passing into the cylinder from *H*, causing the gas in it to expand. In (2), the cylinder is isolated, the gas continues to expand, and its temperature drops. In (3), heat is passing out of the cylinder to *C*, causing the gas to contract.

unlimited supplies of mechanical work, at the expense of just cooling down, say, the oceans. This would be almost perpetual motion (although in a weaker sense than if energy were not conserved).

Carnot took it as axiomatic, and we should probably agree, that this is impossible. The conclusion is then that *any* ideal engine working between the two given temperatures must have the same efficiency. This efficiency must have a value that depends only upon these two temperatures, and not in any way on the details of the engine (whether it uses steam or air, for example).

Thus we have deduced that

$$\text{efficiency} = \text{(something depending only upon upper and lower working temperatures)}.$$

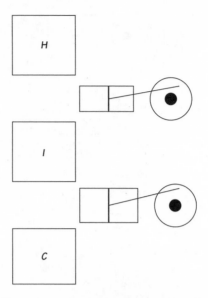

**FIGURE 2.5** Three constant temperature tanks, *C*, *I* and *H* (standing for "cold", "intermediate" and "hot"). Two Carnot engines are shown, one working between *H* and *I* and the other between *I* and *C*. The heat given out to *I* by the first engine is assumed to balance the heat taken in from *I* by the second. Then the combined effect is equivalent to a single engine working between *H* and *C*.

Referring to the definition of efficiency, we can rewrite this as

$$\frac{\text{heat out}}{\text{heat in}} = \text{(something depending only upon the two temperatrures).}$$

We can actually go further and deduce that

$$\frac{\text{heat out}}{\text{heat in}} = \frac{\text{something depending upon lower temperature}}{\text{something depending on upper temperature}}.$$

To see how this last relation comes about, we use another ingenious argument, illustrated in Figure 2.5.

Compare the heat engine, working between an upper and a lower temperature, with a pair of heat engines. The first of these works between the upper temperature and some intermediate temperature. The second works between the intermediate temperature and the lower temperature. The heat out from the first is used as heat in by the second. Now let us verify that our last equation is consistent when applied to the pair of engines working in tandem. For the first, it gives

$$\frac{\text{heat out at intermediate temperature}}{\text{heat in at upper temperature}}$$

$$= \frac{\text{something depending on intermediate temperature}}{\text{something depending on upper temperature}},$$

and for the second engine, it gives

$$\frac{\text{heat out at lower temperaure}}{\text{heat in at intermediate temperature}}$$

$$= \frac{\text{something depending on lower temperature}}{\text{something depending on intermediate temperature}}.$$

Now multiply the left-hand sides of these two equations together. The heat out at the intermediate temperature from one engine is used as the heat in at the intermediate temperature by the other. So these two quantities of heat cancel out. In a similar way, the references to the intermediate temperatures cancel out when the right-hand sides are multiplied together. Thus from these two equations we have deduced the correct result for the single engine working between the upper and lower temperatures.

Carnot's work made little impact at the time of its publication in 1824. It may have been too abstract for engineers, and of too unusual a form to be noticed by scientists. Carnot died in 1832. In 1845 the 21-year-old Scottish scientist William Thomson (later Lord Kelvin) visited Paris and became aware of Carnot's work. He later wrote (quoted in Fox's translation of Carnot's work),

> Nothing in the whole range of Natural Philosophy is more remarkable than the establishment of general laws by such a process of reasoning.

Thomson, and the Prussian Rudolf Clausius, perfected Carnot's ideas and created the branch of science now called *thermodynamics*: the science of heat and its transformation into mechanical and other forms of energy.

Thomson proposed to define the *absolute temperature*, which I will denote by the capital letter $T$, so that for a Carnot engine we have *by definition*,

$$\frac{\text{heat out at lower temperature}}{\text{heat in at upper temperature}} = \frac{\text{lower absolute temperature}}{\text{upper absolute temperature}}.$$

This definition has many virtues. It is a natural definition, independent of the choice of a thermometric substance like mercury or air or hydrogen. It fixes the zero of temperature: we do not have the arbitrary choice whether to call zero the freezing point of water or alcohol or whatever. What it does not do is fix the size of a degree: you could scale $T$ by some constant factor, and that equation would not be affected. In fact the size of a degree on Kelvin's scale (called a *kelvin*) is conventionally defined to be the same as 1 degree Celsius. Temperatures on this scale are written as, for example, 300 K. It turns out that

$$0\,C \approx 273\,K.$$

(The symbol $\approx$ means "approximately equal to".) This is not a fundamental constant of nature. It is just a number that depends on when water freezes and boils. The Kelvin scale of temperature agrees well with that defined by an (ideal) gas thermometer (see the previous section).

The term *entropy* (from a Greek word meaning "transformation", because it is relevant to the transformation of heat into work) was proposed by Clausius. Entropy is defined from Carnot's ideas as follows. First rewrite the last equation as

$$\frac{\text{heat in}}{\text{upper absolute temperature}} = \frac{\text{heat out}}{\text{lower absolute temperature}}.$$

Let us *define* entropy so that, for any object that gains a quantity of heat energy, while remaining at a fixed temperature, the change in entropy is given by

$$\text{increase in entropy} = \frac{\text{heat in}}{\text{absolute temperature}}.$$

This defines entropy as a property that an object has in addition to its temperature. This definition, together with the equation before, tell us that, in a *reversible* process of the type envisaged in Carnot's ideal engines, entropy does not change: the colder reservoir $C$ gains as much entropy as the hotter one $H$ loses.

What happens to entropy in a process that is not reversible? Take a simple example. Take two tanks of water at different temperatures, $T_{hotter}$ and $T_{cooler}$. Put them into contact with each other, so that heat can flow from one to another, but insulate them from everything else, so no other heat can flow in or out. Wait. We know what will happen. Some heat will flow from the hotter to the cooler, until eventually they both settle down at some common temperature $T_{final}$ (which will be in between $T_{hotter}$ and $T_{cooler}$). But what happens to the total entropy? Take the first little bit of heat that flows. This reduces the entropy of the hotter tank by

$$\frac{\text{little bit of heat}}{T_{hotter}},$$

and increases the entropy of the cooler tank by

$$\frac{\text{little bit of heat}}{T_{cooler}}.$$

(Because we have assumed that the heat flow is small, the two temperatures are very little changed during this time.) The increase in entropy of the cooler tank is bigger than the decrease in entropy of the hotter one, so the overall result is an *increase* in the total entropy.

Now we go on to consider the next little bit of heat transferred, and so on. The temperatures will be slightly different now. The hotter tank is not quite so hot and the cooler one is not quite so cool. But the tank that started hotter stays hotter until the process ends and the temperatures are the same. Thus each small change of entropy is greater than zero, and the final total entropy of the two tanks is more than it was originally.

This example is in fact typical: entropy always increases (except in a reversible process, when it remains fixed). This is the basis of the assertion that "entropy increases". One should be a bit careful about this, however. The preceding example is a nice simple one in which we knew what we meant by the entropy at the beginning and the entropy at the end. The two tanks of water at the start, for example, were each calm, undisturbed things, with well-defined temperatures. All we did to initiate the change was gently to push them into contact with each other. To talk about the entropy of a breaking teacup, or a jumping squirrel, would be more problematic.

Entropy is connected with the amount of heat energy that is unavailable (for conversion into useful work). Unlike energy, entropy is not conserved. All we know is that it cannot decrease. Although energy does not disappear, useful energy does.

## 2.7   Entropy and Randomness

In the science of heat ("thermodynamics") there are three key notions: heat, temperature and entropy. If heat is nothing but the random motion of molecules, we should be able to define each of these in terms of that motion. Heat is easy: it is just the total energy in the random motion.

As to temperature, we have already seen in Section 2.5 its interpretation in the simple case of a gas. But we would like a definition that applied to a liquid or a solid or indeed anything. It turns out to be best to define entropy first. So let us do that.

In order to have a definite mental picture, let us take a quantity of some gas, but everything we do should apply to any substance. We suppose that it is in a settled state and, for the moment, isolated from the outside world. What do we know about it? We can suppose that we know the number of molecules in it, the volume it occupies and the total thermal energy it has (that is, the sum of the energies of all the molecules). These things provide a complete "large-scale" definition of our sample of gas. But of course the complete "small-scale" configuration would require an enormously greater amount of information: it would include the position and, at the least, the velocity of each individual molecule (perhaps also something about how each molecule is spinning, for example). There may be something like $10^{24}$ molecules in an ordinary sized sample; the amount of this small-scale information would be enormous. We could never have this information in practice, and we do not want to have it.

The large-scale information, about the volume and energy, that we do have puts some slight restriction on the small-scale information. Now consider the following number

$W =$ (number of possible small-scale configurations consistent with the large-scale information).

I have been a little vague with this definition. The value of $W$ depends upon how precisely we decide to define a configuration. Do we count two configurations as different if the speed of one molecule differs between the two by say 0.001 percent? It turns out that quantum theory (see Chapter 8) cures this vagueness. In the meantime, please accept my assurance that the vagueness does not cause any serious problem.

This number $W$ is a reasonable measure of the randomness when we have only the large-scale information. To illustrate this idea, let us consider a simpler model. Take an ordinary pack of 52 playing cards and shuffle them well. Shuffling is a randomizing process. After shuffling we know nothing more than that we have 52 cards – the "large-scale" information. The number of possible small-scale configurations – that is, actual orders of the cards – is

$$W = 52 \times 51 \times 50 \times 49 \times \cdots \times 3 \times 2.$$

This is a very large number. It comes out to about

$$W \approx 8 \times 10^{67}.$$

So this is our measure of the randomness of a well-shuffled pack of cards. Suppose now we separate a pack into the red cards and the black cards, shuffle each half separately and put them with the 26 shuffled red cards on top of the 26 shuffled black ones. We now have more "large-scale information" (reds are on top). The number of possible configurations is now less. It is

$$W' = [26 \times 25 \times 24 \times \cdots \times 3 \times 2]$$
$$\times [26 \times 25 \times 24 \times \cdots \times 3 \times 2] \approx 1.6 \times 10^{53}.$$

There are two objections to this measure of randomness. The first is a practical one: we get ridiculously large numbers. Even for the playing-card model the numbers are large. For the configurations of molecules in a gas they would be far, far worse. The second objection is more profound. If we have a system divided into two parts, we get the $W$ for the whole system by *multiplying* the $W$s for the two parts. This is illustrated in the last equation, where we multiply the number for the red cards by that for the black ones. But in physics we are used to quantities with which we have to *add* the contributions for the two parts. Energy is an example of such an additive quantity. But we know how to convert multiplication to addition: we do it by taking logarithms. So let us define

$$S = \log W$$

(a formula due to Boltzmann and engraved on his tombstone in Vienna). This is now the definition (almost) of *entropy* (which is traditionally denoted by the letter $S$). For the full pack of cards,

$$S \approx 68,$$

and for the pack with all the reds on top

$$S' \approx 53.$$

In this simple model, we can see how the entropy increases with time. Take the pack with the (shuffled) red cards on top of the (shuffled) black ones. The entropy is 53. Now shuffle the whole pack. Soon it will be just a randomized pack of 52 cards, irrespective of their colour. The entropy has then increased to 68.

This model is very like the example, given at the end of the last section, of two tanks of gas at different temperatures put in contact. Putting them in contact has the effect of "shuffling" the energies of the molecules. Before the tanks were put in contact, we had the information that the molecules of one tank had, on average, more energy than those of the other. After they have been in contact, we have lost that information, so there is a larger number of possible small-scale configurations, and therefore a larger $W$ and a larger $S$.

We can now understand better the law of increase of entropy. If you shuffle an ordered pack of cards, it almost certainly becomes disordered. If you shuffle a disordered pack, it is exceedingly unlikely to become exactly ordered like a new pack. If anyone claimed to make this happen, I would be sure he had cheated. All this is because of the large numbers involved, stemming from the quite large number, 52, of cards in a pack. For a macroscopic piece of matter the corresponding number is something like $10^{23}$. So the chance of entropy's decreasing (for instance, all the molecules of a gas moving by chance into one half of a box) is so small as to be utterly negligible for all practical purposes.

Having now defined entropy in terms of (our information about) the state of the molecules in the substance, we are in a position to define temperature. We simply take over an equation in the last section and write

$$\text{heat energy added} = T \times (\text{increase in entropy}).$$

This is used to define the temperature $T$ (on the Kelvin scale). But we have to explain how the equation is to be used. First of all, we must ensure that the temperature is kept fixed while the changes are going on. We can do this by putting the object we are considering into contact with a large tank of water at the same temperature. Then, whatever we do, the tank will supply or extract heat as necessary to keep the temperature fixed (since it is a very large tank, it can supply heat without significantly cooling down).

How, under these circumstances, can we arrange for the entropy of the substance to change? Take an example. Suppose the substance is a gas in a cylinder with a piston that we can pull out or push in, as represented in Figure 2.6. Allow the cylinder slowly to move out a little. The gas has a pressure, and so is pushing on the cylinder and could do mechanical work (propelling a car, or whatever). Where does the energy come from? It must come from heat supplied by the big water tank. So the "heat energy added" (to the gas in the cylinder) is some positive quantity. Therefore the entropy must have increased. It is easy to understand this. The volume of the gas in the cylinder has been increased. This means that $W$ has increased, because there are more positions that each molecule can be in. This is why $S$ has increased.

The last equation tells us that, provided the temperature $T$ is held fixed, heat is required in order to lose information about the state of the molecules. The higher the temperature, the greater the amount of heat for a given loss of information. This is the way in which temperature is defined.

The definition of temperature may seem rather indirect and difficult to visualize, compared to our preliminary explanation in Section 2.5 for an "ideal" gas. The point about the new definition is that it is absolutely general and applies to gases or liquids or solids

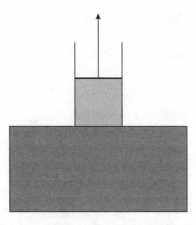

FIGURE 2.6    A cylinder containing gas in contact with a heat reservoir. When the piston is pulled out a short way, a little heat is drawn from the reservoir. The molecules of the gas then occupy a slightly larger volume, so their entropy increases. The reservoir serves to keep the temperature fixed.

or anything. It can be shown that our new definition agrees with the old one in the special case of an ideal gas.

## 2.8    Chaos

In Chapter 1, we met Newton's simple, predictable "clockwork" Solar System. In this chapter, we contemplated a gas of some $10^{23}$ molecules performing a random dance that we could never possibly follow in detail but that can nevertheless be characterized by some statistical properties.

The simplicity of Newton's Solar System arose from two fortunate circumstances. First, the Sun is so much heavier than any of the planets that the gravitational force of one planet on another is much smaller than the force of the Sun on a planet. So there is a quite good approximation in which the orbit of each planet is calculated independently of the others. Similarly, the orbit of each moon round its planet can be approximately computed using only the force due to the planet (because the moons are near to their planets). Second, the inverse-square law has the property (see Appendix A) that each of these planetary orbits closes up and repeats itself periodically.

Starting from this simple approximation, one can try to determine the small corrections due to forces between planets, or between the Sun and the Moon. This is not easy. Even Newton was defeated by the Earth-Sun-Moon system.

One cause of difficulty is due to possible *resonances*. This word comes originally from acoustics. The body of a cello is designed so that its natural frequencies of vibration are similar to the frequencies of the vibrating strings when it is played. Because of this, energy is readily transferred from the strings to the body of the cello, and hence to the air. Similarly, the best way to get a child going on a swing is to push in time with the natural frequency of the swing (that is, the frequency with which it swings freely). Then each push, going with the swing, transfers the maximum energy to it.

The word *resonance* is used to refer to any situation like this, in which a force applied to a system is repeated at a rate related to the natural frequency. Then a small force can eventually have a large effect.

Can there be resonances in the Solar System? I have already described, towards the end of Section 1.6, how gaps in the asteroid belt may have been created where the periods resonate with that of Jupiter.

Throughout the eighteenth and nineteenth centuries, the possibility of resonances made accurate calculations of the Solar System difficult. Mittag-Leffler, a Swedish mathematician, proposed a mathematical contest to mark the sixtieth birthday in 1889 of Oscar II, king of Sweden and Norway. The prize went to the great French mathematician Henri Poincaré for work eventually published as *Les Méthodes nouvelles de la Mécanique céleste*. Poincaré's great new insight was to apply geometrical methods to mechanics.

Let me explain something of Poincaré's approach by means of a simple model. Start with a trivial case: a single pendulum. In order to picture the motion, it is sometimes convenient to think of a two-dimensional space in which the displacement of the pendulum is plotted along one axis and its velocity along the other. Such a space, with velocities as well as positions, is called *phase space*. A point in phase space represents a *state* of the pendulum. Given any point, the equations of motion determine the subsequent states of the pendulum, that is, a curve in phase space. In the case of a

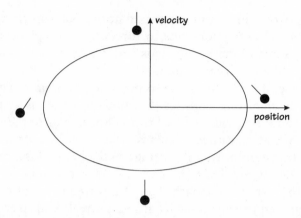

FIGURE 2.7   Phase space for a pendulum. The displacement of the pendulum bob is plotted along the horizontal axis, and its velocity along the vertical one. The closed curve represents one complete period in the motion of the pendulum. If the pendulum keeps swinging (without friction), the representative point in phase space keeps going around this curve. The positions of the pendulum at four points on the curve are indicated.

pendulum (assuming it does not have enough energy to swing right over the top), the curves are closed ones, as in Figure 2.7.

Now go to a less trivial case. Take two pendulums, which can each swing backwards and forwards, and between which we can arrange small interactions, by means of an elastic string or something like that. Now phase space is four-dimensional: two positions and two velocities. Even for this simple model, we need four dimensions, more than most of us can imagine. The Sun-Earth-Moon system requires a 12-dimensional phase space.

Assume that the total energy of the two pendulums is conserved. This energy consists of the kinetic energies of both pendulums together with their gravitational potential energies and the potential energy in the connecting string. It is thus something depending on the four variables of phase space. The condition of energy conservation (for a given total energy) puts one restriction on the allowed values of the four variables. This brings down the four dimensions to three, which we have more hope of picturing. Unfortunately these three dimensions are not in a simple flat space,

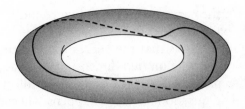

FIGURE 2.8    Phase space for two unconnected pendulums. The figure is meant to lie in the three-dimensional part of four-dimensional phase space, which corresponds to a fixed total energy. The curves representing the successive states of the system lie on a torus (a two-dimensional surface) and curl around it both ways. In the example shown, the two periods have been taken to be equal, and the curve loops just once around both ways, getting back to its starting point. Note: this figure should not be taken too literally, because the three-dimensional space is curved and so its representation on a flat page may mislead.

but something curved, a bit like the surface of a sphere (but generalized from a two-dimensional surface to a three-dimensional one).

Take now the specially simple case in which there is no interaction between the two pendulums. What do typical motions look like in the three-dimensional space? The answer is that the curves describing possible motions all lie on a *torus*. A torus is a surface like the inner tube of a bicycle tyre. One can draw loops on a torus in several ways. One way is to draw a small loop once round the tyre. Another is to draw a large loop, on say the side of the tyre nearest to the hub. The motion of the pair of pendulums is represented by a sort of combination of these two types of path, going round both ways at once. An example is illustrated in Figure 2.8.

This figure shows a very special case. The periods of the two pendulums have been taken to be equal, so the curve in phase space goes exactly once around the torus both ways, then gets back to its starting point. If one period had been twice the other, the curve would have wound twice round the torus one way as it went round once the other. If the ratio of the two periods is any fraction, the curve eventually gets back to its starting point. On the other hand, if the ratio is a decimal like $\sqrt{2}$ or $\pi$ (which is not exactly equal to any fraction), the curve goes on winding round the torus, never getting exactly back where it started.

The important thing is that the curves representing the motion of the two pendulums lie on *two*-dimensional surfaces and do not stray into other parts of the *three* dimensions allowed by energy conservation. The question is, What happens when there is an interaction between the pendulums? Are the motions still restricted to something like a torus, or do they roam all over the phase space allowed by energy, something like a loosely wound ball of wool? In the 1950s and 1960s, three Russian mathematicians, Kolmogorov, Moser and Arnold, proved an important theorem that, for our example of the two pendulums, says something like this:

- If the ratio of the two periods is not a fraction, and the interaction between the two pendulums is "small enough", the torus is slightly distorted but still exists. How small is "small enough" depends upon how near is the ratio to fractions with small denominators (that is to say, it depends upon the absence of important near *resonances*).

When these conditions are not satisfied, the curves may begin to wander about in other regions of phase space, which regions grow as the strength of the interaction increases. This wandering off the torus is called *chaotic* motion. The use of the word *chaos* in this (fairly precise) sense was started in 1975 by James Yorke. This word, like the phrase *black hole*, has fired the public imagination.

The detailed study of chaotic motions was made possible by the invention of calculators and computers. One of the most famous such studies was made by Edward Lorenz in 1963. His equations were for a very simplified model of the rising of hot air in the atmosphere. Part of his aim was to find out how predictable weather could be. He found chaotic solutions, but there were patterns in the chaos and in the onset of chaos. Much of the excitement in chaos studies has been concerned with identifying and explaining the patterns in it.

A key property associated with chaos is that curves that start close together in phase space quickly wander far apart. Any specification of the initial state of a dynamical system (the Solar System, for instance) must inevitably have some imprecision. Measurements

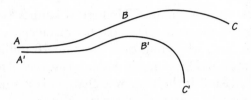

FIGURE 2.9  Predictability in chaotic motions. The initial state of the system could be as well represented by $A'$ as by $A$, because of measurement errors. The paths in phase space through each of these two points are followed into the future. In chaotic motions they may diverge fast, with the deviation doubling (for example, from $BB'$ to $CC'$) in some fixed time interval. The predictions are subject to fast increasing uncertainties.

of initial positions and speeds are subject to errors. Suppose we choose two points in phase space that are close enough together to be equally good approximations to what is known of the initial data. Then we follow the two curves, one through each point, which represent the subsequent states of the system. The separation between these two curves shows the limitation upon our powers of prediction. Figure 2.9 illustrates the idea.

In chaotic motions, there is a time, which I shall call the *error-doubling time*, such that the deviation doubles with every passing of this time interval. Let us illustrate the idea with some made up numbers for the two-pendulum system. Suppose we measure the initial state with an accuracy of one part in a million(!). Suppose, in a chaotic regime, the error-doubling time is 1 minute. After 10 minutes the uncertainty will have increased to about one part in a thousand (because $2^{10} = 1024$). After 20 minutes, there will be really no prediction at all (because $2^{20} = 1,048,576$).

For the Earth's weather, computer studies indicate that the error-doubling time is about two days.

Thus chaos entails inescapable practical limitations on the predictability of determinate systems (determinate because the laws of motion do uniquely determine the motion). Probably we had no right to expect anything else. But before the advent of computers our mathematical experience was necessarily limited to rather special cases of non-chaotic motion.

Is the Solar System chaotic? Numerical studies (see Peterson's book) have given evidence of chaos, for instance, in the orbit of Pluto, with an error-doubling time of some 10 million years. By comparison, the Solar System as a whole has survived for more than a billion ($10^9$) years. If the system is chaotic, is it stable? Could a planet, by chance, find itself moving fast enough so that it escaped? Maybe any chaotically unstable orbits have already been emptied in this manner.

In these brief remarks about chaos, I have limited myself to closed systems (as is, very nearly, the Solar System). Many practical examples are not closed: there may be friction removing energy and external driving forces supplying energy. It is in these cases that chaos reveals some of its most interesting features, like the existence of "strange attractors" in phase space.

Chaos is no doubt a ubiquitous phenomenon in the universe. A blanket statement like this does not get one very far. The interesting, and difficult, questions are ones like, What are the error-doubling-times for various systems? Are there average properties that are more predictable than details? What are the features (like the Giant Red Spot on Jupiter) that persist in spite of the chaos? Can we attain an extra day of reliable weather forecasts? What are the patterns in the chaos?

What has chaos got to do with the subject of this chapter, heat? The idea of phase space can be applied when the system in question is a quantity of, say, gas. The number of dimensions of phase space is then huge, some $10^{24}$, give or take a few powers of 10. We are incapable of imagining such a space, but it is well defined mathematically. The state of motion of all the molecules of the gas at any time is represented by a single point in the phase space. The history of the gas is the curve traced out as this point moves in phase space.

In previous sections, I have used words like *random* and *average*. What do these mean exactly? How can anything determined by laws of motion be called random? Average over what? The more chaotic is the motion, the easier it is to glimpse answers to these questions. Suppose the curve in phase space (the curve that represents the history of the gas) winds in a tangled way in all parts of the space. Then it is reasonable that averages should be taken over all of phase space. Suppose that curves that start near together soon totally diverge from each other. Then the word *random* is appropriate.

## 2.9   Conclusion

Heat is "nothing but" the energy of the motion of tiny particles, in obedience to the laws of motion. But to understand heat and temperature, a further, non-mechanical concept is necessary: entropy, which is defined in terms of our ignorance of the microscopic state of a macroscopic object. The difficult notion of randomness is involved.

The "law of increase of entropy" raises profound questions, since it seems to give a direction to time, a direction which is absent from the microscopic laws of motion. We return to this enigma in Section 14.6.

# ELECTRICITY AND MAGNETISM

*How electricity and magnetism are different aspects of one thing.*

## 3.1 Electric Charges

William Gilbert (1504–1603), physician to Queen Elizabeth, coined the adjective *electric* from the Greek word for amber. It had been known in antiquity that a piece of amber rubbed with a cloth acquired the power to attract small objects. Many other electrically insulating substances, like glass and plastics, behave similarly. An inflated rubber balloon, after being rubbed on clothing, will stick to the ceiling. Metals and damp substances are unsuitable, because any electricity generated on them leaks away immediately.

During the course of the seventeenth century, people realized that electrified objects can repel as well as attract one another. One may easily perform the following experiment at home. Cut two pieces of cooking foil about one centimetre square. Glue each of them to the end of a piece of cotton and hang them up so they are next to each other. Rub a pen on wool and bring it up to the pieces of foil. As soon as the pen touches them they jump away, then left to themselves they hang a little apart. Now rub a sherry glass on the wool and move it near the foils. It will attract them (perhaps rather weakly).

Charles-François du Fay (1698–1739), superintendent of gardens to the king of France, discovered that electric charge made by rubbing resinous material (we would use plastic nowadays) attracts that made by rubbing glass. He inferred the existence of two kinds of electricity, which we now call "positive" and "negative". Glass rubbed with wool or cat's fur gets a positive charge. Plastics so rubbed get a negative charge. From observations like the ones described, we infer that like charges repel, but opposite charges attract.

Benjamin Franklin (1706–1790), scientist, inventor and statesman from Pennsylvania, did experiments with electricity, including making sparks pass from person to person, and he demonstrated that lightning was nothing but an electric spark. He thought that there was only one kind of electricity, and "positive electricity" was an excess of it, "negative electricity" a deficit. In some respects, it is a matter of words whether we adopt Franklin's view or alternatively think of positive and negative as two different types of electricity. But the latter way of thinking is probably better. The use of the words *positive* and *negative* reminds us of Franklin's idea, but it is really entirely arbitrary which we call positive and which negative.

The positive–negative usage conveniently allows us to state a law, which is now extremely well tested:

- The total quantity of electricity, taking account of whether it is positive or negative, does not change.

For example, if two objects with equal but opposite electric charges are brought together, the total electric charge is, and remains, zero.

We now know that some of the fundamental particles of matter, for instance, electrons and protons, have electric charge. Electrons have negative charge, and protons (some 2,000 times heavier) have exactly equal and opposite (i.e., positive) charge. Electrifying a rubber balloon by rubbing with a cloth works by displacing some electrons from the cloth onto the balloon (by a process I do not understand in detail). The balloon then sticks to the ceiling because it repels nearby electrons in the ceiling, driving them away, so that the consequent slight local positive excess in the ceiling attracts the balloon.

Although the production of electric charges by rubbing (and even just by contact) has been known for so long, it is still not very well understood (see Cross's book). The maximum number of extra electrons that can be attached to a surface is only a few for each million surface atoms, so the process depends delicately on fine details of the surface. Frictional electricity can be a nuisance, attracting dust, and a danger, causing explosions. Surface electric charges are used in devices such as photocopiers.

The next question is, What is the law of electric attraction and repulsion, analogous to Newton's law of gravitational attraction (see Section 1.7)? In 1767, Joseph Priestley (1733–1804, a dissenting minister who left England for Pennsylvania; he discovered soda water, oxygen and carbon monoxide) showed that, if a metal cup was charged, all the charge was stored on the outside, none on the inside. He wrote (quoting from Whittacker's book):

> May we not infer from this experiment that the attraction of electricity is subject to the same law with that of gravitation, and is therefore according to the squares of the distance; since it is easily demonstrated that were the earth in the form of a shell, a body in the inside of it would not be attracted to one side more than another?

Newton had indeed proved this property of the inverse-square law. A closely related theorem is demonstrated in Appendix A. Michael Faraday (who will appear prominently later in this chapter) used to give a spectacular popular demonstration in which he himself sat in a highly charged metal cage and came to no harm.

These are indirect demonstrations of the inverse-square law. It was also verified by direct comparisons of the forces between small charged spheres, by, for instance, the Frenchman Charles Augustin Coulomb (1736–1806), who is commemorated in the name of the law of electric force (Coulomb's law) and in the name of one unit of electric charge (a coulomb).

These properties of electric force are summarized in the equation (*Coulomb's law*)

$$\text{attractive force} = -(\text{a constant}) \times \left[ \frac{\text{charge} \times \text{charge}}{(\text{distance})^2} \right].$$

(Compare this with Newton's law of gravitation in Section 1.7.) Note that each of the two charges may be positive or negative. If they are both positive (or both negative), the force has a minus sign: that is, it is a repulsive force. If one charge is positive and the other negative, the minus sign gets cancelled out, and the force is attractive.

So far, we have not given a definition of the magnitude of an electric charge, so the charge terms in the equation remain vague. One possibility is to define *charge* so as to make Coulomb's law hold when the "a constant" in front is set equal to $\frac{1}{4\pi}$. (The reason for making the definition with this factor is just convenience; $4\pi$ is the area of a sphere of radius 1. If you do not put the factor in here, it crops up somewhere else.)

So, we define a unit electric charge to be such that, when put 1 metre away from an identical charge, it repels it with a force of $4\pi$ newtons. This is called the *electrostatic unit of charge*, and I will use it in this book. The official SI unit, the coulomb, is (for historical reasons) equal to about 33,547 electrostatic units.

It is interesting to note the differences between Coulomb's law of electric forces and Newton's law of gravitation.

- Electric charges may be positive or negative so electricity may attract or repel, but gravity always attracts.
- In electricity, like charges repel, but in gravity "like masses" (and all masses are "like") attract.
- The magnitudes of the charges in Coulomb's law have no other definition independently of that equation, but the "gravitational masses" in Newton's law of gravitation turn out to be the same as "inertial masses", which do have an independent definition (in Newton's third law of motion, Section 1.7).
- There is a real sense in which gravitation is very, very weak compared to electricity. The electrical repulsion between two protons is $10^{36}$ as great as the gravitational attraction. (To check this, one needs to know that the mass of the proton is about $10^{-27}$ kg and the charge is about $10^{-14}$ electrostatic units, as well as knowing the value of the gravitational constant.)

In view of this last remark, how is it that the effects of gravity (holding us on the floor, and so on) are so much more obvious than those of electricity? It is because most ordinary sized bits of matter are almost exactly electrically neutral: the numbers of protons and electrons are almost exactly equal. In fact the electrical repulsion makes it very hard to create an imbalance in these numbers. (Inside an atom the electrons manage to avoid falling onto the protons: quantum theory is needed to explain why this is so.) It is electrical forces that explain the structure of atoms, molecules, liquids and solids, and indeed electrical forces are behind all of chemistry.

I shall now explain some ideas, dating from the nineteenth century, that will be useful in the rest of this chapter.

First there is the concept of a *field*, particularly an *electric field*. Few words are used more often than *field* in modern physics, and with repeated use *field* calls to mind a concept with which physicists are very much at home. Yet it is not easy to say precisely what it means. I will begin by quoting a quite conservative definition given by Maxwell (in his *Treatise*):

> The Electric Field is the portion of space in the neighbourhood of
> electrified bodies, considered with reference to electric phenomena.

The reader may well feel dissatisfied with this legalistic-sounding definition.

It is probably more useful to say what we would have to do to find out about "the electric field", and what data would be needed to describe it. We would put a small "test charge" at a point in the space concerned and measure the force on it (the electric force), then repeat this at lots of other points. So the data would be a list of values of the force at lots of points. In principle, the description of the field entails knowledge of the force at "every point", but of course this is just a mathematical idealization. A physicist who claimed to know the electric field near an electrified pen, for example, would mean that it could be estimated (to some degree of accuracy) at any point where it was required. (Actually, the electric field is defined by dividing the force by the magnitude of the electric charge on the test charge.)

The electric field is conventionally denoted by the letter E. The bold print is used because a force is a vector (see Appendix B), having a direction as well as a magnitude associated with it. Since we need E at each point of space, and the position of a point in space may be denoted by a "position vector" x, we may write the field as $E(x)$. This just means that we have a sort of slot machine into which you insert a value of $x$, that is, the position of a point, and out comes a value of $E$, which in turn tells you the force on a test charge at the point $x$.

This rather abstract explanation of what is meant by a field will get fleshed out later on, when we talk about things like the "energy in the field".

The other thing I want to define is a *line of force* or a *field line*. This idea was much used by Faraday. It provides a good way of visualizing a field and making this abstract notion more concrete. Electric field lines are a collection of curves drawn in space, with the following properties:

- At any point in space, the direction of the line gives the direction of the electric field there (i.e., the direction of the force on a test charge). The lines also carry arrows to indicate which way they point.
- The strength of the electric field is proportional to the number of lines going through a unit area: that is, the strength is determined by how closely packed the lines are.
- Field lines do not terminate except on an electric charge. They flow out of a positive charge and into a negative one.

As explained in Appendix A, these rules are consistent because of the inverse-square law.

In order to try to sketch the field lines for a given array of charges, there is one further rule to be observed:

- If one imagines a small test charge being moved round in a closed loop, the total work done on the charge by the electric field should be zero.

This rule implies that any part of the loop along which the field is helping to push the charge should be balanced by another part where the field acts against the motion. The truth of this rule is the condition for potential energy to exist, as explained in Section 2.3 (a hypothetical counterexample is shown in Figure 2.1).

Figure 3.1 shows two examples of field lines. The first of these shows two equal but opposite electric charges. The field lines go from one to another (or go out of the picture). The second is for

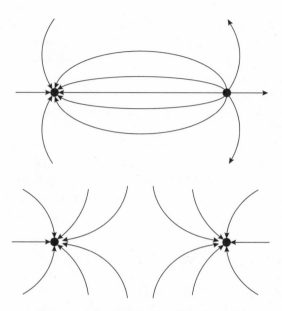

FIGURE 3.1   Two examples of field lines. The first is for two opposite charges and the second for two like charges.

two equal and like charges. All the field lines go off indefinitely. The
lines from the two charges act as if they repelled one another.

Figure 3.2 shows a set of field lines that is incorrect. This pur-
ports to show two equal and opposite charges, with all the lines
concentrated into a "tube", instead of being spread out as in the
first diagram in Figure 3.1. The lines shown are consistent with the
four listed rules, but inconsistent with the further rule about work

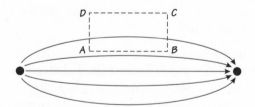

FIGURE 3.2   A wrong configuration of field lines for two opposite
charges, with all the lines concentrated into a "tube". A test charge
taken round the circuit *ABCD* has nonzero total work done on it
by the field.

done on a test charge. If the test charge is taken round the circuit *ABCD*, work is done on it along *AB* (where there is an electric field in the direction of *AB*). Along *BC* and *DA*, the motion is at right angles to any field, so no work is done. Along *CD* we have assumed there is no electric field, so again no work is done. Thus nothing balances the work done along *AB*, so this configuration is impossible. Arguments such as this are sufficient (though this is not obvious) to prove that the field lines look like Figure 3.1.

## 3.2  Magnets

Magnetic (the word comes from the town Magnesia) iron ore was known in antiquity. With it, an iron needle could be magnetized. The magnetic compass was used in the Middle Ages and had been known to the Chinese much earlier. It was natural for believers in astrology to attribute the action of the compass to the influence of the Pole Star. But Gilbert studied magnetism systematically and showed that the magnetic needle actually pointed to a region in the Earth. In fact, the Earth behaves as an enormous magnet.

At first sight, magnets can attract or repel, just as electric charges do. But there is an important difference. Consider first the behaviour of a magnet in the Earth's "magnetic field". The magnet as a whole is not attracted to either of the Earth's magnetic poles. The Earth does not cause the magnet to move along: what it does do is cause it to rotate, until one end points to the Earth's north magnetic pole. We can understand this if we assume that it is not the magnet as a whole that is analogous to an electric charge; rather the magnet has two opposite "poles" and each of these is analogous to an electric charge. One of the magnet's poles is attracted to the Earth's north pole, and the other is attracted to the Earth's south pole. This makes the magnet swing round until it is aligned but does not tend to make the centre of the magnet move (since the forces on the two poles of the magnet are in opposite directions).

What about the force between two nearby magnets? A particular case is illustrated in Figure 3.3. At first glance, it is not obvious whether the repulsive forces overcome the attractive ones, or vice versa. Let us assume the inverse-square law applies to magnetic poles. As an example, take each magnet to be one inch long, and

FIGURE 3.3   Two nearby magnets. The sum of the repulsions between the north poles and between the south poles is greater than the sum of the attractions between the unlike poles.

the gap between the magnets to be one inch. Then the two south poles are one inch apart, the two north poles are three inches apart and each north pole is two inches from a south pole in the other magnet; so the total repulsive force between the two magnets is proportional to

$$\frac{1}{1^2} + \frac{1}{3^2} - \frac{1}{2^2} - \frac{1}{2^2},$$

and this is positive. Whatever the dimensions, two magnets aligned as in Figure 3.3 repel each other.

Now suppose the distance between the magnets is large compared to their lengths. Then there is a lot of cancellation between the attractions and the repulsions. In fact the net force goes approximately as

$$\frac{(\text{magnet strength})^2}{(\text{separation})^4};$$

that is, it decreases with distance much faster than the inverse-square law.

Given that it is magnetic poles not whole magnets that have to be compared with electric charges, the analogy between magnetism and electricity becomes fairly close. We can draw "magnetic field lines" describing the "magnetic field", and we can use the same rules as we did in the last section. An example is shown in Figure 3.4. This figure shows the lines outside a magnet.

Magnetic field lines have a more direct physical interpretation than electric ones. They give the well-known pattern revealed by iron filings on a piece of card (placed on top of the magnet in the example in Figure 3.4). The filings fall into this pattern because each filing becomes magnetized in the field of the magnet and is then lined up in that field, just as a compass needle is lined up in

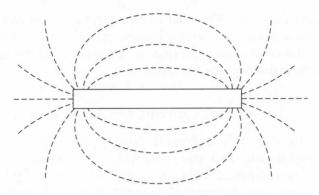

FIGURE 3.4    The magnetic field lines near a magnet.

the Earth's field. One may think of the dashes in the field lines in Figure 3.4 as depicting filings.

Is the analogy between electricity and magnetism complete then? It seems not. No one has ever found an *isolated* magnetic pole, whereas isolated electric charges are commonplace. One can make a very long thin magnet, and then its poles are a long way apart, but still they are joined up by a bar of magnetic material (say iron). What happens if you cut a magnet in half? You do *not* get two magnetic poles. Instead you get two smaller magnets: a new pair of opposite poles appears where the cut is made, so each half has a north and a south pole. (See Figure 3.5.)

Experiments have been done to look for isolated magnetic poles and none has been found to date.

The magnetism of iron (and a few other materials) is due to the individual electrons inside its atoms. Each of these is itself a tiny magnet (with a north and a south pole). In most materials, as many electrons point one way as another, so the magnetic effects of the electrons cancel out. But in iron, there are forces (of complicated origin) between certain of the electrons that tend to make them align

FIGURE 3.5    Cutting a magnet in half. Two new poles appear, so that two half-length magnets result.

in the same direction. What that direction is, may be determined by some outside magnetic field, for instance, the Earth's. Otherwise it is just a matter of chance. At high temperatures, the randomizing effect of heat destroys the magnetization.

## 3.3   Electric Currents and Magnetism

Towards the end of the eighteenth century the study of electricity and magnetism entered a new phase with the production of continuous electric currents. It began around 1780 when Luigi Galvani, an anatomist from Bologna, discovered that metals applied to a nerve in a newly dead frog could make its leg twitch. Galvani was interested in electricity, and he found that linking the frog's leg to a machine generating static electricity (by friction) had the same effect. From these accidental observations, it was deduced that electricity could be produced by metals in contact with moist materials. In 1800, Alessandro Volta, a professor in Pavia, discovered how to increase the effect by constructing a "pile" of repeated layers of copper, zinc and moist pasteboard: the first electric battery. A battery converts chemical energy into electrical energy.

The electricity made in a battery is no different from that generated by rubbing a piece of plastic, but a battery can produce a continuous current rather than the momentary discharge of static electricity. For this reason, batteries made possible many new experiments.

For many years people had suspected some sort of connection between electricity and magnetism. There were rumours that lightning strikes had magnetized iron nails. By 1820, Hans Christian Oersted, a professor in Copenhagen, had shown that, if a compass needle was placed near electric current in a wire, with the axis of the compass parallel to the wire, the needle turned so as to be transverse to the wire. Evidently, the electric current was producing a magnetic field, transverse to the wire. An electric current is nothing but moving electric charges (for example, the moving electrons in a metal wire). So moving electric charges produce magnetism. This is the first step in the unification of electricity and magnetism. Further experiments showed that the magnetic field decreased inversely as

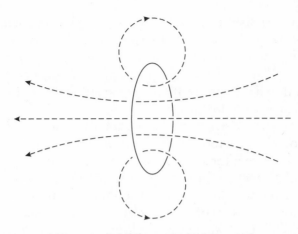

FIGURE 3.6   Magnetic field lines (dashed lines) due to a current circulating in a wire loop (continuous line).

the distance from the wire: that is, doubling the distance halved the field, and so on.

Figure 3.6 shows, as an example, magnetic field lines due to a circular current-carrying wire. There might be a battery (not shown in the figure) attached to the wire loop to produce the current.

The magnetic field of the Earth is due to circulating electric currents in the molten iron core, so Figure 3.6 can be regarded as a very rough description of the Earth's magnetic field, both outside and inside the Earth's surface (with the north pole to the right of the centre). Note that the field lines are closed loops (or else go out of the diagram).

What would the field lines look like inside a magnet: that is, how could we complete Figure 3.4 inside the magnet? Since the magnetism is due to electrons, each of which is a tiny magnet, one could draw the field lines inside the magnet with each line beginning and ending on an electron. Electrons of course have electric charge, and they are also known to be spinning on an axis. If the charge is spread out, the spin means that there are circulating electric currents. One way of thinking of the electron's magnetism is as being due to these circulating currents, again roughly like Figure 3.6. If we adopt this point of view, the field lines pass through inside each electron, a little like the lines in Figure 3.6. Then the field lines in Figure 3.4 would be completed inside the magnet to form closed loops. This description should be taken with a small pinch of salt. As I will explain in Chapter 8, quantum theory shows that statements about what happens at the very small scale of electron size cannot be taken too literally.

Since an electric current produces a force on the poles of a magnet, it follows by Newton's law of action and reaction that a current should experience a force in a magnetic field. Since the force on a pole of the magnet is at right angles to the direction of the wire, the force on the wire must also be at right angles to it. A current is nothing more than a lot of moving electric charges (electrons in the case of a metal wire), so a moving electric charge must experience a force in a magnetic field. This fact gives one way to define the strength of a magnetic field. I will state the definition in a special case (naming it, as is customary, after the great Dutch physicist H. A. Lorentz):

- If a charged particle is moving at right angles to the magnetic field, the force on the particle is given by (field) × (charge) × (speed).   (Lorentz force)

Of course, in addition to being a definition, this rule makes some assumptions: that the force is proportional to the charge and to its speed. This is plausible, because the current carried by charges increases with their speed. If the charge is not moving at right angles to the magnetic field, the force is less, reduced by a factor given by the sine of the angle between the velocity and the field. In particular, a charge moving exactly in the direction of the magnetic field experiences no force.

The direction of the force has not been stated. It is at right angles both to the field and to the velocity of the charge. The sign of the force has also to be stated. It is such that, if one looks along the force, the rotation from the velocity towards the magnetic field appears to be clockwise.

There is a very important property of this force exerted by a magnetic field on a moving charge: it does no work. This is because the force is at right angles to the motion, and such a force can do no work (compare Figure 2.2). For example, if a marble is rolling in a smooth bowl, the bowl certainly exerts a force on the marble, but it does no work on the marble. (By contrast, gravity does work on the marble and so changes its speed.)

(In parenthesis, I remark that magnetic field *could* have been defined like electric field, by starting with the force between two magnetic poles. This is not done in practice, partly because isolated magnetic poles probably do not exist.)

As electric currents produce magnetic fields and experience forces in magnetic fields, it follows that two current-carrying wires should in general exert forces on each other. Very shortly after Oersted's discovery, this expectation was confirmed by André-Marie Ampère. He demonstrated that two long parallel wires carrying currents are attracted or repelled according to whether the currents are in the same or opposite directions to each other. The force per unit length on one of the wires decreases inversely as the distance between the wires.

The magnetic field is transverse to the wires. The force on one of the wires is at right angles both to the wire and to the field and is therefore directed towards the other wire, as illustrated in Figure 3.7.

The magnetic effects of electric currents are of course the basis of the operation of electric motors, electric bells, and so on. In 1823, soon after the discoveries of Oersted and Ampère, the electric telegraph, using the same principles, began to be developed.

Ampère went on to construct a complete mathematical description of the forces between currents. He did it by imagining the

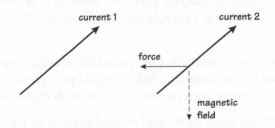

FIGURE 3.7 Two parallel electric currents in directions indicated by the large arrows. The magnetic field due to current 1 at current 2 is shown by the dashed arrow. The direction of the force on current 2 is also shown.

wires to be made up of a lot of very short segments and writing down the force between two such small segments. The force for ordinary lengths of wire could then be deduced by adding the forces on the separate segments. Ampère's rule for the force between two short segments is quite complicated, depending upon the directions of each segment relative to the line joining them. But it does have one simple feature: it decreases as the square of the distance between the segments. In this respect, it is like Newton's law of gravity and Coulomb's law (Section 3.1). In fact Ampère's work was in the Newtonian spirit, being content to have a mathematical description of the forces and not asking for any mechanism in the intervening space. Forces regarded in this way are described by the phrase *action-at-a-distance*. Much more work was done in this spirit during the nineteenth century, especially by German mathematicians.

But, here, I will continue to give the field lines point of view, partly because it is simpler to explain without mathematics and partly because it inspired the great discoveries of Faraday and Maxwell later in the century. Our first question then is, What is the shape of the magnetic field lines near a straight wire carrying an electric current? Since their direction is everywhere transverse to the wire, the lines must go round the wire in circles (with the wire going through the centres of the circles). How close together are the lines? Since the force decreases inversely as the distance, the spacing must increase proportionally to the distance from the wire, that is, proportionally to the radii of the circles. Such field lines are sketched in Figure 3.8.

For any system of electric currents, there is a general rule that controls the field lines. This rule may be stated as follows (Ampère's law):

- Take a unit magnetic pole round any closed loop. The work done by the magnetic field is equal to a constant times the total electric current threading through that loop.

Several things need to be said in explanation of this rule. First, it should be compared with the corresponding rule for an electric field, stated in Section 3.1. Second, magnetic poles probably do not exist, so the pole mentioned should be thought of as an imaginary one, invented in order to formulate the rule. (Alternatively, one

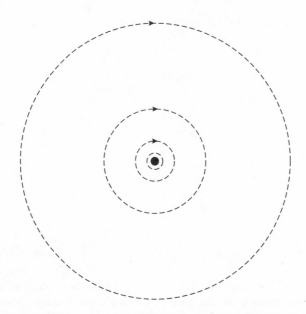

FIGURE 3.8   The magnetic field lines (circles) due to a straight current-carrying wire. The diagram shows a cross section; the wire is the black dot in the centre. The current is taken to be directed away from the reader, into the paper. The spacing between the field lines decreases with their radii.

could use one pole of a very long thin magnet, whose other pole was held a long way away.) Third, the rule allows for a general configuration of electric currents, in straight or curved, narrow or thick wires. The important thing is that only those wires count that pass *through* the loop. Wires that do not are irrelevant. There is also a rule to say what counts positively and what negatively. To state this rule, imagine you are in the wire looking in the direction in which the current is flowing. If the loop round which the pole is carried appears to you to be clockwise, then that current counts positively. In the opposite case, the current counts negatively (that is, it is to be subtracted rather than added). This rule is illustrated in Figure 3.9.

I have already mentioned the work done in carrying particles around closed loops. In Section 2.3, I said that this work should be zero in order that potential energy could be defined (and

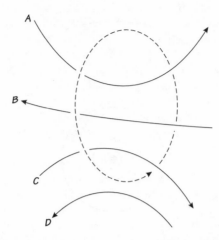

FIGURE 3.9   An illustration of the rule about magnetic fields produced by electric currents. The dashed line represents the closed loop
around which a magnetic pole is carried. The continuous lines represent four current-carrying wires. Currents *A* and *C* count positively,
current *B* negatively. Current *D* does not count at all.

conservation of energy maintained). (See Figure 2.1.) In Section 3.1,
I said that the work was indeed zero for an electric charge carried
round in an electric field. Now, for magnetic poles, we have a case
when the work is not zero. How does this square with conservation
of energy? A short answer would be to say that magnetic poles don't
exist, so it is a hypothetical question anyway. But I think this is not
the whole story. Even if one supposes that magnetic poles do exist,
energy is still be conserved, in spite of there being no way to define
potential energy. The energy of the motion of the electric charges
in the currents has to be included. Taking the magnetic pole round
a loop causes an electric field that alters the current (unless there is
some external source of energy, like a battery). Thus the energy of
the magnetic pole *and* the electric charges is constant.

Finally, I mention a very important thing: the "constant" in
Ampère's law. This constant is not something we can choose at
will. Magnetic field and electric current have already been defined
(by Coulomb's and Lorentz's laws, respectively). There is no freedom of choice left. So this constant is something to be measured,

and we don't know what answer we shall get until the experiments are done. We can, however, work out what dimensions the constant must have. This is done in Appendix D, and the answer is that the dimensions are 1 over the square of a speed. Call the constant $\frac{1}{c^2}$, where $c$ has the dimensions of a speed. Measurement gives

$$c = 2.99792458 \times 10^8 \text{ metres per second}$$

very nearly. (But see Appendix D for the modern status of this number.)

Let me stress that, in principle at any rate, the determination of $c$ depends on just two sorts of measurement: fixing a standard charge by the force between two charges (from which current strength can be defined) and measuring the force on a moving charge in a magnetic field. In Chapter 4, the quantity $c$ will gain a completely new meaning.

It may be surprising that such a large number as $3 \times 10^8$ should crop up in the measurements of everyday things like currents and charges. The following example illustrates the sort of thing that happens. Take two parallel wires, one centimetre apart, each carrying a current of 1 ampere (using SI units, for the moment). The magnetic force on either of them is

$$2 \times 10^{-7} \text{ newton per centimetre.}$$

(By comparison, it takes a force of about 10 newton to lift a 1-kilogram weight.) Now consider the total electric charge that passes when a current of one amp flows for one second. Imagine that we could take two such charges and place them one centimetre apart. The electric force between them would be about

$$9 \times 10^{13} \text{ newton.}$$

We see that the ratio of the last two numbers is very large (in fact it is $\frac{1}{2}c^2$ with $c$ in centimetres per second).

Incidentally, the magnetic field one centimetre from a wire carrying one ampere is comparable in strength to the Earth's magnetic field. What about the electric charge mentioned earlier – the total charge passing in one second in a current of one ampere? To put this (positive) charge on 1 kilogram of copper would mean roughly removing one electron for every million atoms of copper. From the last paragraph, it follows that carrying such a large concentration of charge on a small object is totally unrealistic, because of the huge forces that would blow it to pieces.

## 3.4   Faraday and Induction of Electricity by Magnetism

The Royal Institution of Great Britain was founded in 1799, following a proposal by Rumford (he of the cannon-boring experiments on heat; see Section 2.4). Humphry Davy was the first professor of chemistry at the Institution. Michael Faraday, attracted by Davy's public lectures, applied to him for a post and became an assistant in 1813. On Davy's death in 1829, Faraday succeeded him as director. His lectures were popular in Victorian London, and he started the Christmas lectures at the Royal Institution, which are given to this day (and televised).

Faraday had been a bookbinder's assistant. He taught himself science and never mastered much mathematics. He was an experimenter of unsurpassed genius, and he had a wonderful scientific intuition. There are some great scientists, Newton for one, for whom it is difficult to feel much human warmth. Faraday, from what one reads, was a man to be admired for his goodness as much as for his genius.

One idea that motivated Faraday was that there should be a unity in the physical forces, notably electricity and magnetism. (He would have liked to include gravity too, but this was not to be.) People knew that electric currents produced magnetism, so Faraday wondered whether magnetism could produce electricity. Around 1831 he made the crucial discovery that a *changing* magnetic field produces a flow of electric current in a conductor. He saw that the essential thing was that the magnetic field lines and the conductor should move relative to one another. Here are three examples.

Move a magnet near a wire. Since the magnet moves, its magnetic field lines change, and a current is "induced" in the wire. Move a wire in the magnetic field of a fixed magnet. The wire moves relative to the field lines, and again a current is induced in it. Change the current in one wire. The associated magnetic field changes. If another wire is near, a current is induced in it. These effects underlie the operation of dynamos and transformers.

There is an effect (called *self-induction*) even with a single electric wire (wound into a coil, let us say, since this makes the effect more marked). Suppose the current in it is changed. Then the associated magnetic field changes. According to Faraday, this produces an electric field, which itself tends to change the current in the wire. I have not said anything yet about the direction of this effect, but in fact for self-induction the tendency is to counteract a change in current. An increase or decrease in current that would have been sudden is slowed down by self-induction. This property of a coil is used to protect electrical devices from too sudden changes.

In brief, then, Faraday's discovery was that a magnetic field changing with time produces an electric field (which in turn can cause current to flow in a conductor). I will now give a more precise formulation of Faraday's law of induction. First it is necessary to define *magnetic flux* through a surface. Take any surface like a disc or a hemisphere (an "open" surface with a boundary, not a "closed" surface like the whole of a sphere). Then, if it is in a magnetic field, the flux through it is defined simply to be the number of field lines that go through it. Here I am assuming that the field lines are so defined that the number of field lines crossing a unit area (which is at right angles to the field) gives the strength of the magnetic field.

There is a very important property of flux defined in this way: the flux is the same for any two surfaces that share the same boundary rim. The reason for this is that magnetic field lines have no ends (since I am assuming that isolated magnetic poles do not exist), so if a line of force goes through one surface with a given rim it must go through another. An example is shown in Figure 3.10.

I can now state *Faraday's law*:

- The rate of change with time of the magnetic flux through a surface is equal to minus the work done in carrying unit electric charge around the rim of that surface.

Here are some comments:-

- The rule makes sense because of the property just stated: we can choose *any* surface with the given rim.

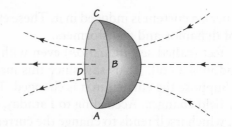

**FIGURE 3.10**   Two surfaces, a circular disc and a hemisphere, with the same rim, the circle *ABCD*. Any line of force that goes through one surface goes through the other. This diagram also illustrates the convention connecting the positive direction for field lines to the direction round the rim (the direction defined by the order *ABCD*).

- If some work is done carrying an electric charge round the rim, there must be an electric field present: the field "induced" by the changing flux.
- The "minus" sign in Faraday's law means something only if we have some convention about the direction in which the charge is carried round the loop. The convention is similar to one used in the last section. Place a clock on the surface and count magnetic field lines as positive if they go in the direction in which the clock faces. Then the electric charge is to be carried round clockwise as defined by this clock. Notice that a rule like this, requiring one to know what "clockwise" means, has appeared three times: twice in the previous section and now once again. Each time, a magnetic field is involved.

Faraday's rule begins to unite electricity with magnetism. It also begins to give the *dynamics* of electromagnetism, because it tells us a time rate of change (of the magnetic field), just as Newton's gives the dynamics of motion of matter because it tells us the rate of change of velocity.

Faraday thought of field lines as having physical reality. He thought of them as being in tension, and also repelling one another. With a little bit of imagination, one can see roughly how the patterns in Figure 3.1, for example, can be understood as a consequence of balancing these two properties. Faraday's outlook was in complete contrast to the action-at-a-distance methods that had triumphed from Newton to Ampère. Perhaps one of the factors that influenced Faraday's way of thinking was his lack of mathematical training. Ampère's action-at-a-distance equations cannot be understood in such an intuitive way as field lines. If this is so, Faraday's comparative mathematical ignorance may have aided the progress of science.

Faraday was as famous as a chemist as for his work on electricity and magnetism. He was in demand as a consultant, on such matters as the possible use of sulphur dioxide as a poison gas in the Crimean War and the protection of the pictures in the National Gallery against pollution.

## 3.5   Maxwell's Synthesis: Electromagnetism

James Clerk Maxwell (1831–1879) was a Scotsman, educated at Edinburgh and Cambridge. He was a professor at, successively, Aberdeen, King's College, London, and Cambridge (where he was the first head of the Cavendish Laboratory).

Maxwell was certainly not deficient in mathematics, but he had a great respect for Faraday. Here are Maxwell's own words on the subject (from the preface to the first edition of Maxwell's *Treatise*).

> Before I began to study electricity, I resolved to read no mathematics on the subject till I had first read through Faraday's *Experimental Researches in Electricity*. I was aware that there was supposed to be a difference between Faraday's way of conceiving phenomena and that of the mathematicians, so that neither he nor they were satisfied with each other's language. I had also the conviction that this discrepancy did not arise from either party being wrong. I was first convinced of this by Sir William Thomson. . . .
>
> As I proceeded with the study of Faraday, I perceived that his method of conceiving the phenomena was also a mathematical one, though not exhibited in the conventional form of mathematical symbols. I also found that these methods were capable of being expressed in ordinary mathematical forms, and thus compared with those of the professional mathematicians.

For instance, Faraday, in his mind's eye, saw field lines traversing all space where the mathematicians saw centres of force attracting at a distance: Faraday saw a medium where there was nothing but distance: Faraday sought the seat of the phenomena in real actions going on in the medium.

(The "mathematicians" Maxwell refers to were mostly German.)

Maxwell imagined a mechanical model of electric and magnetic fields that satisfied the laws that Faraday had discovered. Here is a little of Maxwell's description of it (from a letter to W. Thomson in 1861, quoted in Whittaker's book):

> I suppose that the "magnetic medium" is divided into small portions or cells, the divisions or cell walls being composed of a single stratum of spherical particles, these particles being "electricity". The substance of the cells I suppose to be highly elastic, both with respect to compression and distortion; and I suppose the connection between the cells and the particles in the cell walls to be such that there is perfect rolling without slipping.

(This "rolling" is identified with the magnetic field.) We need not bother to try to understand Maxwell's scheme: it has only historical interest. It was a return (temporarily for Maxwell) to mechanical model building, slightly reminiscent of Descartes's vortex model of gravity (see Section 1.6). However, this model did lead Maxwell to discover how to complete the rules of electromagnetism. He found that his model was consistent only if a "displacement current" was added to the ordinary electric current. The word *displacement* (which is still used in this context) is a reminder of the historical origin in a mechanical model.

When Maxwell came, in the early 1970s, to write his *Treatise*, he did not mention the mechanical model, but gave only the equations. In some respects, this was a return to the tone of Newton's *Principia*, being content with equations and not seeking to imagine a mechanism. There is an important difference though: Newton dealt only with the motion of bodies; Maxwell included electromagnetic *fields* and their changes with time.

The search for mechanical models, whether as teaching aids or analogies or attempts to find the real structure of the aether, continued after Maxwell's death. People eventually found that this was not a useful strategy. As Poynting wrote (quoted in Hunt's book):

> [Such explanations] are solely of value as a scaffolding enabling us to build up a permanent structure of facts. . . . And inasmuch as we may at any time have to replace the old scaffolding by new, more suitable for new parts of the building, it is a mistake to make the scaffolding too solid.

It is easy for us to see with hindsight that Poynting was right. But at much the same time there was doubt about how literally to take the molecular theory of gases (see Section 2.5). In *this* case "the ridiculously simple hypothesis that a gas consists of an immense number of small particles in motion" (Heaviside's words, quoted in Hunt's book) did prove to be the literal truth.

I will now explain what Maxwell added to Faraday's rules, as described in the last section. We will then be in possession of a complete set of rules, equivalent to "Maxwell's equations", as the equations of electromagnetism are universally called.

We start from Ampère's law, which connects the work done taking an electric charge round a loop to the current flowing through that loop. This law makes sense provided that the "current flowing through the loop" is unambiguous. We can choose a surface of which the loop is the rim and find the total current flowing through that surface. But does it matter what surface we choose? Take the example in Figure 3.11. Here I have shown two different surfaces,

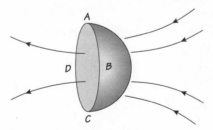

FIGURE 3.11 An example of two surfaces, one a disc and one a hemisphere, which both have the loop *ABCD* as rim. More current is flowing through the hemisphere than the disc. This is because, in the region between them, the quantity of electric charge is increasing with time. (This diagram is similar to Figure 3.10, but that illustrates that magnetic flux *is* the same through the two surfaces, whereas this figure shows that total current need *not* be the same.)

one a hemisphere and the other a disc, with the same rim *ABCD*. Does the current flowing through the hemisphere have to be the same as through the disc? Suppose more current is flowing through the hemisphere than the disc, as indicated in the diagram. Then the amount of electric charge in the region between must be increasing. This would not be the case if we were assuming everything to be unchanging with time. But if we want to allow time changes, then there is nothing to rule out the situation in the diagram. Then Ampère's law makes no sense.

What Maxwell saw was how to fix up the rule. We want to add something to the current so as to get a new quantity that *is* independent of which surface we choose. If the charge in between is increasing (as would be the case in Figure 3.11) then the number of electric field lines coming out must be increasing. This suggests the following. Work out the electric flux (that is, the number of electric field lines) through one of the surfaces. Work out how fast this is increasing with time. Add the result to the current through that surface. The total *is* independent of the surface chosen. For the example in Figure 3.11, the rate of change of electric flux through the disc must be different from that through the hemisphere, because the electric charge in between is increasing.

We are now in a position to state *Maxwell's laws* of electromagnetism. The first is the new one; the other three recall rules we have already met.

(i) The work done taking a unit magnetic pole round any closed loop is equal to the current (times $\frac{1}{c^2}$) plus the rate of increase of the electric flux through that loop (times $\frac{1}{c^2}$). (Remember that $c$ is a constant of nature, with the dimensions of a speed, whose value was given in the previous section.)

(ii) (Faraday's law of induction): The work done taking a unit electric charge round any closed loop, times $-1$, is equal to the rate of increase of the magnetic flux through that loop.

(iii) Magnetic fields may be represented by field lines that have no ends.

(iv) Electric fields may be represented by field lines that end only on electric charges, and the number of field lines coming out of a charge is proportional to the magnitude of that charge.

Here are some comments on these rules.

- The first two each hold for *any* closed loop. It is this generality that makes the rules powerful enough to control the fields. It is actually sufficient to take the rules just for very small loops, because big loops can be built up by stitching together lots of small ones. In this case, the rules become what are usually called "Maxwell's equations".

- The first two rules have somewhat similar forms. Take the special case in which there are no electric charges and no currents. (Even then there can be interesting fields, as we shall see in the next chapter.) In this special case, if we replace the electric field by $c$ times the magnetic field, and replace the magnetic field by $-1/c$ times the electric field, then the first two rules are interchanged with each other.

- The first two rules are *dynamical*: they tell us about changes with time. In this respect, they are like Newton's laws of motion, which tell us about the rate of change of velocities in terms of the gravitational force, this force in turn being determined by the positions of the bodies concerned. In fact, if we think of electric field as being *analogous* (and it is no more than an analogy) to velocity, and magnetic field as analogous to position, then the second rule gives rate of change of electric field ("velocity") in terms of magnetic field ("position"). In the first rule, it is the other way about.

  Suppose there is a system of moving electric charges with associated electric and magnetic fields. Newton's equations give the rates of change of the velocities, and Maxwell's equations give the rates of change of the fields. These equations are sufficient to determine everything at all times, provided we have sufficient information at some starting time ("initial time"). Sufficient information would be all the positions, all the velocities and all the electric and magnetic fields everywhere.

- The last sentence points up a vital difference between fields and particle motions. Given a number of particles, let us say seven, there is a limited amount of information needed to describe their motion at any given time: seven positions and seven velocities. But with fields, we need their values at *every* point of space – an unlimited amount of information. In practice, we might be content to know the average values of the fields over little regions of size, say, one millimetre, throughout a total region of, say, size one metre. But, if more accuracy were required, we would have to take a finer mesh of little regions.

So electromagnetic fields vary with time in a dynamical way just as the positions and speeds of particles do. What happens to conservation of energy? It turns out that energy can be assigned to the fields in such a way that the total energy of particles *and* fields does not change with time. What is more, the field energy can be thought of as distributed over space, as the fields are. The rule for field energy is this:

- Take a little region of space. Work out the quantity

$$\frac{1}{2}[(\text{electric field})^2 + c^2(\text{magnetic field})^2]$$

where the fields are evaluated at that region of space. Multiply this by the volume of the little region. Then this is the amount of electromagnetic energy assigned to the little region. The total energy in the fields is found by adding up contributions like this for all regions of space.

There is an analogy (only an analogy) with particle energy. The preceding electric term is analogous to kinetic energy. Similarly, the magnetic term is analogous to the potential energy. Here also the field appears squared. This is in fact like the gravitational potential energy of a pendulum, which is proportional to the square of the angle to the vertical (as long as that angle is small).

As well as energy, we would like to be able to assign momentum to the electromagnetic field so as to make the total momentum

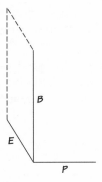

FIGURE 3.12   How to find the momentum in an electromagnetic field. The lines $E$ and $B$ represent the electric and magnetic field strengths at a little region of space, and the parallelogram is drawn having these lines as two of its sides. The magnitude of the momentum is the volume of the little region of space times $c^2$ times the area of this parallelogram. The direction is at right angles to $E$ and $B$, as shown by the line $P$.

constant in time. The rule for so doing is illustrated in Figure 3.12. This rule was discovered by John Henry Poynting in 1884, five years after Maxwell's death.

Earlier in this chapter, I remarked upon the similarity between Coulomb's inverse-square law of electrostatic forces and Newton's law of gravitation. But by the end of the chapter electromagnetism appears very different from action-at-a-distance gravity, with the electromagnetic fields operating in the space between the charged particles. The difference occurred at the point where we allowed positions and fields to vary with time. What, then, happens to gravity when fast time variations are included? The answer to this question is left to Chapter 7.

## 3.6   Conclusion

In the present chapter, I have explained how electricity and magnetism are unified. This does not mean that they are the same thing. It means that they are different aspects of something more general: electromagnetism. In all the principles we have encountered, the electric and magnetic fields appear inextricably bound up with one another. This is a pattern of "unification" that we will encounter again in later chapters.

Maxwell's equations carry the seed of Einstein's unification of space and time (see Chapter 5) and also provide the model, when suitably modified, for the modern theory of nuclear forces (Chapters 11 and 12).

# 4

# LIGHT

*How light is a wave-like electromagnetic field.*

## 4.1 Waves

This chapter is about the nature of light. Section 4.8 describes one of the great unifications of physics: the demonstration that light is just part of electricity and magnetism.

The first thing to be explained is the wave nature of light, so I begin by saying what is meant by a wave. We are all familiar with water waves, but I will define a wave in a general way.

In a wave, a shape propagates over a long distance, but matter (or whatever the wave is "in") moves only locally. For example, if a stone is dropped into the middle of a sizeable pond, waves may be propagated to the edge of the pond. But the actual water is only moving locally. For example, the surface goes up and down.

We need to define one or two terms. The simplest sort of wave is what is called a *simple harmonic wave* (the name comes from the connection with musical notes). Here the shape is like that of a corrugated surface. To define it mathematically, we can imagine doing the following. (See Figure 4.1.)

Take a wheel with a peg on its side. Take a vertical pen with a slot in its stem and with the peg in the slot. Let the wheel rotate at a steady rate, so that the pen is moved from side to side (remaining always vertical). Let the nib of the pen rest on a long sheet of paper (under the wheel), which is being pulled at a constant speed in the direction of the wheel's axis. Then the curve traced out on this paper has the form of a simple harmonic wave. Some readers may know it as a "sine curve".

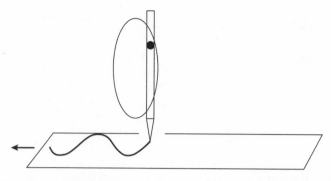

**FIGURE 4.1**  Tracing out a waveform. The point of a pen rests on a horizontal strip of paper, which is moving steadily to the left. A rotating wheel, with a peg engaging into a slot, causes the pen to move from side to side.

We can use Figure 4.1 to define some important terms. The speed with which the curve is moving from left to right on the paper, which is the same as the speed with which the paper is being pulled along, is called the *speed* of the wave. The time for the wheel to rotate all the way round once is called the *period*. The distance the paper moves while the wheel goes all the way round once is called the *wavelength*. One over the period is called the *frequency*. From these definitions, we see some relations:

$$\text{wavelength} = \text{speed} \times \text{period},$$

$$\text{speed} = \text{frequency} \times \text{wavelength}.$$

At any given position, the period is the time for a complete swing from side to side and back again. After one period, the wave repeats itself. At any given time, the wavelength is the distance for a complete swing from side to side and back again.

The radius of the wheel is called the *amplitude* of the wave. The swing between a "crest" and a "trough" is twice the amplitude.

There is a very important property, which many types of wave have to a good approximation, called *superposition*. This property is that, if we have two possible wave motions, we can construct another possible wave motion by just adding together the amplitudes of the first two. The two waves carry on independently of each other, but the total amplitude (at any given place and time) is got by adding together the two components.

For some types of wave, the superposition property holds, and the speed is independent of the wavelength (or at least these assertions are true to good approximation). When this is so, things are very simple. One can superpose several simple harmonic waves, and thus build up a waveform of a more complicated shape. Because the components all move with the same speed, this more complicated shape moves along without changing.

A simple example of a wave is got by taking a stretched rope (preferably a slightly elastic one) and wiggling one end up and down. The speed is determined by two factors: the tension with which the rope is stretched and the mass per unit length. To a good approximation, this example possesses the superposition property. For instance, if two people simultaneously begin wiggling opposite ends of the rope, two waves moving in opposite directions just pass through one another. While they are in the process of passing through, the shape of the rope looks more complicated than it would with either wave separately.

Note that the rope moves just up and down (or from side to side). It does not move in the direction of the wave (that is, along the rope).

Waves on the surface of water are easy to see, but their properties are rather complicated. The speed in general does depend on the wavelength. For example, take waves on a lake when the wavelength is more than a few centimetres and a lot less than the depth of the lake. For such waves, the speed increases with the wavelength. When the wash of a boat reaches the shore, long waves arrive before short ones.

In surface waves, the motion of a molecule of the water near the surface is approximately circular, rotating so that it moves in the direction of the wave at the surface and back in the opposite direction beneath the surface.

When a large stone stands in a fast-flowing river, wave-like shapes are visible on the surface, but they are not moving relative to the stone (that is, relative to the observer on the bank). They *are* moving relative to the water in the river. Their wavelengths get adjusted so their speed relative to the river is just equal to the speed of the river, with the result that they stay with the stone.

## 4.2 Sound

It is not hard to see that sound is a type of wave motion, or at least that it involves vibrations. For loud, low notes the vibration can be felt. One can see the vibration of a string that produces sound. The connection between musical pitch and the length of a vibrating string was studied in antiquity, reputedly by Pythagoras (c. 560–480 B.C.).

For sound in a gas or liquid, the quantity that varies in a wave-like way is the pressure. This was known by the first century A.D. By the seventeenth century, the connection between pitch and frequency was understood, and the first measurements of the speed of sound had been made. Newton calculated the speed (in terms of the pressure and density of the air) but did not get it quite right. His calculation was corrected by Laplace in the early nineteenth century, by allowing for the fact that temperature, as well as density and pressure, varies along a sound wave.

Sound waves in air do (to a good approximation) satisfy the two properties mentioned in the last section: superposition and the independence of speed on wavelength.

The speed of sound in air is about 340 metres per second. The frequency of the A above middle C (a′) is now defined to be 440 per second, so the corresponding wavelength in air is about 77 centimetres. For each rise of one octave, the frequency is doubled (or the wavelength halved).

The superposition property is crucial to music. The movement of air caused by a string quartet is no doubt very complicated. But it consists of a superposition of the simpler vibrations caused by each string of each instrument (each of which in turn is made up of a fundamental and several harmonics). Each of these simple waves is superposed in the vibrations of the air and of the listener's eardrum. The human ear has the marvelous ability to analyze the complicated vibration into its component frequencies (to some extent, at least).

Sound provides an example of a general phenomenon associated with wave motion: *beats*. Suppose the A above middle C on a piano is ill tuned, so that one of the strings vibrates at 441 per second and another correctly at 440. Suppose at some instant the waves from both strings produce a maximum of pressure, so the loudness is maximal. Half a second later the second string has lagged behind the first by half a period, so the second produces a maximum of pressure when the first produces a minimum. The total pressure then is just average, as if there was no sound at all – that is, the loudness is a minimum. Thus moments of maximum loudness occur at one-second intervals, alternating with moments of quiet. Beats like this are unpleasant to the ear, at least if their frequency is low. The simplest theory of consonance is that two musical sounds are consonant when there are few (slow) beats.

Beats are an example of the more general phenomenon of *interference*: the superposition of two waves to produce variations of intensity, either in space or in time.

In a sound wave in air, the motion of molecules of the air is backwards and forwards in the direction of the propagation of the wave. Where the pressure is a maximum, the molecules move out, that is, forwards in front of that region and backwards behind. A wave in which the motion is, like this, in the direction of propagation is called *longitudinal*.

In solids, different kinds of sound waves are possible. There are longitudinal, compression waves, just as in a gas or liquid. But in addition there can be waves in which the solid vibrates from side to side, being displaced alternately to one side and then the other half a wavelength along the wave. The elasticity of the solid is necessary to make it spring back from a displaced position. (A fluid displaced in this way would just stay displaced.) Such waves are called *transverse*. For a wave moving in a given direction, the transverse vibrations can take place in any direction at right angles to the direction of the wave. The direction of vibration is called the direction of *polarization* of a transverse wave.

Sound waves traveling through the Earth, produced by an earthquake, for example, are called seismic waves. The longitudinal waves travel faster than the transverse ones. The centre of the Earth is made of liquid metal, and the transverse waves cannot go through it.

## 4.3   Light

The following properties of light are all more or less obvious:

- In a uniform medium, like air, light seems to travels in straight lines. We cannot see around corners. Straight "rays" of sunlight may be visible slanting through a hole in the clouds on a rainy day.
- Light is *reflected* from smooth surfaces, as if it bounced off (and like sound echoing off a large wall).
- When passing from one transparent medium to another, light is generally *refracted*, that is, bent. Refraction causes a pencil partly immersed in water to look bent.
- Light can be white or it can be coloured.

Any theory of the nature of light must at least explain these facts.

The third property, refraction, is the basis of the operation of lenses, in, for example, spectacles, telescopes and microscopes (and within the eye itself). Spectacles are thought to have appeared in the thirteenth century. The telescope and microscope date from the start of the seventeenth century (as I mentioned in Section 1.1).

From antiquity on, detailed laws about light have been expressed in terms of the concept of a *ray*. We may define a ray as a very thin beam of light, produced, for example by letting the light from a small source pass though a very small hole (which is not too near to the source). The assumption is that such a pencil of light will remain very narrow (at least for a reasonable distance) and could in principle be made as thin as we liked by making the hole and the source smaller. A broad beam of light could be regarded as made up of a lot of "rays".

In terms of rays, the law of reflection says that the reflected ray makes the same angle with the mirror as does the incident ray. This is illustrated in Figure 4.2.

There is a law governing refraction, which was found by Willebrord Snel (1580–1626) of Leyden and by Descartes. This law is explained in Figure 4.3. Here there are meant to be two transparent media with an interface between them. For example, there might be air above the surface and glass below. Suppose a ray of light $I$ comes through the air and hits the surface of the

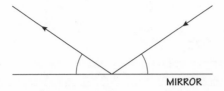

FIGURE 4.2    The law of reflection. The two angles indicated are equal.

glass at $P$, then is refracted as the ray $R$. The ray $I$ might meet the surface at any angle, but the law states there is something that is independent of that angle and depends only on the two media (air and glass). To construct this something, draw a circle with its centre at $P$, and let it intercept the incident and reflected rays at $A$ and $B$. Then find two points $A'$ and $B'$ on the surface so that the lines $AA'$ and $BB'$ are each at right angles to the surface. The law states that

$$\frac{PA'}{PB'} = \text{a constant.}$$

Here the "constant" is independent of the slope of the incident ray.

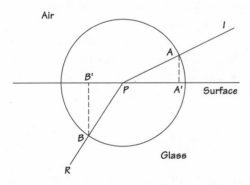

FIGURE 4.3    The law of refraction. A light ray $I$ is incident from air onto a sheet of glass. It passes through the interface at $P$, and the refracted ray is $R$. A circle centre $P$ cuts the rays at $A$ and $B$. $A'$ and $B'$ are points in the interface such that $AA'$ and $BB'$ are each at right angles to the surface. The law says that the ratio $PA'/PB'$ (the ratio of these two lengths) is the same whatever the angle between $I$ and the surface.

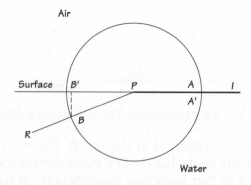

FIGURE 4.4   The limiting case of a ray *R* in water incident upon the surface, at such an angle that no light gets through into the air above.

But it does depend upon the two media. For air and flint glass, the constant is about 1.65. For air and water, it is about 1.33. It is generally greater than 1 when the lower medium is denser than the upper one.

The direction of the light rays in Figure 4.3 can be reversed, so that the light is incident from below in the glass and is refracted on passing into air. The diagram then remains the same. But then a question arises. What happens if we make the angle between *R* and the surface smaller, until *I* lies in the surface, the situation illustrated in Figure 4.4? Then the light ray does not penetrate into the air above. If *R* makes an even smaller angle with the surface, no ray in the air is possible. What then happens to the incident light energy? The answer is that it is *reflected* back into the water, just as if the surface were a mirror. In fact there is always some reflection back into the denser medium, as well as refraction. In the limiting case in Figure 4.4, the reflection becomes *total*.

A fish looking up from beneath the water sees the sky only if it looks within about 50 degrees of the vertical. Outside that angle it sees just reflections from beneath the surface (assuming the surface is exactly level).

Now about colour. It is a common observation that sunlight (which is white) can produce rainbow colours if it passes through a fluted glass window or a cut glass vase, for example. Newton investigated this with characteristic thoroughness (quoted in Ronchi's book):

> In the beginning of the year 1666 ... I procured me a triangular glass-Prisme, to try therewith the celebrated *Phaenomena of colours*. And in order thereto having darkened my chamber, and made a small hole in my window shuts, to let in a convenient quantity of the Sun light, I placed my Prisme at its entrance, that it might be thereby refracted to the opposite wall. It was at first a very pleasing divertisement, to view the vivid and intense colours produce thereby; but after a while applying myself to consider them more circumspectly, I became surprised to see them in an *oblong* form; which according to the received law of Refraction, I expected should have been circular.

Newton expected a circular image on the wall because the hole in his "window shuts" was circular.

Newton showed by his experiments that

- The light that we perceive as white is a mixture of different colours.
- If one colour is selected out of the spectrum produced by a prism, it is not altered by refraction through a second prism.
- Light that appears white can be made by mixing suitable pure spectral colours.
- Light of different colours is refracted by different amounts: that is, the "constant" in the law of refraction depends upon the colour of the light (as well as on the two transparent media). The blue/violet end of the spectrum is refracted more than the yellow/red end.

By some analogy with the musical scale, Newton identified seven colours in the spectrum (in the rainbow, if you like). Of course, the colours merge continuously into each other, and the choice of colours to which we choose to assign names is a bit arbitrary (and varies from language to language).

Unlike the ear, which can analyze a sound into notes of definite pitch, the eye is not able to analyze white light into its component colours. The eye's sensitivity to colour is rather complicated. For example, a mixture of pure green and pure red lights can give the same visual experience as pure yellow light, yet we know the first case is different because a prism can resolve it.

These discoveries of Newton's may not seem surprising to us now, but they were far from obvious to Newton's contemporaries. We make a white piece of paper red by painting it with red paint. Did not the prism somehow "paint" white light different colours, that is, *add* something to the white light? The mathematician Barrow, senior to Newton at Cambridge, described Newton's optical experiments as "one of the greatest performances of Ingenuity this age hath afforded".

Newton's researches, as described previously, did not explain the physical difference between, say, red and blue light, nor why the second is refracted more than the first. To do so would require a theory of the nature of light, and I have not yet addressed this question.

Finally, in this section, there is the question of the speed of light. Light certainly travels very fast. Aristotle and Descartes, for example, believed it to be transmitted instantaneously (as is gravitational attraction, according to Newton's theory). In 1676 the Danish astronomer O. Rømer deduced the speed of light from observations of the satellites of Jupiter. The diameter of Earth's orbit is about 600 million kilometres, so the distance of Jupiter from Earth varies by this amount according to whether the two planets are on opposite sides of the Sun or the same side. Light takes, as we now know, about 2,000 seconds to travel 600 million kilometres. Therefore, a satellite of Jupiter, in its orbit around Jupiter, appears to be 2,000 seconds "late" when Earth is at its maximum distance compared to when Earth is at its minimum distance. By observing this "lateness", Rømer deduced the speed of light. He got about three-quarters of the correct value (which I have used in the preceding numbers). His conclusion was not accepted by everyone, but it was accepted by Newton and Huygens (whom we will meet in Section 4.5).

## 4.4   The Principle of Least Time

Before returning to the question, What is light?, I will take up a topic of enormous significance in physics. This is a principle (*Fermat's principle*) that I will state in the form

- Between any two given points, light takes the path that it can traverse in the least time.

This idea goes back to antiquity, for example to Hero of Alexandria in the first century A.D., but it was the great French mathematician Pierre Fermat (1601–1665) who realized that the principle could apply to refraction as well as to, for example, reflection.

The principle is so different from anything we have encountered before in this book that some explanation is called for.

First, Fermat's principle presupposes that light travels with some finite speed (not instantaneously), and, as we shall see, that the speed is different in different substances, in air and water, for example.

The principle refers to two different points: a point from which the light starts and a point that it is assumed to reach. Given these two points, and given the nature of the intermediate substances, the principle says that, out of all imaginable routes it might take, the ray of light "chooses" the one along which it takes least time.

Contrast this with, for example, Newton's treatment of the motion of a particle. In this case, one starts with the particle at a given position with a given velocity. Then Newton's laws tell us the change in the velocity in a very short time, so the position and velocity are deduced after that very short time. Then one may repeat this process, and thus work one's way step by very short step along the orbit of the particle. Newton's laws are stated in terms of very small changes in very small times, but they enable us to deduce motion over a long time. I will call physical laws like Newton's *local*. By contrast, Fermat's principle refers to the complete path of the light right from the start. I will call this type of principle *global*.

You might love or hate the principle of least time according to your philosophical principles. A materialistically minded person might say, How can a light ray "choose" anything? A religious person might applaud the principle as showing the perfection of God's creation. Fortunately, we need not exercise our minds about this, since I shall shortly explain that the "global" principle of least time can be deduced from "local" considerations.

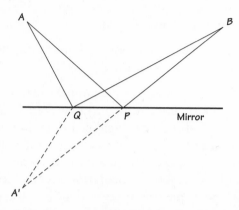

**FIGURE 4.5** The principle of least time applied to reflection in a flat mirror. We want the path of a light ray from the point *A* to the point *B*. *APB* is the true path; *AQB* is another imaginable route. *A'* is the "reflection" of *A* in the mirror.

But for now, I will show how Fermat's principle can be used to give three of the properties of light mentioned: straight line propagation (in a homogeneous medium), reflection and refraction.

In a homogeneous medium, like air (when there are no big temperature differences), the speed of light is the same everywhere. Therefore, the path that takes least time is the one of least length – the shortest path. But everyone knows that the shortest path between two points is a straight line. Therefore, by the principle, light should travel in a straight line.

Now take reflection, which is not quite so trivial. In Figure 4.5, *A* and *B* are the two given points between which we want to find the path of a light ray. Two imaginable paths, *APB* (in fact, the true path) and *AQB*, are shown. We want to show that *APB* is shorter than any other, and so shorter than *AQB*. We can do this by a trick. Let *A'* be the "reflection" of *A* in the mirror, and draw *A'P* and *A'Q*. Clearly the length of *A'PB* is the same as *APB*, and the length of *A'QB* is the same as *AQB*. But *A'PB* is a straight line and *A'QB* is not, so the former is the shorter. Therefore, *APB* is shorter than *AQB* for any *Q* (as long as *Q* is different from *P*).

Now for refraction. In Figure 4.6, there are two transparent media, say air and glass, with a flat surface between them. A ray of

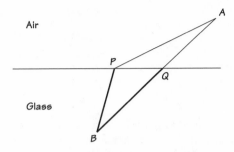

**FIGURE 4.6** Refraction. *APB* is the true path of a light ray from *A* to *B*, satisfying Snel's law. *AQB* is another imaginable path. The paths of the rays in glass are drawn as thicker lines to suggest that light moves slower there.

light goes from *A* to *B*. The true path is *APB*. Suppose that *AQB* is some other imaginable path (which I happen to have chosen to be a straight line). According to Fermat's principle, *APB* should be the path of least time. I do not know a simple way to prove this and will be content with qualitative remarks. How can *APB* take less time than the shorter path *AQB*? It can, provided that the speed of light in glass is *less* than in air. This is because the path *APB* has more of its length in air than does *AQB* (*AP* is longer than *AQ*). The reason is the same as the reason that the quickest car route between two towns may not be the shortest: it may utilize more motorway than the shortest one does.

Thus we see, roughly at least, that the principle of least time can explain three of the main properties of light, but only if we assume that light travels faster in air than in glass (or water, for example). In fact the "constant" in Snel's law (Section 4.3) has to be equal to the speed in air divided by the speed in glass. We shall see in the next section that different theories of the nature of light made different predictions about the speeds in different substances, so the principle of least time is consistent with some theories and not others.

We can use the principle of least time to understand, at least qualitatively, some optical phenomena. Figure 4.7 shows a simple lens producing an image at *B* of a point source of light at *A*. The rays from *A* to *B* should minimize the time of travel. Examples of

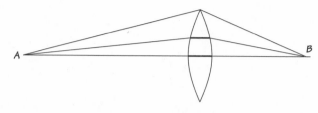

FIGURE 4.7   A lens bringing rays from *A* to a focus at *B*. The three rays shown all take the same, minimum time, because the light travels slower in the glass of the lens.

three rays are shown, all with the *same* minimum time. It is assumed that the light travels slower within the glass lens (as indicated by the thicker lines). Then, provided the lens has a suitable shape, the rays shown will all take the same time from *A* to *B* – although the straight path is the shortest, it has a bigger portion within the glass.

The final example is one with an inhomogeneous transparent medium, that is, one whose properties vary continuously from place to place. Figure 4.8 shows a road surface on a very hot day. The sun heats up the road, which in turn makes the air just above the road warmer than the air higher up. This temperature variation causes the density of the air to decrease near the road surface. There is no abrupt change, just a continuous variation. As one might guess, light travels slower in denser air, so the speed of light is faster nearer the road. What is the path of minimum time from the point *A* to the point *B*? It will take advantage of the faster light speed nearer the road to dip down below the straight line from *A* to *B*, as shown in the figure. So the ray behaves a bit as if it were reflected from a mirror just above the road surface. Thus one may sometimes see a portion of the sky "reflected" from the road: a mirage.

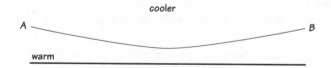

FIGURE 4.8   A ray of light from *A* to *B* following a curved path because of the faster light speed in the warm, less dense air near the hot road surface. This effect can cause a mirage.

## 4.5 What Is Light?

During much of the seventeenth and eighteenth centuries there were two competing theories about the nature of light. One was that light consisted of a stream of particles; the other, that light was a wave motion in some medium (often called the *aether*), analogously to sound's being a wave in air. Each of these theories faced serious difficulties. If light is a stream of particles, how is it that two light beams pass through one another without deflection? If light is a wave, why do light rays seem to travel in straight lines without spreading out sideways?

The most influential proponent of the particle theory was Newton. He thought that the difficulty of explaining straight propagation was fatal to the wave theory (quoting from Query 28 of the *Opticks*):

> Are not all Hypotheses erroneous, in which light is supposed to consist in Pression or Motion propagated through a fluid Medium?... For Pression and Motion cannot be propagated in a fluid in right [i.e., straight] lines, beyond an Obstacle which stops part of the Motion, but will bend and spread every way.

(Newton's second great book, the *Opticks*, was first published in 1704, some thirty years after his experiments on the spectrum. The style of the *Opticks* is very different from that of the *Principia*. The former is in English, the latter in Latin. The *Principia* reads a bit like a mathematics book, with a series of "propositions". The *Opticks* has more "observations" than "propositions", as well as some 30 "queries" in which Newton allows himself to speculate, but also, as in the example just quoted, makes clear his own very definite opinion.)

The particle theory of light easily explains reflection as being due to the particles "bouncing" off the mirror. Newton showed that the theory can also explain refraction, as follows. Suppose that the matter in the glass attracts the particles of light approaching from the air, when they get very near to the surface of the glass. Then the light will be bent towards the surface, as illustrated in Figure 4.9. The light is thus refracted in the right direction (compare Figure 4.3), and indeed Newton showed that Snel's law could be deduced.

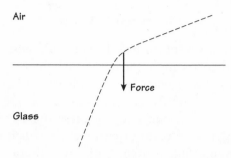

Air

Force

Glass

FIGURE 4.9   A light ray assumed to consist of a stream of particles being bent as it approaches the surface of a piece of glass. The particles are assumed to experience an attractive force directed towards the surface of the glass.

There is an important consequence of this explanation of refraction. Since the light is attracted towards the glass, it accelerates and is moving *faster* by the time it has entered the glass, contrary to what had to be assumed for Fermat's principle of least time to be valid. This does not disprove the particle theory. Appealing as the least time principle is, it might be wrong. It was not until 1850 that the speeds of light in air and water were measured. Until then, there was no direct test of Newton's assumption.

Newton could also explain the production of colours in refraction, by supposing that the particles of blue light (for example) were attracted to the glass more strongly than those of red light.

An important insight about the wave theory of light was made by Newton's near contemporary Christaan Huygens (pronounced roughly "Hoygens", with a hard *g*) (1629–1695). Huygens was a Dutch Protestant, born in the Hague and educated at the University of Leiden. He invented the pendulum clock and the balance-spring timekeeper and discovered the rings of Saturn. One of his great contributions to wave theory is an idea that I will call *Huygens's wavelet principle*. Let me try to explain Huygens's idea.

First imagine a wave moving along a stretched rope, from left to right. Take any point, *P*, on the rope. All that the portion of rope to the right of *P* knows about the portion to the left is the motion of *P* itself. If the rope to the left of *P* were removed, but *P* were caused to move up and down in exactly the correct way, the wave to the right of *P* would be just the same. This seems more or less obvious. So we can if we want consider the motion of the rope to the right of $P$ as being "radiated" by the moving point $P$.

Now go from this one-dimensional example to a two-dimensional one, say, waves on the surface of water. Instead of the point $P$, now imagine a line $L$ drawn on the surface, with waves moving across it (say, from left to right). Just as before, we may think of the wave to the right as being generated by the motion of the surface of the water all along the line $L$. We may go further and imagine this line to be made up of a lot of little drops of water, $P, Q, R, \ldots$, as many and as small as we please. Then the wave to the right is due to the motions of the drops $P, Q, R, \ldots$. Now we invoke the *superposition* principle, explained in Section 4.1. According to this, we may imagine first the wavelet produced by the motion of $P$ alone, then the wavelet produced by $Q$, then that by $R$, and so on; then the actual wave to the right is just got by adding (at any point to the right) the displacements in all these separate wavelets.

The extension to a wave moving in three dimensions, like a sound wave in air, is pretty obvious. The line $L$ is replaced by a surface, with waves moving across it. This surface must be imagined as made up of many small drops of air, and the wavelets due to each drop must be added.

To show that this principle is at least self-consistent, let us see how, according to it, a simple wave maintains its form. In Figure 4.10, a simple plane wave is moving from left to right. The thick lines represent the crests of the waves. The line $L$ is just any line that the wave crosses. According to Huygens's principle, we should imagine wavelets radiated by lots of points on $L$. I have shown just two points, $P$ and $Q$. The crests of the wavelets emanating from these are indicated by semicircles. If the principle is right, all these wavelets should add up to give simply the plane wave continuing to the right of $L$. It is a complicated matter to add all these wavelets everywhere. But we can verify one thing. That is, that the crests of the waves expected to the right all lie on a series of crests of the wavelets: in the figure, the semicircular wavelet crests all touch a line representing a crest of the plane wave. Note that the spacings between the line and between the semicircles are all the same, all equal to one wavelength.

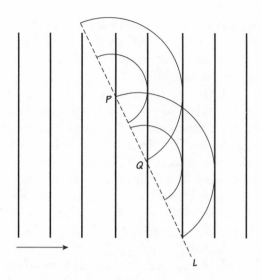

**FIGURE 4.10** Huygens's wavelet principle applied to a simple wave. The wave to the left is incident on the line $L$. The thick lines represent the crests of waves. The wavelets emanating from the two points $P$, $Q$ are represented by the semicircles.

Take a different example: a plane wave incident on a small hole in a barrier. By "small" I mean small compared to a wavelength. In this case, it is enough to draw just one wavelet, and the picture is as in Figure 4.11. To the left of the barrier there is a plane wave, but to the right there is a circular wave spreading out from the hole.

This illustrates what Newton meant when he wrote that the wave should "bend and spread every way" (in the quotation at the beginning of this section). He was right, but only for a *small* hole. What happens for a large hole (one large compared to the wavelength) is harder to see.

Now let us see how Huygens was able to explain refraction. Figure 4.12 shows a plane wave in air incident upon the surface of a piece of glass. Think of the surface as being like the line $L$ in Figure 4.10. Wavelets emanate from the points of the surface into the glass. But now we have to take account of the wavelength of the wavelets. All the waves and wavelets must have a common frequency, in order to keep in step with one another. Huygens assumed

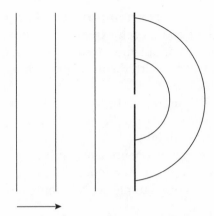

FIGURE 4.11    A plane wave incident on a small hole in a barrier. It is good enough to take just one Huygens wavelet beyond the barrier.

that the speed of light was *less* in glass than in air. Then the wavelength must also be less in glass than in air, as indicated in the figure. In that case, the wavelets in the glass add up to give a plane wave, but it is tilted with respect to the wave in air, in the way indicated in Figure 4.3, and in agreement with experiment.

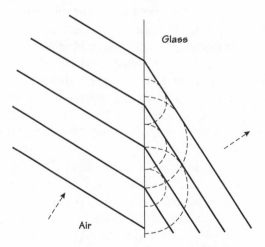

FIGURE 4.12    A plane wave in air incident on the surface of a piece of glass. Within the glass, wavelets of *shorter* wavelength are drawn. They add up to a refracted plane wave.

It is remarkable that Newton's particle theory and Huygens's wave theory of light could each explain refraction, but by making *opposite* assumptions about the speed of light in glass. Huygens's assumption *is* consistent with the principle of least time (described in the last section).

Huygens visited London in 1689 and met Newton. They both addressed the Royal Society on the same day, Huygens speaking on gravity and Newton on light.

There are two types of phenomena, well known at the time of Newton, that I have not yet mentioned. The first is the colours observed in thin films. For example, as Newton wrote in his *Opticks*:

> If a Bubble be blown with Water first made tenacious by dissolving a little soap in it, it is a common Observation, that after a while it will be tinged with a great variety of Colours.

Newton goes on to describe these colours in great detail. If one washes up a milk bottle in a detergent solution, and fails to rinse it out, it is easy to see the colours in the bubbles inside. The colours vary slowly as the thickness of the bubbles changes. Newton tried to explain these colours by an ingenious elaboration of the particle theory.

The second observation that I have not mentioned is that light does *not* quite travel in straight lines. This was demonstrated by Francesco Maria Grimaldi (1618–1663). He passed a very thin pencil of light through a very small hole and saw that the image on a screen was bigger than the hole. This type of effect is called *diffraction*. The reason it is not observed much in everyday life is that it requires such small apertures. With a bit of luck, one can see diffraction as follows. Score a slit about 1 cm long, and as narrow as you can make it, in a piece of kitchen foil. View a ceiling light bulb through the slit, holding it about 10 to 20 cm from your eye (it helps to close the other eye). You may see an image bigger than the slit and traversed by some dark lines parallel to the slit.

As a matter of fact, diffraction effects are quite common, for example, when a small bright light source, like the Moon, is viewed through moist air. Water droplets of size about 10 to 20 times the wavelength of the light produce diffraction maxima at angles of a degree or two from the direction of the source. The diffracted light may appear as a "corona" round the Moon, of apparent size a few times the Moon's diameter. (This is not to be confused with the much larger lunar "halo" caused by *refraction* by ice crystals.)

## 4.6   Light Waves

Throughout most of the eighteenth century, the particle and wave theories of light each had adherents. Then the wave theory triumphed, mainly as a result of the work of two men, Thomas Young (1773–1829) and Augustin Jean Fresnel (1788–1827).

Young trained as a physician in London, Edinburgh, Göttingen and Cambridge. He studied the physiology of the eye and propounded what is essentially the correct theory of colour vision (that the eye contains receptors of three different kinds, each sensitive to light in different bands of colour, one red, one green and one violet: we judge colour from the different degrees of stimulation of these three types of cell). Later in his life he deciphered the Egyptian scripts on the recently discovered Rosetta stone (which bears its inscription in Greek as well as in two Egyptian scripts), thus beginning the study of ancient Egyptian languages.

Young discovered and explained the phenomenon of *interference*. He made light (from a single source) pass through two very thin, nearby parallel slits in a screen and produce an image on a screen beyond. He observed that the image consisted of a succession of alternate light and dark stripes, with a bright stripe in the middle (i.e., beyond the line midway between the two slits). This utterly contradicts the particle theory, which predicts two bright images of the slits and darkness in between.

Young's observation is very easy to explain by the wave theory, using Huygens's wavelets. Figure 4.13 shows a cross section through the experiment, with the two slits in the screen at $X$ and $Y$. Light of a definite wavelength is incident directly onto the screen from below. The widths of the slits are assumed to be small compared to the wavelength. The semicircles indicate the "crests" of Huygens's wavelets coming from the two slits. Of course, there are "troughs" in between the crests. At some points in the region above the screen, two crests lie on top of one another. There, they reinforce one another, producing a double-size wave, that is, maximum

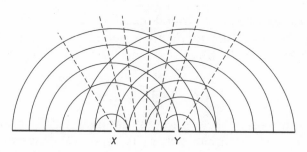

**FIGURE 4.13**    Interference. *X* and *Y* show the positions of the two slits in the screen (only a cross section through this screen is drawn). Light is incident from below, and wavelets emerge from the two slits. The dashed lines indicate directions in which the wavelets add up constructively; in between they tend to cancel out.

brightness. At other points, a crest of one wavelet lies on a trough of another, thus canceling each other out and producing darkness. The points of maximum brightness lie on lines, shown dashed in the figure. In between these lines lie the points of darkness. Thus, if a screen were placed at the top of the diagram, there would be alternate bright and dark stripes, the former where the dashed lines met the screen.

This is an example of the phenomenon of "interference". It is characteristic of waves. Especially noteworthy is the feature of *destructive interference*, the production of darkness where the two wavelets cancel each other out. Particles cannot behave like this.

Young himself expressed the idea of interference very accurately (quoted in Ronchi's book):

> The law is, that whenever two portions of the same light arrive at the eye by different routes, either exactly or very nearly in the same direction, the light becomes more intense when the difference of the routes is a multiple of a certain length [the wavelength], and least intense in the intermediate states of the interfering portions; and this length is different for light of different colours.

This last fact, that wavelength is related to colour, Young deduced by carrying out his interference experiment with light of different colours. It is clear from Figure 4.13 that the spacing between the interference "fringes" (the bands of maxima and minima) depends upon the wavelength (that is, the spacing between the semicircles in the figure), so the wavelength can be worked out from the interference pattern. In fact, Young deduced that the wavelength of red light was about $7 \times 10^{-4}$ millimetre and that of violet about $4 \times 10^{-4}$ millimetre. The production of the colour spectrum by a prism is explained if the speed of light in glass increases with wavelength. Then, on entering glass, blue light changes its wavelength more than red, and so (by Figure 4.12) is refracted more.

The production of beats in sound (see Section 4.2) is a related effect. Beats are alternations of maximum and minimum intensity in *time*. Interference involves alternations in *space*. Of course, interference can be observed with sound too: it is just an accident that beats are more familiar.

Interference also explains the appearance of colour in soap bubbles and other thin films. Take the case of light reflected off a soap film. Some of the light is reflected off the top surface of the film and some off the bottom, the latter therefore traversing a greater distance and so having crests and troughs in different places from the former. If the crests of the two reflected waves happen to coincide (for a given thickness of film, angle of reflection and wavelength of light), maximum reflection will result. If crest happens to coincide with trough, there will be minimum reflection. Since these conditions depend on the wavelength, that is, on the colour, red light is reflected in some directions and blue in others: hence the appearance of the colours.

Instead of the two slits in Figure 4.13, an array of many equally spaced slits produces similar interference effects. Such an array of equally spaced narrow slits is called a "diffraction grating". An audio compact disc (CD) behaves as a diffraction grating. A CD has a spiral track about $0.6 \times 10^{-3}$ millimetre wide with a spacing of $1.6 \times 10^{-3}$ millimetre. These distances are similar to the wavelengths of visible light, so the light reflected from successive tracks can interfere, producing maxima and minima, depending on the colour of the light and the angle of view. These effects are easily observed.

One might think that Young's work would have convinced everyone of the truth of the wave theory of light. In fact, Young was violently attacked in some quarters. But the victory of the wave theory was finally assured by the work of Fresnel. Augustin Fresnel was born of strict Jansenist parents in 1788. He became a French government civil engineer. He died of tuberculosis at the age of 39, having been awarded a medal by the Royal Society in the last year of his life. Fresnel solved the last essential problem of the wave theory: to explain why to a good approximation, light usually travels in straight lines, and why in special circumstances it deviates, that is, diffraction occurs. Remember that it was the straight propagation of light that Newton thought to be inexplicable by the wave theory.

I will give one example of the type of argument that Fresnel invented. Let light be incident on a slit whose width is comparable to the wavelength of the light (not, as in Figure 4.11, much less). To be definite, choose the width to be equal to five wavelengths. If we are to use Huygens's wavelet method, we must first imagine the slit divided up into a lot of segments (a lot more than five) and take a wavelet emerging from each segment. The difficult problem then is to find out how all these wavelets add up. It is not simple as it was in Figure 4.13, where there were only two wavelets.

To simplify matters, I will assume that we are observing the slit from a long distance (much longer than the width of the slit). Then the rays of light from the slit to the eye are almost parallel. The situation is shown in Figure 4.14. A cross section of the slit is shown. For the sake of the clarity of the drawing, I have divided it into only three segments, each of which emits a Huygens's wavelet, but please imagine a lot more segments and wavelets, say 10 or 20. Light is incident on the slit directly from below. It is viewed from a long distance, out of the top right-hand corner of the diagram, in a direction defined by the angle $a$, as indicated. We only need portions of the wavelets in that direction, drawn as the three small arcs (representing crests of wavelets). We need to add up the contributions from these wavelets. This is the difficult problem (especially remembering that we really need many more than the 3 drawn), which Fresnel showed how to solve.

I will show how we can use a graphical method to get a general idea of the result. We need to treat the wavelets in a little more detail than just thinking about the crests, as we have done up to now. To do this, go back to the construction of a simple waveform

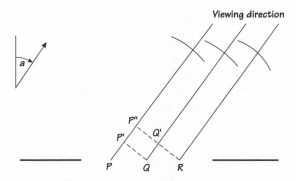

FIGURE 4.14   Diffraction by a slit, viewed from a long distance in a direction defined by the angle *a* shown. For the purposes of the drawing, just three Huygens wavelets are shown, emanating from *P*, *Q* and *R*. Small portions of the wavelets are drawn. The distances *PP″* and *QQ′* show the delay of two of the wavelets compared to the third.

given in Figure 4.1. There we used a point going round on a wheel. The position of this point tells us where we are in the wave, on a crest or a trough or something in between. Draw a little arrow pointing from the axis of the wheel to the position of the point in question. These little arrows are a way of representing the position on the wave.

In Figure 4.15, one wavelength of a wave is drawn. Up above it, little arrows represent the position on the wave. In the first position, the wave is at neither a crest nor a trough, but is "going up". In the

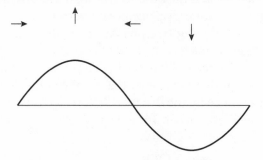

FIGURE 4.15   Phasor arrows representing four positions on a wave: "going up", "up", "going down" "down".

second, it is at a crest. In the third it is "going down", and in the fourth it is at a trough. Of course the direction of the arrow varies continuously between the four representative points shown. Such arrows are sometimes called *phasors*, because the position along a wave within one wavelength is called the *phase*.

At any fixed time, as one moves along the wave direction, the phasor arrow goes round and round, making one complete turn per wavelength. At any one point, as time elapses, the phasor arrow also goes round and round, making one complete turn per period.

Now we can use these arrows to add up wavelets. The rule is this. To add up two wavelets, draw the phasor arrow for one, start at the head of that arrow and draw the phasor for the second. Then the resulting total displacement represents the combined wave; in particular the length of the total displacement gives the strength of the combined wave. The intensity of the combined wave is then got by squaring this total length (this is because energy usually involves the square of something).

The reader who has looked at Appendix B may recognize that the rule for combining phasors is just the rule for "adding" vectors.

We can at least check that this rule makes sense in two special cases. Suppose each wavelet is at a crest. Then the two arrows are parallel (both upwards, in the convention of Figure 4.15). Combining them gives a line of double the length, and the intensity is a maximum (in fact four times the intensity of each wavelet). If one wavelet is at a crest and the other at a trough, the arrows go up and down again, so the total displacement is zero (as is the intensity).

The rule for combining two phasors generalizes in a fairly obvious way to three or more.

Now we can go back to Figure 4.14. The wavelet emanating from $Q$ is in advance of that from $R$ by the distance $QQ'$. Similarly, that from $P$ is in advance by the distance $PP''$, which is twice $QQ'$. This means that the phasor arrow for the $Q$ wavelet has rotated through an extra angle, which is given by

$$\frac{\text{length } QQ'}{\text{wavelength}} \times 360 \text{ degrees.}$$

Similarly, the phasor for wavelet $P$ is in advance by twice this angle.

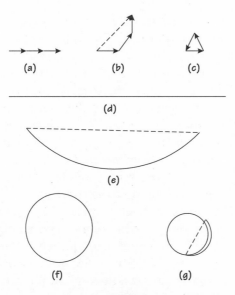

FIGURE 4.16    The first three diagrams show the combination of
the three phasors of the wavelets in Figure 4.14. In (a), the angle
between successive phasors is 0; in (b) it is 45 degrees; in (c) it is
120 degrees. The other four diagrams represent the combination
of many small phasors. The dashed line in each case represents the
resultant of all the phasors.

From Figure 4.14, you can see that $QQ'$, and therefore the angle,
increases as the angle $a$ increases.

Figure 4.16 shows some combinations of more phasors. In (a),
the three phasors for the three wavelets in Figure 4.14 are shown
for the case when the angle $a$ is zero, and hence the angle between
phasors is zero. The total displacement has its maximum value, 3,
so the intensity, 9, is maximal. In (b), the angle between phasors is
taken to be 45 degrees. The intensity comes out to be about 6.8. In
(c), the angle is taken as 120 degrees. Then the phasors just bend
round and get back to where they started, so the intensity is zero.

Now we can see how to treat many wavelets, that is, many phasor arrows. The figures will be like (a), (b) and (c) but have many
little sides instead of just 3. In any such diagram, the angles between any two neighbouring sides are all the same. When there are
many little sides, the figure is quite well approximated by a smooth

curve, which is in fact part of a circle, or alternatively something winding round more than once. The total length along the curve must always be the same (given by the number of wavelets). This means that, the more it winds round, the more compact the figure becomes. In Figure 4.16, (d) is the special case in which the angle between phasors is zero, and the curve becomes a straight line, giving maximum intensity. (e) is a case in which the arc curls round a bit. The intensity is given by the square of the dashed line and is less than maximum. (f) is the case in which the curve just closes up (like [c]). The intensity is zero. (g) is an example in which the curve has gone round more than once (actually somewhat less than one and a half times). The intensity, given by the square of the length of the dashed line, is then about 5 percent of the maximum given by (d). In general, whenever the curve goes round a whole number of times and gets back to its starting point, the intensity is zero. Between any two of these zeros there is a maximum, but these maxima get smaller and smaller as the curve gets more and more curled up.

Figure 4.17 shows the dependence of the intensity, plotted vertically, on the viewing angle (called $a$ in Figure 4.14), plotted horizontally. The example chosen here is for a slit of width about 10 wavelengths. There is significant intensity up to about five degrees, and even little bumps outside that angle. This is the phenomenon of diffraction. If we make the slit wider (compared to the wavelength), the angular spread gets less and less. The position of the first zero is at an angle of about

$$\frac{\text{wavelength}}{\text{width}} \times 57 \text{ degrees.}$$

For example, for a slit of width 1 mm and blue light of wavelength $5 \times 10^{-4}$ mm, the first zero is at about three-hundredths of a degree.

The upshot of all this is that (contrary to what Newton thought) the wave theory *can* explain why light goes in straight lines (very nearly) so long as very small distances are not involved.

We can also answer the question posed in Section 4.4: how does light "know" to find the path of least time? The principle of least time is only true when the wavelength is small compared to all other distances concerned. Really, Huygens's wavelets spread out in *all* directions, but they add up to a large maximum, as in the centre

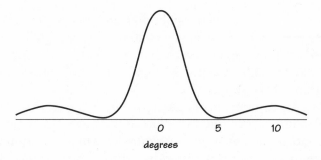

degrees

FIGURE 4.17   The light intensity in diffraction by a slit. The intensity is plotted vertically against the viewing angle, called *a* in Figure 4.14. The example shown is for a slit 10 wavelengths wide.

of Figure 4.17, only near the path of least time. This is because the phasors add up to a maximum, as in Figure 4.16(a) and (d), near the path of least time. For other paths, the phasors combine to something negligibly small, as in Figure 4.16(c) or (f).

## 4.7   Waves in What?

So light has some of the distinctive properties of a wave, especially interference and diffraction. Water waves and sound waves share these properties. But what is light a wave *in*? What property of what medium varies in a wave-like way? The simplest guess is that, like sound in air, light consists of pressure/density waves in some medium. This hypothetical medium was called the *aether* (or *ether*), from a Greek word meaning "the heavens". The aether would have to be very fine or thin, so as to have no other appreciable effects, but at the same time very "springy", so as to explain the very high speed of light (300,000 kilometres per second). Suppose we think all this to be possible: it still turns out that this picture of the aether is too simple.

We can show this simply at home. Take a pair of Polaroid sunglasses. Hold them up and view through one of the lenses a patch of blue sky, in a direction at right angles to the direction of the Sun. Try two positions, the first where the glasses are horizontal, the second where one lens is vertically above the other. The sky should look brighter in the first position.

An alternative experiment can be done indoors on a dull day. Let a lamp be reflected in a pane of glass and view it through the Polaroid sun-glasses. Choose a position so that the light from the lamp strikes the glass at an angle of about 30 to 40 degrees. Compare different orientations of the sunglasses.

Why do these experiments prove that light waves are more complicated than compression waves like sound waves? Imagine a sound wave is moving towards you horizontally. There is nothing about the wave to distinguish the up-down direction from the left-right direction. The wave just consists of motions of air to and from you, and all directions at right angles to this to-and-from direction are equivalent so far as the wave is concerned. The preceding experiments show that light (from the blue sky or reflected from the glass) does, on the contrary, "know" about directions at right angles to its direction of motion. How can this be?

We have in fact already, in Section 4.2, met an example of a wave that could have this property. That was a sound wave in a *solid*, which can vibrate transversely as well as longitudinally. The transverse vibrations do pick out a direction perpendicular to the direction of motion. An even simpler example is given by waves on a stretched rope. If the rope is held horizontally, for example, the vibrations could be up and down or sideways, or somewhere in between. This property of a transverse wave is called its *polarization*.

It can be shown that the polarization of light is never longitudinal, but *always* transverse (in empty space, at least), so in this respect light is different from sound in a solid. The challenge then is to imagine some sort of "solid" aether, which can support transverse vibrations and not longitudinal ones, *and* at the same time not be observable in any other way. In the nineteenth century, the study of the deformation of elastic substances was being developed, and many attempts were made to devise an aether with the right properties. But, as we shall see in the next section, such attempts eventually proved unnecessary.

Phenomena connected with polarization had puzzled people since the seventeenth century, when crystals of calcite (calcium carbonate) were taken back to Europe. Objects viewed through such crystals appear double. A beam of light is split into two, refracted through different angles. Huygens realized that this was because the two kinds of light had different speeds in the crystal. Young recognized that the components with different speeds had different polarizations. A crystal is able to have this effect because the atoms within it are arranged in a regular, lattice pattern. Polarized light can go at different speeds, depending on its direction of vibration relative to the crystal lattice pattern. A homogeneous substance (that is, one with no special pattern within it), like water, could not possibly have such a property.

## 4.8   Light Is Electromagnetism

I have described how, by the time of the publication of Fresnel's *Mémoire sur la diffraction de la lumière* in 1819, it was understood that light was a transverse wave motion, with very high speed and short wavelength (compared with everyday standards of length). The speed of light in water was measured around 1850 and found to be about three-quarters of the speed in air, as required on the wave theory to explain the refraction of light passing between air and water.

In 1862, James Clerk Maxwell (for whom see Section 3.5.) showed that there must exist time-varying patterns of electric and magnetic fields that had all the properties (including the right speed!) of light. This was one of the most remarkable unifications of science: light follows as a sort of by-product of something else apparently completely unconnected. So far as I know, this had not been anticipated, though Faraday (ever seeking for unifications) had suggested that light might consist of vibrations of his lines of force.

If light is electromagnetism, why does not light affect, for example, a magnetic compass needle? The reason is that the electric and magnetic fields in ordinary light vary very fast with time, oscillating about $10^{15}$ times per second, and a compass needle could not possibly respond so fast.

In order to explain Maxwell's idea, I must describe the pattern of electric and magnetic fields in a light wave and verify that they obey Maxwell's rules of electromagnetism (Section 3.5). Figure 4.18 shows how the fields are. The wave is moving in the direction shown by the big arrow. The electric ($E$) and magnetic ($B$) fields are always at right angles to each other and to the direction of the wave. The diagram represents the fields at a particular instant of time, and at two places along the wave that are separated by half a wavelength. These two places are chosen to be ones where both fields have maximum strengths but opposite directions (like a crest and a trough in a water wave). Halfway in between, the fields are both zero. The fields change in a wave-like way along the direction of motion and (at any place) oscillate in time.

The strengths of the electric and magnetic fields are always and everywhere related by

$$E = cB,$$

where $c$ is the constant of nature discussed in Section 3.3.

We must now check that the fields represented in Figure 4.18 do obey Maxwell's rules governing electromagnetism. First take rule (ii) in Section 3.5. This rule refers to a closed path (around which an electric charge is taken). As an example of such a path, take the rectangle shown by a dashed line in the figure. If you go round it clockwise, you go in the direction of the electric field along both of the vertical sides. The electric field is at right angles to the two horizontal dashed lines, so no work is done taking the charge along these. Thus the work done on the charge is determined by the magnitude of $E$.

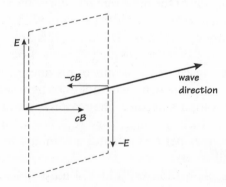

FIGURE 4.18    The electric ($E$) and magnetic ($B$) fields in a simple, plane light wave. The wave is moving in the direction of the thick arrow. The fields are shown (at one instant of time) at two points along the wave, separated by half a wavelength. Halfway in between, both fields are zero.

The other quantity needed to check rule (ii) is the rate of change of the magnetic flux through the dashed loop in Figure 4.18. The magnetic field is everywhere at right angles to the area inside the loop, and it varies from one side to the other. But we need the rate of change of the magnetic field with time. This is zero on the two edges, is everywhere positive inside the loop and has a maximum in the middle (just as the rate of rise of water in a water wave is maximum in between a trough and a crest). The rate of change is determined by the frequency of the wave. The width of the dashed loop in Figure 4.18 is half the wavelength. Putting these things together, one finds, when factors of 2 are worked out, that rule (ii) is satisfied provided that

$$\text{frequency} \times \text{wavelength} = c.$$

Remembering the properties of waves mentioned in Section 4.1, this means that the speed of the electromagnetic wave in Figure 4.18 has to be given by the constant $c$. This constant is found by comparing the (magnetic) force between electric currents with the (electric) force between electric charges, and its value from these measurements was found (see Section 3.3) to be very nearly 300,000 kilometres per second. Thus Maxwell's theory predicts that waves with this speed should exist. And indeed this is just the measured speed of light.

Maxwell's other rules can be verified from Figure 4.18 in just the same sort of way.

Maxwell's electromagnetic wave is *transverse*, in the sense that both the electric and magnetic fields are at right angles to the direction of motion of the wave. The *polarization* of the wave is conventionally defined to be the direction of the electric field (the magnetic field is then fixed to be at right angles to that).

The wave in Figure 4.18 is a wave propagating by itself and filling all space. Of course, this is unrealistic. In practice, a wave has to be generated by something. Electromagnetic waves are generated whenever electric charges oscillate back and forth, or, what is the same thing, whenever oscillating electric currents are present. Visible light is usually generated by the motion of electrons within atoms, because it is only these tiny charges that can oscillate at the necessary high frequencies (about $10^{15}$ times per second).

Although all these conclusions follow from Maxwell's work, they were not immediately accepted by everyone: not, for example, by the prestigious William Thomson (Lord Kelvin). After Maxwell's death, two men in Britain especially developed Maxwell's work. These were George Francis FitzGerald (1851–1901) of Trinity College, Dublin, and Oliver Heaviside (1850–1925). Heaviside came from a poor home and worked as a telegraph operator from age 18 to age 24. After that he had no regular paid employment, though he earned a little from publishing articles in *Electrician* – exceptionally mathematical articles for that journal. He was contemptuous of authority and prone to be sarcastic. He never attended scientific conferences. Among many other things, he first stressed the partial symmetry between magnetism and electricity (mentioned in Section 3.5), and he was one of the inventors of *vectors* (briefly explained in Appendix B).

Apropos of the electric telegraph, I should mention its connection with electromagnetic theory. The telegraph began in the 1830s, and the first successful undersea cable was laid in 1851. Britain led in this technology, which linked up the empire. The current conveying Morse signals has to vary with time, so it produces varying magnetic and electric fields outside the conducting part of the cable. There is a sort of wave running along close to the cable. The signal was liable to be distorted. Heaviside used Maxwell's theory to predict that the distortion would be reduced if self-inductances (mentioned briefly in Section 3.5) were inserted into the cable, but for several years the authorities of the Post Office Telegraph Department (especially the chief engineer, William Henry Preece) refused to accept his conclusion, and animosity resulted.

Electromagnetic waves, as in Figure 4.18, should be able to exist for *any* frequency, as long as the wavelength is related to that frequency in the correct way. Oscillating currents such as can be produced with electric circuits should also produce electromagnetic waves, but it was not clear how to detect these. This prediction of Maxwell's theory was realized by the brilliant Heinrich Hertz (1857–1894) in 1887/8. He excited oscillations in one electric circuit by means of a spark, which set the current flowing to and fro momentarily, rather as the hammer in a piano makes a string vibrate. The oscillating current in the circuit, in turn, excited electromagnetic waves (with wavelengths in the region of one metre) in his laboratory. These he was able to detect by means of a second wire loop with a small spark gap in it. If the loop was of the right size, its natural frequency for current oscillations could be the same as the frequency of the electromagnetic radiation, and a spark in the loop was visible. This is analogous to a piano string's *resonating* to a sound of the right pitch.

It is a bit strange that this discovery should have been made in Germany, where there had been other theories of electromagnetism than Maxwell's. But Hertz was at first assistant to Helmholtz (then much more prestigious than Maxwell), who had developed a variation of Maxwell's theory. Hertz's discovery was announced by Fitzgerald at a meeting of the British Association at Bath in 1888 and aroused great interest. (Edison's phonograph was displayed at this meeting, and Bernard Shaw addressed the economics section.)

In 1894, at a another meeting of the British Association in Oxford, Oliver Lodge (1851–1940) demonstrated transmission of electromagnetic waves from one building to a neighbouring one. He had an improved method of detecting the currents in the receiving circuit, a device called a "coherer" (a tube containing metal powder that lost its electrical resistance after a high-frequency current had gone through it).

In 1896, Guglielmo Marconi (1874–1937) arrived in England and patented his "invention for Improvements in transmitting electrical impulses and signals and in apparatus therefor" (radio signals, that is). In 1899, Marconi transmitted across the English Channel, and in the early 1900s he and others were transmitting farther and farther, using larger aerials (and longer wavelengths) and more electric power.

Returning now to light, let us see how Maxwell's theory accounts for just two of the properties of light mentioned earlier in this chapter: that light moves slower in water than in air, and that light from the sky is polarized.

There are two characteristics that underlie both properties. The first is that electrons, even when bound in atoms, can be made to vibrate to some extent, and the oscillating electric field in a light wave generally has this effect on the electrons in any transparent substance through which it passes. The electrons vibrate at the same frequency as the light wave.

The second characteristic is that an oscillating electric current emits electromagnetic radiation (again at the same frequency as the oscillations). This is what happens in the aerial of a radio or television transmitter. An electric current produces magnetic lines of force circling about it (see Figure 3.8). As the current alternates, the lines of force keep flipping direction. These lines of force, in first one direction, then the opposite one, are radiated outwards, just like the changing magnetic field in the electromagnetic wave in Figure 4.18. It is not very surprising, then, that a wave is emitted. Of course, electric lines of force will be present too, because of the oscillating electric charges.

Now take light passing through a thin sheet of glass. As it goes through, it sets some electrons oscillating, and therefore radiating. The total wave on the far side of the glass will be made by combining the original wave with the extra radiation from the electrons. The extra radiation lags behind the original wave. In fact it turns out that the phasor arrow (see Figure 4.15) for the extra radiation is at 90 degrees to the phasor for the original radiation. These phasors combine as in Figure 4.19. The total wave beyond the glass sheet is represented by the thick arrow in the figure. Remembering that

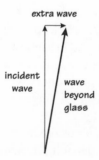

FIGURE 4.19   The phasors for a light wave passing through a thin sheet of glass and for the extra radiation emitted by oscillating electrons in the glass. The result is to rotate slightly the phasor for the original radiation. This is just as if the wave had arrived at the far side of the glass a little late, that is, had travelled slower within the glass.

FIGURE 4.20  The light radiated by a patch of sky is polarized in the direction of the dashed line, that is, at right angles to the direction of the Sun.

the phasor arrow goes round anticlockwise as one moves forward along the wave (see Figure 4.15), one sees that the phasor for the total wave lags a little behind. This is exactly what is expected if the light had moved slower through the glass.

Now about the polarization of the blue light from the sky: this light is emitted by electron in atoms in the air oscillating in response to the electric field in sunlight. (It is blue, because the faster oscillations in the blue part of the spectrum radiate more than the slower ones in the red part.) As one would expect, the light emitted by an oscillating charge is polarized in the direction of that oscillation. Take a patch of sky in a direction at right angles to the direction of the Sun, as in Figure 4.20.

The sunlight is a mixture of all possible polarizations (i.e., it is "unpolarized"). Consider two possible polarizations, the first in the direction of the line from the *SKY* to the *OBSERVER*. This sets an electron in the sky oscillating in that direction, but then none of the radiation emitted would go towards the observer (because the electric field in a light wave is at right angles to the direction of the wave). For the opposite extreme, take light from the Sun polarized in a direction at right angles to all the solid lines in Figure 4.20. This sets the electron oscillating in the direction of the dashed arrow in the diagram. In this case, radiation is emitted towards the observer, and it is polarized in the direction of the dashed line. Thus, that radiation that the sky does emit towards the observer is polarized in the direction perpendicular to the direction of the Sun.

If one takes a patch of sky in a direction not exactly at right angles to the Sun's direction, then the light from that patch is partially polarized (that is, there is more with one polarization direction than with the other).

Finally, how does the Polaroid in sunglasses work? The plastic contains long thin dye molecules, all aligned in the same direction. Light polarized in this direction sets electrons moving along the dye molecules, and this absorbs the light by taking energy out of it. Light polarized at right angles to the dye molecules is not so absorbed. Thus the Polaroid lets light of one polarization pass through but tends to absorbs the other.

The type of polarization described previously, and illustrated in Figure 4.18, is called *plane polarization*. The electric field is always in a plane (the plane of the dashed rectangle in Figure 4.18), and the magnetic field is always in another plane at right angles. There is another simple type of polarization called *circular polarization*. This is formed by superposing two plane polarized waves with electric fields in two directions at right angles, and that are "out of phase" so that the fields of one are maximum where the fields of the other are zero. The result is that the strengths of the electric and magnetic fields are constant along the wave, but their directions change: they rotate steadily about the direction of propagation of the wave. The wave is like a screw: it turns as it moves along. The rotation can be clockwise or anticlockwise (viewed from the point of view of the direction of propagation), and these are called *right* and *left* circularly polarized waves.

## 4.9   Conclusion

Light exhibits the effects of interference and diffraction, which mark it unmistakably as a wave motion. Like the radio wave, it is a wave-like variation of the electromagnetic fields. The speed of light (in vacuum) $c$, which makes its first appearance in Maxwell's equations, is an important constant of nature.

# SPACE AND TIME

*How Einstein found space and time to be different aspects of a unified space-time, with the speed of light built into its structure.*

## 5.1 Electrons

By the end of the nineteenth century the theory of electromagnetism had reached a state of maturity and completeness similar to that of gravitation theory in the eighteenth century. Yet the whole edifice concealed at least three subtle internal contradictions, which were not to be resolved for another quarter century. I shall explain these contradictions, and their resolutions, one by one, in this chapter and then in Chapters 7 and 9. But let me begin by describing how things looked at the turn of the century.

The *electron* was discovered in 1897 by Joseph John Thomson. The apparatus he used was essentially the same as the cathode ray tube in television sets, and so on: a high-vacuum glass vessel with a heated wire at one end. The wire emits some electrons that pass through a small hole, are steered by electric and magnetic fields and produce a bright spot on a fluorescent screen at the other end. Thomson showed that the position of the bright spot was as expected if the "cathode rays" were a stream of negatively charged particles deflected by electric and magnetic fields according to the laws of electromagnetism. He could deduce for the electron the value of the ratio

$$\frac{\text{electric charge}}{\text{mass}},$$

but not the values of the charge and mass separately. This is because the acceleration is, according to Newton's third law (Section 1.7), the force divided by the mass, and the force is proportional to the charge.

Assuming that the charge was the same as the minimum charge of particles (ions) carrying current in solutions (in electrolysis), Thomson deduced that the mass of the electron was some thousandth of the mass of a hydrogen atom (the lightest atom). We now know the fraction to be $\frac{1}{1837}$. He argued also that electrons were universal constituents of all atoms.

Before Thomson's result, there had been doubt whether "cathode rays" were particles or waves of some kind. Some German physicists, including Hertz, had believed they were waves, because they had not been able to detect any deflection by electric fields. Thomson succeeded because he had a better vacuum, so that there were fewer other charged particles in the tube to cancel out the applied electric field.

For a long time there had been speculation whether there might be an "atom" of electricity, and Thomson's experiments finally proved this to be the case. The name *electron* had been coined by Stoney in 1891. Thomson himself did not use the word until 1915.

Since matter is electrically neutral there must also be some positive charge. In 1907, Rutherford discovered that this positive charge was concentrated in a very small region in the centre of the atom, the atomic *nucleus*.

The important ingredients in electromagnetic theory at the turn of the century were the electric and magnetic fields themselves (see Chapter 3) and electrons. The electrons were the sources of the fields, and the fields exerted forces on the electrons and thus affected their motions. (The atomic nuclei, being less mobile than the electrons, were less important, except in providing the background positive charge.) This is the language that I have used, anachronistically, in Chapter 3. Most of the electrical and optical properties of matter were then to be explained, in principle at least, by the motion and vibration of electrons. Mobile electrons in metals accounted for the conduction of electric currents. Oscillating currents in wires produced radio waves. Oscillating electrons in atoms produced light. Moving electrons in the surface of a metal accounted for reflection. In Section 4.8, I have explained how light is slowed down in substances like glass by its interaction with electrons. The man who did most to develop electron theory was the great Dutch physicist Hendrik Lorentz (1853–1928).

In all this, Maxwell's equations were used alongside Newton's laws of motion. Also, as far as gravitation was concerned, Newton's law of gravity remained unchanged. Two of the contradictions in this world view resulted from a mismatch between Maxwell's equations and Newton's. In the next section, I will explain the first contradiction.

## 5.2   Is the Speed of Light Always the Same?

Galileo (and Descartes) realized that motion with constant velocity can persist without any force being applied. For instance, a block of ice sliding on a frozen lake slows down only in so far as there is a frictional force exerted by the lake on the block. This was the cornerstone of the mechanics of the seventeenth century, which distinguished it from Aristotle's. I have explained this in Section 1.5, but it is important enough to warrant repetition. In a steadily moving aeroplane, with the window blinds down, one cannot tell that one is moving (although one may guess from the engine noise and vibration and so on). A game of billiards in the aeroplane (provided the table were properly horizontal) would look just like a game on the ground. This is not so if the pilot suddenly increases the power in the engines, and so accelerates the plane: then the billiard balls would seem to curl towards the back.

This idea is reflected in Newton's laws of motion (Section 1.7): they do not mention the velocity but only its rate of change (i.e., the acceleration). The extra velocity of the aeroplane, if it is constant, does not alter the acceleration of the billiard balls.

We can enshrine this in a principle, which we may call *Galileo's principle*:

- All observations in a closed laboratory are the same, whatever constant velocity that laboratory might have.

Think of the "laboratory" as being the aeroplane mentioned and the observations as being on the movement of the billiard balls. The laboratory has to be "closed"; otherwise the observer could look out of the window and see whether the plane was moving relative to the ground.

This principle is of a type called an *invariance principle*, because it states that observations do not *vary* with the velocity of the laboratory. Invariance principles play a vital role in modern physics. We will meet many more examples.

Galileo's principle might also be called a principle of *relativity*. This is because it says that what count are the velocities (of the billiard balls, for example) *relative* to each other and the billiard table and the aeroplane, not the *absolute* velocities. It may come as a surprise to use the word *relativity* about a principle enunciated some 300 years before Einstein, with whom the word is popularly associated. But as we shall see, what Einstein did was subtly to modify Galileo's principle.

The question now is, How does Galileo's principle square with Maxwell's equations and the theory of electromagnetic waves (like light)? Maxwell predicted that electromagnetic radiation (in a vacuum) should have a speed given by $c = 3 \times 10^8$ metres per second, a constant whose value is deduced from measurements of forces between electric charges and between electric currents (see Section 3.3). Suppose a steward on the aeroplane shines a torch beam towards the front. If speed of the beam is $c$ relative to the plane, surely an observer on the ground will think the light has speed given by

$$c + \text{(speed of the plane)}.$$

But this contradicts what the observer on the ground would deduce from Maxwell's equations. So who is right, Galileo or Maxwell?

Compare with sound. The speed of sound in air can be predicted to be about 340 metres per second, but this is the speed *relative to the air*. A swallow flying towards a source of sound would think that the sound approached it faster than 340 metres per second. We cannot apply Galileo's principle to observations made on sound by the swallow, because that bird is not a *closed* laboratory: the air around has an effect coming from outside the swallow's "laboratory".

Is something similar true of light? Is Maxwell's prediction really only correct relative to some hypothetical medium, the *aether*, analogous to the air in the case of sound? This aether would be something that electric and magnetic fields are properties of, as most nineteenth-century scientists thought. If so, the Earth is flying through the aether, just as a swallow flies through the air. Then the speed of light as measured on the Earth would not be exactly Maxwell's value.

The motion of the Earth is complicated. The Earth is moving round the Sun at about 30 kilometres per second. The Sun is moving round in its galaxy (the Milky Way) at 250 kilometres per second. The Milky Way is moving at some 600 kilometres per second towards the Virgo cluster, which is itself in motion. Fortunately we do not need to add all these motions up. It is enough that the velocity of the Earth is certainly different in, say, December and June, because it is going oppositely relative to the Sun in these two months. So the speed of light should appear to vary through the year. The speed of the Earth (relative to the Sun) is about $10^{-4}$ of the speed of light, so the effect, although small, might be measurable.

But it is not so easy. To measure the apparent speed of light, one has to send the light off *and* get it back again. One can send the light in the direction of the Earth and reflect it off a mirror so that it comes back again. But then the apparent speed is greater on the way out but less on the way back. For the total time taken, these two effects almost cancel out, but not quite. The total time taken turns out to be greater than it would be if the Earth were at rest by a fraction of about $10^{-8}$, that is, one part in a hundred million (this is the square of the fraction $10^{-4}$ mentioned earlier).

In the last year of his life, 1879, Maxwell mentioned the possibility of making such a measurement, but he thought the effect must be too small to be measured. An American physicist, Albert Michelson (1852–1931), realized that it could be done. His idea was to use interference (see Section 4.6). He split a beam of light into two parts and sent one part on the back and forth route described and the other part on a back and forth route at right angles to the Earth's motion. He then combined the two parts, so that they interfered with each other. The times taken along the two paths would (if motion relative to the aether were observable) differ by one part in about $10^{8}$, that is, by about $3 \times 10^{-16}$ second (the light's path length was about 10 metres). How can anyone measure such a small time interval? The answer is that, although very small by ordinary standards, this time is comparable to a period of a wave of the yellow light used. The interference pattern depends on the time interval expressed as a fraction of the period: for instance, if there were no time difference the two waves would add up, but if the time delay were exactly half a period (so that a crest came with a trough) they would cancel out.

Now rotate the whole apparatus through 90 degrees, so that the two paths are interchanged. The interference pattern should change. The advantage of an experiment like this, which measures the *difference* between two small effects, is that all sorts of unknown corrections cancel out.

Michelson carried out his difficult experiment in 1881 and repeated it in 1887 with Edward Morley (1838–1923). No effect was observed. The apparent speed of light did *not* depend upon the speed of the Earth. This "null result" was one of the most crucial observations in the history of physics.

(Although the speed of the observer does not affect the *speed* of light, it does affect the *colour*. The effect of the recession of an observer, for example, is to stretch out the time interval between the arrivals of successive wave crests. Thus the frequency of the light is perceived to be reduced: it is "red-shifted".)

The result of the Michelson-Morley experiment would be understandable if the aether near the Earth were somehow "dragged along" with the Earth, much as the atmosphere goes along with the Earth. But this possibility had been ruled out long ago (just after Newton's death) by Bradley's observation of the change in apparent direction of a star at different times of year (mentioned at the end of Section 1.7). The explanation of this effect assumes that rays of light move in straight lines from the star to the Earth. If light is a wave motion in the aether, and if the aether were dragged along with the Earth, then rays would be bent as they entered the region of moving aether, and this would tend to cancel out Bradley's effect.

To prevent possible confusion, note that Bradley's observation was of a change in the apparent *direction* of light, not in the apparent *speed*. Also, note that Bradley measured something of the order of one part in $10^4$, not one part in $10^8$, as in the Michelson-Morley experiment.

With "aether drag" ruled out, then, the Michelson-Morley result was a great puzzle at the end of the nineteenth century. One ingenious explanation was suggested independently by Fitzgerald in 1889 and Lorentz in 1892. This explanation was to assume that any object (including the apparatus used by Michelson and Morley) moving through the aether is somehow *shortened* in the direction of motion. It is easy to arrange for this shortening to cancel the expected effect in the experiment exactly. The shortening need be only very small, about one part in $10^8$.

This explanation may seem far-fetched. But if matter is made of atoms, and these consist of electrons and positive nuclei held together by electromagnetic forces, and electromagnetism is something to do with the aether, then it is quite possible that motion in the aether might alter the shape of an atom. We shall see in the next section that this *Fitzgerald-Lorentz contraction* does indeed occur in Einstein's theory, though the reason for it is in some sense much more basic than Fitzgerald and Lorentz supposed.

## 5.3   The Unity of Space and Time

In the year 1905 (the same year in which he explained Brownian motion, as described in Section 2.7), Einstein cut through the apparent contradiction between Galileo's principle and Maxwell's theory of electromagnetic waves.

First, Einstein proposed it to be a fundamental principle that (*Einstein's principle*):

- All observers, with whatever constant velocity they may be moving, find the same value for the speed of light (and other electromagnetic radiation) in vacuum. And nothing can travel faster than the speed of light.

It is natural to suppose that it was mainly the Michelson-Morley experiment that led Einstein to propose this principle. Curiously, Einstein denied this (see Pais's book, p. 172).

Einstein also retained Galileo's principle, in the form I have stated it in in the last section. There then seems to be a stark contradiction: surely the speed of an observer affects the speed she finds for light. Einstein removed this contradiction by formulating new rules about how moving observers define space and time. In order to explain this, I must first describe how Galileo and Newton (and nearly everyone before 1905) would have done this.

## 5.4   Space, Time and Motion

If we want to describe the motion of a particle, it is very natural to plot a graph of position (in metres say) against time (in seconds say), that is, with the position measured along one axis and the time along another. In such a diagram, a point represents a particular position at a particular time, for instance the position and time at which two billiard balls collided. Such an occurrence, with a particular position and time, is usually called an *event*.

If we follow the history of the motion of a particle, we get a curve in the space-time graph. In particular, if the particle is moving at a constant velocity, we get a straight line. Such curves are called *world-lines*.

Take Figure 5.1, in which I have only shown one dimension of space. Here time is plotted along the vertical axis and position along the horizontal one. Slices corresponding to successive times are represented by the (horizontal) dashed lines.

Take the top diagram in Figure 5.1 first. *OA*, *OB* and *OC* are the world-lines of three particles. The first is at rest, the second is moving forward with a constant speed and the third is also moving forward but with double the speed. According to Galileo's principle, it is only relative velocities that are important, so we can equally well represent the situation from the point of view of *OB*. This is done in the lower diagram in Figure 5.1, where *OB* is at rest and *OA* is now moving backwards. The relative distances (at any given time) of the three particles are the same in the upper and lower diagrams.

The process of getting from the top to the bottom diagram is called a *Galilean transformation*. The lines of constant time (the dashed lines) are exactly the same in the two diagrams. This may be expressed by saying that time is an *invariant* under Galilean transformations. The important point is that the upper and lower diagrams in Figure 5.1 represent identical physical situations, but from different points of view.

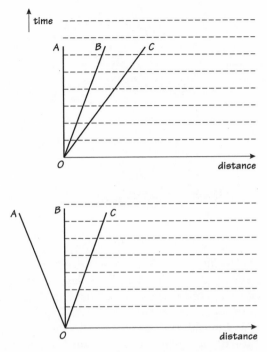

FIGURE 5.1    A space-time diagram (with only one dimension of space). Slices corresponding to different times are represented by the (horizontal) dashed lines. *OA*, *OB* and *OC* represent the world-lines of three different particles, each moving with constant velocity. The top diagram shows the point of view of *OA*, taking this particle to be at rest. The lower diagram gives the point of view of the particle *OB*. The relative velocities are the same in the two diagrams.

In order to reconcile Galileo's principle with the constancy of the speed of light, Einstein realized, space and time must have different properties from those assumed in Figure 5.1. The speed of light must play an essential role. So we must put into the diagrams something representing light. Suppose there is a flash of light (as from a camera flash) at time 0. The light will move out at the speed of light, $c$. So after some time $t$ has gone by, the light will have reached out to a sphere of radius $ct$, that is, a sphere whose radius grows with time. We would like to show this expanding sphere on a space-time diagram. Unfortunately, this would require a four-dimensional figure,

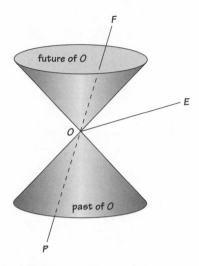

**FIGURE 5.2**   The *light cone*. A space-time diagram (with two space dimensions) showing an expanding light flash from the event *O*: the light cone. This cone divides space-time into three regions: the future, inside the light cone at later times than *O*; the past (inside at earlier times) and outside the light cone. Also shown are the world-line *POF* of a particle moving at a constant velocity, and the line *OE* connecting *O* to an event *E* outside the light-cone.

three dimensions of space and one of time, which is impossible to draw. We can do better if we limit ourselves to two dimensions of space, say the surface of the Earth (taken to be flat). Instead of an expanding sphere, we then have an expanding circle. This can be represented in a perspective drawing, as in Figure 5.2.

This figure shows time increasing vertically upwards, and two dimensions of space, which we may think of as plotted in horizontal planes. At the event *O* a light flash occurred. At times later than *O*, the light from the flash illuminates a circle of steadily increasing radius. The history of these expanding circles is represented by the upper part of the cone in the figure. If we include the lower part of the cone, the complete (double) cone is called the *light cone* at the event *O*. (Note that the lower part of the cone does not represent an expanding light flash: it would correspond to an "imploding" light circle, if such a thing were possible.) Individual light rays correspond to straight lines on the surface of the cone, emerging from $O$. There is nothing special about $O$. Any event, that is, any point in the space-time diagram, has a light cone associated with it.

The shape of the cone (whether it is tall and thin or short and squat) depends upon the units used for time and distance. In the figure, I have used units such that the angle of the cone is 45 degrees. For example, I could use seconds for time and *light seconds* for distance, where a light-second is the distance moved by light in one second, that is, $3 \times 10^8$ metres. Of course, this is an enormous unit of distance, but with any ordinary unit, like metres or kilometres, the light cone would be so flattened that it would not look like a cone at all in my drawings.

People use the word *light cone* whether they are talking about three, two or one spatial dimension. In three spatial dimensions, cross sections of the cone are spheres. In one spatial dimension, the "cone" consists of just two lines. For the remainder of this section, I will use just one space dimension, so that space-time diagrams are two-dimensional (as in Figure 5.1).

Figure 5.2 shows also an example of a world-line, $POF$, representing a particle moving with a constant velocity. According to Einstein, that velocity must be less than the speed of light, that is, less than 1 in our units. This means that $OF$ must lie "inside" the light cone, as shown in the figure. I have also put in an event $E$ "outside" the light cone. The line $OE$ *cannot* represent the world-line of a particle, for if it did, the particle would be moving with a speed greater than the speed of light. We will return to the question of what lines outside the light cone, like $OE$, do represent.

## 5.5  The Geometry of Spacetime

The light cones are the first features needed to build up Einstein's account of space-time. They tie together space and time in a way that is absent from diagrams like Figure 5.1. In fact we must learn to think of space-time as an entity having its own *geometry*. To emphasize this point, I will henceforth write *spacetime*, without a hyphen.

This use of the word *geometry* needs some explanation. We are used to the idea of geometry as being a part of mathematics that

deals with lines, distances, angles, circles and so on. It is the part
of mathematics useful for drawing maps, surveying and working
out areas of carpets, heights of pyramids and so on. This familiar sort of geometry has certain laws, such as "The angles in a
triangle add up to 180 degrees" and Pythagoras's theorem. In the
nineteenth century, mathematicians realized that they could invent
other mathematical systems with different laws (but self-consistent
ones, of course). They would therefore speak about *geometries*, in
the plural. Our familiar geometry is then called, more precisely,
*Euclidean geometry*, after the Greek of about 300 B.C. who systematized its laws. Euclidean geometry, the mathematical system,
seems to represent very well the physical, three-dimensional space
in which we live.

As soon as we start to "do geometry" by drawing diagrams on
flat pieces of paper, we automatically find ourselves doing Euclidean
geometry. This fact makes it difficult to visualize other geometries.
I ask the reader to keep this warning in mind in what follows. The
diagrams that follow represent some things faithfully but misrep-
resent other things.

Einstein's theory requires us to think about a spacetime having its
own geometry, which is different from Euclidean geometry. Spacetime with this geometry is called *Minkowski* spacetime, after Hermann Minkowski, one of Einstein's mathematics teachers.

The first point that strikes one about Minkowski spacetime is
that it has *four dimensions* (three of space plus one of time), unlike the three dimensions of ordinary Euclidean space. But in some
ways this is not such an important difference. We can think about
one-dimensional Euclidean space (restricted to a line), and two-
dimensional space (restricted to a plane) and ordinary three-
dimensional space; it is not too difficult to extrapolate this series to
think about four-dimensional Euclidean space (though it is difficult
to draw it). But this would not be the same as Minkowki spacetime.
Equally, we can omit one space dimension from Minkowski spacetime, and to get a three-dimensional Minkowski spacetime (as in
Figure 5.2), and still be left with the special Minkowski features.
Indeed, we can omit two space dimensions and still have an interesting two-dimensional Minkowski spacetime. This is what I shall
usually do in the following. Then we can draw diagrams, on the

two-dimensional pages of this book, to represent two-dimensional spacetime. But in so doing we must be constantly on guard against being misled: we are trying to represent Minkowski spacetime by drawings on the Euclidean page.

What, then, are the laws that make up Minkoskian geometry, analogous to the laws like Pythagoras's theorem that constitute Euclidean geometry? One of these laws has already been introduced: it is the existence of the light cone – a sort of defining element of the geometry as points and lines are defining elements of Euclidean geometry.

To go further, we must think how physical measurements about spacetime can be made. The simplest device for making such a measurement is a *clock*, because we are talking about space*time*, and clocks measure time. More precisely, if we have the world-line of any particle, we can imagine a clock moving along with that particle, and thus giving a measure of time along the world-line. We must imagine a lot of identical and reliable clocks, so that we can select one to accompany any particle we like. Thus the rate of progress of time is defined along any world-line. For example, in Figure 5.2, we can define time along the world-line OF.

Note the contrast with the diagrams in Figure 5.1. Here time was assumed to be defined, "from outside", the same for all particles and all motions. In Einstein's theory we now have a "time" for each separate world-line. This is often called *proper-time*, because it is "proper" to a particular particle (the German word was *eigen*, which is translated as "proper" or "own"). In Einstein's theory, it is simpler to give a physical interpretation of proper-time than of distance.

The question now is the following. Suppose we have two particles, starting at a common event but moving with different velocities, each carrying one of a pair of identical clocks. The progress of these clocks allows us to mark out time steps along each of the world-lines. How are these two sets of time steps related? Do equal time intervals correspond to equal segments in our diagram? How can we compare times on the two world-lines?

The obvious method we have at our disposal is to send light flashes from one particle and receive them at the other. In Figure 5.3, $l$ and $l'$ are the two world-lines. A light flash is emitted at an event

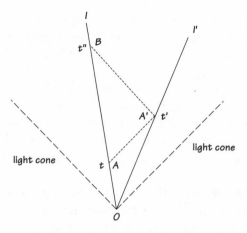

FIGURE 5.3  A diagram used to state a law of Minkowski geometry (showing only one dimension of space). $l$ and $l'$ are the world-lines representing the histories of two particles, which were together at the event $O$ and which each move with constant velocity. The events $A$, $B$ on $l$ and $A'$ on $l'$ are such that a light flash from $A$ gets to $A'$, and a flash from $A'$ gets to $B$. The clock on $l$ reads 0 at $O$, $t$ at $A$, $t''$ at $B$ (so $t''$ refers to the time elapsed between $O$ and $B$); that on $l'$ reads 0 at $O$ and $t'$ at $A'$. (The light cone at $O$ is drawn for reference.)

$A$ on $l$ and received at the event $A'$ on $l'$. Let the clock on $l$ (having started at zero at $O$) record a time $t$ at $A$, and the clock on $l'$ record $t'$ at $A'$. It does not look as if $t' = t$. What is the connection between the two sets of clock readings?

Let us assume that the world-lines $l$ and $l'$ are on an equal footing: neither particle has any preferred status. Then the relation of $t'$ to $t$ is the same as the relation of $t''$ to $t'$. This makes it plausible to assume that

$$\frac{t'}{t} = \frac{t''}{t'}.$$

We can rewrite this relation (which I will call *Minkowski's law*) as

$$t \times t'' = t'^2.$$

This is the fundamental law of Minkowski geometry. In some ways, it is the counterpart of Pythagoras's theorem in Euclidean geometry. The latter relates measurements of distances in different directions. The last equation relates measurements of times on different world-lines. Pythagoras's theorem is about squares of lengths: the last equation refers to one square, at any rate.

I must immediately warn that the three times $t$, $t'$ and $t''$ are *not* faithfully represented by the lengths of the lines drawn on the paper in Figure 5.3. If you try to verify Minkowski's law by measuring the lines in the diagram with a ruler, it will not in general work. Minkowski's law is a statement about the readings of clocks, not about lengths of lines in the picture. (There is one thing that the picture gives right: the ratio of the times $t''$ and $t$ is equal to the ratio of the lengths of the lines $OB$ and $OA$. Along any *one* world-line, lengths can represent times, but not if we compare two different world-lines.)

Note that Minkowski's law means, in particular, that $t \times t''$ is independent of the direction of the world-line $l$ through $O$: that is to say, it is independent of the velocity of the clock carried through $O$. The product $t \times t''$ depends only upon $O$ and $A'$.

Figure 5.4 shows an example of the use of Minkowski's law. This figure is a spacetime diagram showing the world-lines of two particles, moving with (different) constant velocities. Each of the particles carries a similar clock. There is some initial time when the two particles coincided (that is, the three world-lines all pass through one event). At that initial time, the two clocks were synchronized at a time that we may take to be zero. The figure shows some subsequent readings (in seconds, say) of the two clocks. These have been chosen to be consistent with Minkowski's law. For each world-line, the length on the figure that corresponds to one second is different. The faster the particle (that is, the nearer the world-line is to the light cone), the longer the length representing one second appears in the figure.

## 5.6  Lorentz Transformations

The next thing I want to explain is the idea of an observer's description of spacetime. Take the example of a map of a city. The real

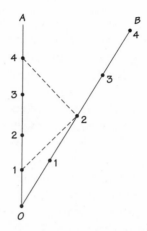

**FIGURE 5.4** The world-lines of two particles moving with constant velocities, each carrying a clock. The two particles were initially together at the same point (O), and their clocks were both then set to zero. Some subsequent clock readings are shown. They are consistent with Minkowski's law. The dashed lines represent light flashes, as required in the statement of Minkowski's law (compare Figure 5.3).

information conveyed by the map are facts like "The bus station is twice as far from the swimming pool as from the library". This is, let us assume, a *fact*, and any accurate map must agree about it. But it is often convenient, for purposes of reference, to divide a map up into squares, perhaps labelled by a letter together with a number, like D8. This is a *convention*, and equally good maps might use quite different conventions, though there should be a way (perhaps rather complicated) to translate from any one convention to any other.

In a somewhat similar way, diagrams we draw to represent space-time inevitably make choices that are really irrelevant. This is illustrated in Figures 5.5 and 5.6. Take Figure 5.5 first. This is a spacetime diagram with only one space dimension shown. I have drawn a whole "grid" of light cones (or parts of light cones) as thin lines. For example, the light cone at the event O is represented by the two lowest thin lines, that at P by the two thin lines going through the P. We may suppose that the lines OP and OQ are world-lines of two particles moving with constant speeds. The

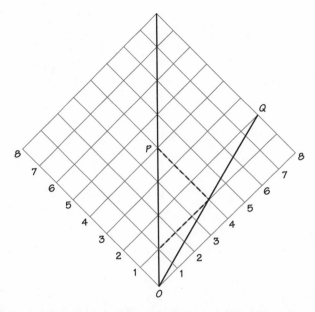

**FIGURE 5.5** A spacetime diagram with a grid of light cones drawn in. The sides of the light cones are numbered in a regular way. *OP* and *OQ* are two portions of world-lines.

particle represented by *OQ* is moving to the right relative to that represented by *OP*.

I have labelled the lines of the light cones uniformly from 1 to 8. This provides a means of labelling the events. For example, *O* is (0, 0), *P* is (4, 4) and *Q* is (2, 8). With this labelling, there is a rule for working out the time elapsed along world-lines. One multiplies the two numbers in the labelling and takes the square root. For example, the time from *O* to *P* is $\sqrt{4 \times 4} = 4$, and the time from *O* to *Q* is $\sqrt{2 \times 8} = 4$. One can verify that this rule is consistent with the Minkowski's law (and with Figure 5.4). As an example, the time (from *O*) at the event (1, 1) is 1, at (4, 4) is 4 and at (1, 4) is 2. These numbers are connected by Minkowski's law as they should be ($1 \times 4 = 2^2$).

So, to determine measurable times as registered by clocks, one multiplies together the two sorts of label and then takes the square root. If one stretches one set of labels and shrinks the other set

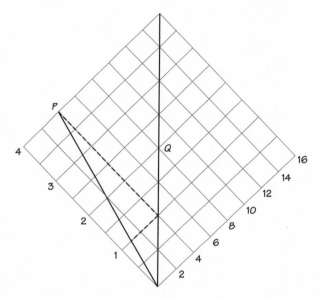

**FIGURE 5.6** A space-time diagram representing exactly the same physical situation as Figure 5.5. The labeling along one side has been stretched by a factor of 2, and that along the other has been shrunk by the same factor. The events and world-lines represented are identical to those in figure 5.5.

in the same ratio, nothing physical changes. This is illustrated in Figure 5.6. Here one set of labels has been stretched by a factor of 2 and the other set shrunk by 2. The events called *P* and *Q* here are the same as those called *P* and *Q* in Figure 5.5. In the diagram, they appear in different places on the page, but the *same* thing is being represented. We may, if we wish, say that Figure 5.5 is drawn from the point of view of an observer moving along the world-line *OP*, but Figure 5.6 from the point of view of one moving along *OQ*. But they are only different points of view, not different things. It is an unfortunate fact that if we try to represent spacetime by a diagram we must inevitably make an arbitrary choice of point of view.

The change that turns Figure 5.5 into Figure 5.6 is called a *Lorentz transformation*. As we have seen, this change consists of a stretching along one direction on the light cone and an equal shrinking along the other.

## 5.7 Time Dilation and the "Twin Paradox"

Time plays a strikingly different role in Einstein's theory than it did in Newton's. In the Newtonian picture, as in Figure 5.1, there is an *absolute time*, somehow given from "outside" and shared by all observers. In Minkowski spacetime, all that is given are the proper-times registered by different observers' clocks. There is no common "time" given from outside. In particular, there is no special meaning to saying that two events are simultaneous.

There is a consequence of all this that is so notorious it deserves special mention. It is often called the "twin paradox", but as I shall explain, it is really no paradox at all. Take two twins. One of them, "he", remains on the launch pad. The other, "she", goes off in a rocket at constant speed on a long journey, reverses and returns at constant speed. He sees the time of her journey as dilated, so, on her return, he sees more time as having elapsed than she does. She is now *younger* than he.

The situation is illustrated in a spacetime diagram in Figure 5.7, in which the speed of the rocket is taken to be $\frac{4}{5}$ the speed of light. Take diagram (a) first. The line $OB$ is the world-line of the twin who remains at rest. The world-line $OAB$ represents the astronaut twin. $OA$ corresponds to the outward journey and $AB$ to the return journey. The proper-times recorded by the twins' clocks are shown in the diagram. If we think first about the outward journey, $OA$, and compare with the examples in Figure 5.4, we see that the scales of proper-time along the two lines have been indicated correctly (that is, consistently with Minkowski's law in Section 5.5).

Having reached $A$, the astronaut begins her return journey with her clock starting at a reading of 3. Because the outward and return journeys are assumed to be at the same speed, the same proper-time elapses along $AB$ as along $OA$. So finally at $B$ his clock reads 10 but hers only reads 6. This sort of effect is called *time dilation*. The astronaut has aged less than her twin.

(Space travel is not a way of attaining a longer life. True, the astronaut is younger than her twin on her return, but she has lived four years less during the journey. Her heart has made $\frac{6}{10}$ of the beats of her twin [neglecting the effects of the excitement and stress of rocket travel].)

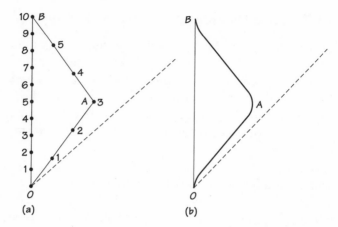

FIGURE 5.7   The "twin paradox". The twin remaining on Earth is represented by the line $OB$, the astronaut twin by the world-line $OAB$. The proper-times recorded by the twins' clocks are indicated. Diagram (a) depicts an unrealistic situation in which the spacecraft accelerates and decelerates instantaneously; (b) allows for short periods during which these changes of speed occur (represented by the curved portions of the world-line).

Why is this sometimes thought to be a "paradox"? The argument goes like this:

> Einstein's theory is a theory of "relativity". This means that motion is relative, and only relative motion has any meaning. Hence we can just as well think of the astronaut as being at rest and the other twin as being the one who moves. If we do this, we find the opposite conclusion: the astronaut ages faster than the twin on Earth. Thus the two supposedly equivalent ways of reasoning lead to opposite, mutually contradictory, conclusions.

The fallacy here lies in the second sentence. Einstein's theory is a theory of relativity, but only in the precise sense of Galileo's principle (Section 5.2). That is, there is relativity for motions with *constant* velocity. But the astronaut certainly does not move with constant velocity. She starts at rest and has to accelerate to the speed of her outward journey. When she reaches her destination, she must slow down to land, then accelerate again for the return journey. Finally, she must slow down to land on Earth. These periods of acceleration are ignored in diagram (a). Diagram (b) is more realistic, showing the accelerations as small portions of curved world-line. So there is *not* symmetry between the two twins. One definitely accelerates and the other does not. The astronaut is in no doubt about that: at blast off, for instance, she feels pressed back against her chair.

But, you may object, how do we know how the astronaut's clock behaves during the accelerations, that is, along the curved portions in Figure 5.7(b)? The arguments in this chapter are silent on this point. Could her clock run faster during acceleration so as to cancel out the slowing effect that we derived earlier? The answer to this is that the difference in ages of the two twins on return goes up as the length of the journey. Keeping the curved portions in (b) the same, we can make the straight portions as long as we please. Whatever happens along the curved bits, the effect along the straight portions must dominate for a sufficiently long journey.

The whole matter is really very simple. The world-lines $OB$ and $OAB$ are different. The first is straight and the second is not. The proper-times elapsed along the two world-lines are different. Because of the rather surprising properties of spacetime diagrams, the proper-time along a straight line is *longer* than along one that is not straight. Thus the twin at rest ages faster.

(Suppose that, while it is accelerating, the rocket maintains a constant acceleration equal to the acceleration due to gravity on the Earth's surface, that is, about 10 metres per second per second. Then, to attain a speed comparable to the speed of light, the rocket must operate for something like one year – by the astronaut's clock, say. Thus to make a journey like that represented in Figure 5.7(b), the astronaut must be prepared to travel for several years: as far as the nearest stars.)

Time dilation is a very well tested prediction of Einstein's theory. It has been tested by taking clocks in aeroplanes. A less direct but perhaps more striking kind of test is the following. The Earth is bombarded by nuclear particles (coming from elsewhere in the universe), called *cosmic rays*. On striking atoms in the upper atmosphere, these sometimes produce particles called *muons*, moving very fast. The only important thing about muons for present purposes is that they decay spontaneously, turning into three other particles, after an average time (when at rest) of 2.2 microseconds (millionths of a second). If a muon is created 20 kilometres up and is travelling very close to the speed of light, it takes about 67 microseconds to reach ground level, that is, about 30 times the average lifetime. Statistically, very few muons should then survive to reach the ground. But suppose the muon has 99 percent the speed of light. Then the time dilation factor is about 7. So the lifetime according to an observer on Earth is increased to about 15 microseconds, and an appreciable fraction of the muons survive to ground level.

## 5.8   Distances and the Lorentz-Fitzgerald Contraction

Thus far I have described spacetime in terms of clock readings. But what about the spatial properties of spacetime? Surely these should involve distances that are measured with rulers, measuring rods, measuring tapes or what have you. In practice only comparatively small distances are measured directly this way. Longer distances, including all astronomical distances, are measured indirectly by using light or radio waves. But still, we ought at least to have a definition of distance in spacetime.

In order to give such a definition, we first state another law of Minkowski geometry, which is just a variant of Minkowski's law in Section 5.5. Look at Figure 5.8. This is like Figure 5.3 except that the event $P$ is now outside the light cone at $O$ (that is to say, $P$ is in neither the future nor the past of $O$). The line $l$ represents the world-line of a particle moving with some constant velocity. The light cone at $P$ meets $l$ in two events, $B'$ and $B''$, so a light flash from $B'$ goes to $P$ and its reflection goes to $B''$. A clock carried along $l$ registers at $B'$ a time $t'$ before $O$ and at $B''$ a time $t''$ after $O$. (Note the difference from Figure 5.3, where both $B'$ and $B''$ are after $O$.) Then the new law may be stated

$$t' \times t'' \text{ is independent of the choice of } l$$

(for given $P$). This means that, for given $O$ and $P$, one gets the same value for $t' \times t''$ whatever world-line $l$ one chooses through $O$, that is to say, whatever is the velocity of the particle carrying the clock.

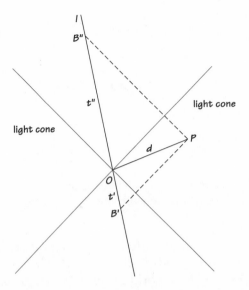

FIGURE 5.8 Defining distance in a spacetime diagram. $P$ is an event outside the light cone at the event $O$. $l$ is a world-line through $O$. The light cone at $P$ meets $l$ at two events, one a time $t''$ after $O$ and the other $t'$ before.

In view of this law, it is natural to define the *distance*, call it $d$, of $P$ from $O$ by

$$d^2 = t' \times t''.$$

Compare this with Minkowski's law, in which $t$ was a time, which could be measured by a clock carried from $O$ to $P$. In the new relation, we are defining a distance, which we do not yet know how to measure.

Figure 5.9 shows examples of distances marked out in accordance with the preceding rule. This figure should be compared to Figure 5.4. The pattern is very similar except that the lines are now "outside" the light cone. They do not represent world-lines of any particle, because nothing can move faster than light. The lines are graduated according to the rule. Nearer the light cone, given increments in distance are represented by longer lines on the diagram.

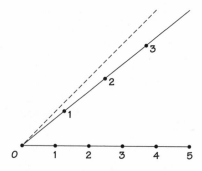

**FIGURE 5.9** Examples of space-like distances in a spacetime diagram. A light flash is used to calibrate the distances in accord with Figure 5.8.

We still have to see how different observers describe the length of a ruler (a measuring rod) in Einstein's theory. The history of a particle is represented by a world-line in a spacetime diagram. What about the history of a ruler? There are world-lines corresponding to its two ends, and the history of the ruler corresponds to the expanse of spacetime in between. We may call this a *world sheet*.

Look at Figure 5.10. Here, the history of a ruler is represented by the strip between the lines $AA'A''$ and $BB'B''$ (these being the world-lines of the two ends of the ruler). The fact that these two lines run straight up the page means that I have drawn the diagram from the point of view of an observer moving with the ruler. There is no doubt about the length of the ruler according to this observer moving with it: it is represented by a line like $A'B'$.

But now imagine a second observer moving with some constant velocity relative to the ruler. Let his world-line be $l$, so that he coincides with one end of the ruler at the event $A'$ (of course, he will pass the other end of the ruler at some later time). How can this observer measure the length of the ruler? He cannot be at both ends at once. The best he can do is to use light signals. Let him send out a light flash at some event $F$ on $l$, which reaches the far end of the ruler at an event $R$ such that the reflected flash reaches $l$ again at $G$. Let $F$ be chosen so that the time along $FA'$ is equal to the time along $A'G$ (these times are recorded by a clock carried by the moving observer). Then it is natural for the moving observer to represent the length of the rod by $A'R$.

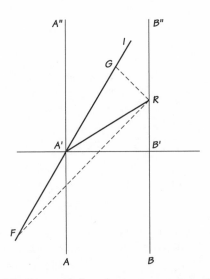

**FIGURE 5.10**   The history of a ruler in a spacetime diagram. $AA'A''$ and $BB'B''$ are the world-lines of its two ends, so the history of the ruler is represented by the strip in between these two lines. The length of the ruler (according to an observer moving with it) is represented by $A'B'$. $FA'G$ is the world-line of an observer moving with constant velocity with respect to the ruler. The dashed lines represent light flashes used by this observer to measure the length of the ruler. A flash is emitted at $F$, reflected at $R$ and observed again at $G$, with $F$ chosen so the time elapsed along $FA'$ is equal to that along $A'G$. This observer represents the length of the ruler by $A'R$.

It remains to compare the lengths represented by $A'B'$ and $A'R$. On the diagram, the latter looks longer. But we must use the definition of distance contained in Figure 5.8. Looking at the examples in Figure 5.9, one sees that $A'R$ in fact represents a shorter distance in Minkowski geometry than does $A'B'$. Thus the moving observer assigns a shorter length to the ruler than does an observer at rest with the ruler. This effect is called *Lorentz-Fitzgerald contraction* and was mentioned at the end of Section 5.2.

The effect comes about because the two observers use different slices of the world sheet of the ruler. The observer moving with the ruler uses slices like $A'B'$ and regards $A'$ and $B'$ as simultaneous events. The moving observer uses slices like $A'R$ and regards $R$ as

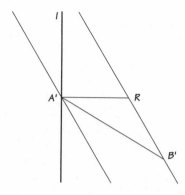

FIGURE 5.11    The length of a moving ruler. The history of the ruler corresponds to the strip between the two sloping lines. The events A', B' and R and the world-line l correspond to those in Figure 5.10.

being simultaneous with A'. In Einstein's theory there is no absolute notion of simultaneity (as there was in the Newtonian diagrams in Figures 5.1 and 5.2).

It is interesting to redraw Figure 5.10 from the point of view of the observer moving along the world-line l. This is done in Figure 5.11. The world-line l is now drawn straight up the page. Relative to this, the ruler is moving backwards and so is represented by a strip sloping to the left. Three of the events marked in Figure 5.10 are also shown in Figure 5.11 and designated by the same letters. The distances to be compared are the ones represented by the lines A'B' and A'R. On the page, the former appears as the longer line. To find the actual distances d in Minkowski spacetime, one needs a comparison calibration, such as appears in Figure 5.9. It is not quite obvious, but in fact the Minkowski distance represented by A'B' is bigger than that represented by A'R.

As one application of the Lorentz-Fitzgerald contraction, consider a cosmic ray muon moving through the atmosphere from the point of view of an observer moving with the muon. This observer ascribes a value of 2.2 milliseconds to the lifetime. How then, from this point of view, does the muon reach ground level? The answer is that, for this observer, the height of the atmosphere suffers a Lorentz-Fitzgerald contraction by a factor of about 7, so that the

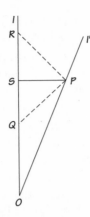

FIGURE 5.12  Minkowski geometry. Two observers, moving with constant velocities with respect to one another, have world-lines *l* and *l'*. They are together at the event O, when they synchronize their clocks. The event P is on *l'*. A light flash emitted by *l* at Q and reflected at P is observed again by *l* at R. The event S, midway in time between Q and R, is judged by *l* to be simultaneous with P.

height appears to be only 20/7 kilometres, and the time to descend through it is about 4 microseconds.

To close this section, I mention another way of expressing the laws of Minkowski geometry. Figure 5.12 shows the world-lines *l* and *l'* of two observers, with P an event on *l'*. The event S is judged by *l* to be simultaneous with P. This means that, if a light flash is emitted at Q, reflected at P and received again at R, then S lies midway in time between Q and R. The observer *l* defines the distance SP according to the rules of Figures 5.8 and 5.9. (In fact, in the figure, that distance is just equal to the time interval QS.) The time that the observer *l'* measures between O and P is related to measurements made by the observer *l* through the following equation:

$$(\text{time } OP)^2 = (\text{time } OS)^2 - (\text{distance } SP)^2$$

(measuring distance and time in units in which the speed of light is 1). This equation is a simple corollary of the laws of Minkowski geometry previously stated, though I shall not give the proof.

The preceding equation relates three sides of a triangle, *OSP*. It looks a little like Pythagoras's theorem in ordinary Euclidean geometry, which relates the sides of a right-angle triangle. The Euclidean condition of being right-angled is replaced by the condition that the observer *l* judges *S* and *P* to be simultaneous. (The angle *OSP* does happen to be a right angle in Figure 5.12 as drawn, but this really has no significance in Minkowski geometry: it occurs only because I have chosen to put *OS* vertically on the page.) Whereas Pythagoras's theorem has the *sum* of two squares on the right-hand side, the Minkowski law has the *difference* of two squares. This is one way of contrasting Minkowski and Euclidean geometry. Note that the time *OP* is shorter than the time *OS* although it looks the other way about in the figure. This is just because equal time intervals on different world-lines look different in the diagrams, as in Figure 5.4.

## 5.9 How Can We Believe All This?

We must now take stock of what Einstein has done. How can it be, for example, that different observers can assign different times to the same event (as in Figure 5.7, for example)? Is not this contrary to common sense?

The first thing to be said is that in all our everyday experiences speeds are very small compared to the speed of light. For example, 1,000 kilometres per hour is only about one-millionth of the speed of light. If we use units in which the speed of light is 1, as I have, the line representing a motion with a speed of 1,000 kilometres per second will be almost indistinguishable from the (vertical) time-axis. The effects of time dilation and so on will be one part in a million million, and so not noticeable. In fact, as we have seen, it took the extraordinary accuracy of the Michelson-Morley experiment to be sensitive to possible effects arising from the Earth's speed (round the Sun) of about 30 kilometres per second. Thus everyday experience is of little value as a guide to what happens when things are moving at speeds comparable with the speed of light.

The second thing to be said is that we should ask ourselves what we (remaining on Earth) might mean by a statement like "The clock in the spacecraft halfway to Mars *now* says 12 noon". How would we check what that clock says? We could have arranged for the astronaut to flash a light when her clock said 12 noon, but it takes perhaps 20 minutes for the light to reach us, so what do we mean by "now"? (Our answer was provided in Figure 5.10, for example, where the observer *l* defines the event *A'* to be simultaneous with *R*.)

## 5.10   4-Vectors

Appendix B explains the idea of a (three-dimensional) *vector*: something like velocity or force or electric field that has a direction as well as a size associated with it. Vector quantities arise frequently whenever we are dealing with things happening in ordinary three-dimensional space. But, as we have seen, Einstein's theory is naturally expressed in terms of spacetime, which has four dimensions, three of space and one of time. Space and time are not the same, but they are bound up together. It is no surprise, then, that the concept of a *4-vector* is natural. This is something that has magnitude and also a direction in *four*-dimensional spacetime. It is called a 4-vector because it needs four numbers to specify it. For example, it might be "resolved" into a "component" in a time direction together with components in each of three space directions.

We have already met examples of 4-vectors in segments of straight world-lines, like OQ in Figure 5.5. (Since two dimensions of space were omitted in these diagrams, the 4-vectors appear as "2-vectors", but we should try to imagine another two dimensions not shown in the diagrams.)

Thus far, we have described the "geometry" of spacetime. We need now to do physics in a manner consistent with Einstein's theory – for example, to state laws of motion for particles and laws of electromagnetism. To do this, the language of 4-vectors is natural.

## 5.11   Momentum and Energy

In Section 1.7, we defined *momentum* as being mass times velocity and wrote Newton's law of motion as, "Rate of change of momentum equals force". Momentum was a 3-vector. In Section 2.3, we introduced the important idea of energy, in particular the kinetic

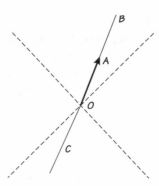

**FIGURE 5.13**  The world-line *COAB* of a particle moving with constant velocity. A clock carried with the particle registers one second between the events *O* and *A*. Then *OA* is defined to be the particle's 4-velocity. (The light-cone at *O* is shown by dashed lines.)

energy of a particle. Energy is not a vector quantity: it has no direction associated with it. How do momentum and energy go over into Einstein's theory?

Take a particle moving with constant velocity, and so having a world-line like *OQ* in Figure 5.5. The history of this particle is represented by a point moving along this world-line. We define the the *4-velocity* to be the rate of movement of this point in spacetime. The rate with respect to what? (Remember there is no one absolute time in Einstein's theory.) The natural choice is with respect to proper-time, as registered by a clock carried along with the particle. The 4-velocity so defined is indeed a 4-vector.

How do we find the 4-velocity in this simple case? Figure 5.13 shows the world-line of a particle moving with constant velocity. Choose the events on the world-line so that a clock carried along with the particle registers 0 at *O*, 1 second at *A*, 2 seconds at *B*, −1 second at *C* and so on. (We could, of course, use a different unit of time. I use seconds in order to be definite.) The 4-velocity is defined to be the displacement in spacetime in 1 second, so it is represented by *OA* (as indicated by the thick arrow). Note that we have no freedom of choice in the length of the 4-velocity 4-vector. By definition, the length must be chosen so that 1 second has elapsed on the clock. Thus Figure 5.14 shows three possible 4-velocity vectors, *OA*, *OB* and *OC*. Given the direction of each of

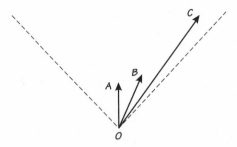

FIGURE 5.14    Examples of possible 4-velocity 4-vectors. The lines drawn represent the motion of the particle in a time of, say, 1 second as measured, in each case, by its own clock.

them, its length is fixed by the 1-second condition. The nearer the light cone such a vector lies, the longer it appears in the diagram, for the reason explained in connection with Figure 5.4. In this example, the particle represented by $OC$ is moving to the right relative to that represented by $OB$, which in turn is moving to the right relative to $OA$. Of course, we may also say that $OA$ is moving to the left relative to $OB$, and so on.

What is the physical interpretation of such a 4-velocity? First of all, it is useful to multiply it by the mass of the particle. Call the resulting quantity the *4-momentum*. How is this 4-vector viewed by a given observer? This observer measures time by his clock and has his own measures of time and distance in spacetime. His view of the momentum 4-vector is shown in Figure 5.15.

The observer may express the momentum 4-vector as a vector sum (see Appendix B) of a piece along the time direction and a piece parallel to space directions. The former is identified with the energy of the particle, the latter with its ordinary 3-momentum. For particles moving slowly compared to light (that is with 4-velocity far from the light cone), Einstein's energy and 3-momentum are closely related to Newton's. For particles moving near the speed of light there are important differences.

In these remarks, I have implicitly been using units in which the speed of light is 1. In ordinary units, the 3-momentum should be multiplied by $c$ and the mass by $c^2$.

The length of the momentum 4-vector is fixed to be $mc^2$. From Figure 5.15, this fact imposes a connection between energy and

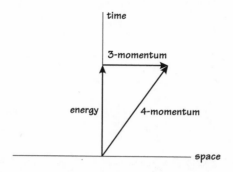

**FIGURE 5.15**  A momentum 4-vector as viewed by a particular observer. The observer uses time- and space-axes in spacetime, as shown. He may then express the 4-momentum as the vector sum of a piece along in the time direction (the energy) and a piece along the space-axes (the 3-momentum).

3-momentum:

$$(\text{mass})^2 c^4 = (\text{energy})^2 - (\text{momentum})^2 c^2.$$

This equation is analogous to the last equation in Section 5.8 (compare Figure 5.15 with Figure 5.12): it is the counterpart in Minkowski's geometry of Pythagoras' theorem in Euclid's, with a difference instead of a sum on the right-hand side. Here, to be more precise, *energy* means "Einsteinian energy as measured by some observer" and *momentum* means "Einsteinian 3-momentum as measured by the same observer". So this is Einstein's relation between energy and momentum.

What happens for a particle at rest (relative to the observer)? The momentum is zero. Then the preceding equation tells us that

$$\text{energy} = (\text{mass}) \times c^2.$$

This is, of course, Einstein's notorious equation. The first conclusion is that we were wrong to identify Einsteinian energy with Newtonian kinetic energy. In fact, it can be deduced from the equation in the last paragraph that

$$\text{Einstein energy} \approx (\text{mass}) \times c^2 + \text{Newton energy},$$

where this is an approximation that gets better the slower the speed of the particle (compared to $c$).

The importance of energy lies in its being conserved (see Section 2.3). In this respect, if one adds to energy some quantity that does not change at all, it makes no difference. So the (mass)$c^2$ term on the right-hand side of the last equation has no effect as long as particles are neither created or destroyed: this term just sits in the equations and is irrelevant to changes of energy. But in more general situations, we learn that mass and energy are really the same thing and may be interchangeable. But the "rate of exchange" between mass and energy is the factor

$$c^2 = 9 \times 10^{16} \text{ (metres)}^2 \text{(seconds)}^{-2}.$$

This is immense in ordinary units, so that, when mass does change, there is a very large change in energy (as we are all too well aware).

Figure 5.16 shows a graph of the relation between energy and momentum according to Einstein.

Einstein's law in Section 5.3 says that nothing can go faster than light. What happens if a force keeps acting on a particle, so that it keeps accelerating? Will it not eventually go faster than light? In a consistent theory, something should prevent this from happening.

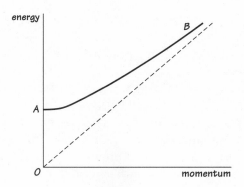

FIGURE 5.16   Einstein's relation between energy and momentum. The graph is drawn using units in which the speed of light $c = 1$. $OA$ represents the rest energy, $mc^2$. The dashed line $OB$ is the zero mass case, $m = 0$. The portion of the curve near $A$ corresponds to the Newtonian approximation. For large momentum, the curve approaches the dashed line (because the mass is nearly negligible compared to the momentum).

Indeed this is the case. Suppose that the speed of a particle approaches that of light. This means that the 4-velocity gets near to the light cone, like $OC$ in Figure 5.14. As this happens, the length of the line representing this 4-vector increases and increases without limit. The same applies to the 4-momentum (which is just the 4-velocity multiplied by the mass), and so by Figure 5.16 the energy and the 3-momentum increase without limit. Thus a continually acting force would, as one would expect, increase the energy and momentum without limit, but it would do this by driving the speed nearer and nearer to the speed of light, never quite attaining that speed.

Note the difference from Newtonian theory, in which a particle moving with the speed of light would have momentum equal to (mass) $\times c$, very large certainly, but finite. Any given force would provide this amount of momentum if applied long enough.

Notwithstanding all this, there is a possibility allowed by Einstein's theory that would make no sense in Newtonian mechanics: that the mass should be exactly *zero*. Looking back to Einstein's equation connecting momentum and energy, one sees that it makes good sense even if the mass is zero, reducing then to

$$\text{energy} = c \times (\text{momentum}).$$

In this case, the particle would be moving along the light cone, that is to say, with the speed of light. Proper-time is not defined along the light cone: it is not possible to imagine such a particle to carry a clock. In a figure like Figure 5.14, a particle with no mass would move along the light cone, but we cannot represent its velocity correctly by a vector like $OA$, $OB$ or $OC$.

Are there particles in nature with zero mass? We shall see in Chapter 9 that quantum theory associates particles, photons, with light. Not surprisingly, photons move with the speed of light and so have zero mass. There are other particles, neutrinos, that have either zero mass or a very small mass.

## 5.12 Electricity and Magnetism in Spacetime

The preceding arguments presuppose the existence of a force describable in a way consistent with Einstein's theory. Is the gravitational force of such a nature? Not in its Newtonian form, certainly.

The very statement of Newton's law of gravitation presupposes absolute time. When we say that the gravitational force of the Sun on the Earth is *now* such and such (calculated from Newton's law of gravitation), we need to know where the Sun is *now*. The force is assumed to be transmitted instantaneously. If a giant hand suddenly moved the Sun, the gravitational force on the Earth would change at the same instant. This is utterly forbidden by Einstein: no influence should travel faster than light.

Can Newtonian gravity be modified to make it consistent with Einstein? If so, maybe the modifications would be very small for bodies, like Earth, that move slowly compared to light. The answer is yes, but the modification requires yet another revolution in our view of space and time. This is the subject of Chapter 7.

Electric and magnetic forces are more promising. After all, Maxwell explained light as an electromagnetic phenomenon, and it was the finite velocity of light that forced Einstein's theory on us. Electric and magnetic forces are not transmitted by instantaneous action-at-a-distance. In Faraday's and Maxwell's formulation, some electric charges produce an electromagnetic *field*, which then propagates to other places, where it can produce forces on other charges. The field appears as an intermediary. It is quite natural that changes in it should not travel faster than light (which is itself such a sequence of changes in the field).

There remains one question. We have seen that quantities fit naturally into Einstein's theory when they can be put together into 4-vectors, like distance and time or momentum and energy. Can this be done with the electric and magnetic fields (at some position and time)? Each of these fields separately needs three numbers to specify it – they are each 3-vectors as regards ordinary space. So there is a total of six numbers in the electromagnetic field at a point. How can we fit these into 4-dimensional spacetime?

In spacetime, we have so far encountered points (events) and lines (world-lines, etc). Another sort of geometrical thing one can imagine in spacetime is a *plane*, that is to say, a flat surface. This is a natural progression, if we think of a point as being zero-dimensional, a line as one-dimensional and a plane as two-dimensional. It is rather difficult to imagine a plane in four-dimensional spacetime, but we can certainly imagine one in ordinary three-dimensional space. We shall be concerned only with the orientation of a plane, not where it is. The next question is, How many *independent* orientations of planes are there in four-dimensional space?

Before answering this, we must pause and think what we mean by independent (the proper mathematical term is *linearly independent*). What do we mean if we say that there are two independent directions in two-dimensional space? Let us talk about the directions north and east on a map, for example. Is north-east an independent direction? We would say not, because we could construct a journey in a NE direction out of one N and then one E. So we can do with two independent directions, not more and not less. Similarly, there are four independent directions in spacetime. We can choose them to be one in the "time" direction (call it $t$) and three along three space axes ($x$, $y$ and $z$). There is nothing special about this particular choice, but it is one possible one.

If we choose any two different directions, that defines a plane. For example, in three-dimensional space, the floor of a house is a plane that contains the north and east directions. A north-facing wall contains the east and "up" directions, and so on. Similarly, in four-dimensional spacetime, two independent directions define a plane. We can now guess the number of independent planes in spacetime. Let us call the plane containing the $x$ and $y$ directions $(x, y)$. Then a set of independent planes is

$$(x, y), \quad (y, z), \quad (z, x), \quad (x, t), \quad (y, t), \quad (z, t).$$

There are six of them, just right to allow a correspondence between the electromagnetic field and directions of planes. If we make this correspondence, the magnetic field goes with the first three planes and the electric field with the last three.

This correspondence teaches us something more. The choice of a "time" direction in spacetime is defined only relative to some particular observer. Therefore the division between the first and last three planes is also made only relative to some observer. It follows that the distinction between a magnetic and an electric field can also be made only relative to an observer. Suppose an observer sees an electric field only (no magnetic one). Another observer moving relative to her will (in general) see a magnetic field as well (and a different electric one). The electromagnetic field must be taken as a single entity, just as spacetime is a single entity in Einstein's theory.

All the laws of electromagnetism in Chapter 3 can be expressed in a way that fits naturally into Minkowskian spacetime and that is consistent with the laws of Einstein's theory.

The electromagnetic wave, described in Section 4.8 and illustrated in Figure 4.18, has electric and magnetic fields related to a direction of motion and are characterized by a frequency and a wavelength. If a given wave is viewed by two observers (in motion relative to each other), they will ascribe to the wave different frequencies, wavelengths and directions of motion and different values of the fields. Yet they must agree about the geometrical configuration of the fields: that the electric and magnetic fields are everywhere at right angles, and each at right angles to the direction of propagation. This geometrical configuration is, although this is not obvious, something compatible with the geometry of Minkowski spacetime.

## 5.13   Conclusion

Galileo and Newton had said that there was no absolute standard of rest or of motion at a constant velocity. Maxwell's electromagnetic theory of light provided a standard speed, the speed of light, $c$. The experiments of Michelson and Morley demonstrated that this speed was not relative to some fixed aether. Einstein resolved this paradox by uniting space and time into spacetime, with its own geometry founded on the light cone. The light cone anchors space and time together but also separates space-like from time-like directions in spacetime.

Newton's dynamics has to be modified to make it fit with the geometry of spacetime. The modification makes a very small difference at slow speeds, but it prevents any particle from attaining the speed of light because energy grows without limit as the speed approaches $c$.

Maxwell's laws are consistent with the geometry of spacetime, so they cannot help but predict that electromagnetic radiation should move at the speed $c$.

Measurements of the speed of light used to be regarded as measurements telling us something about nature. What they really do is to connect three things: light, our chosen unit of time (say the second) and our chosen unit of distance (say the metre). From a modern, post-Einstein point of view, we would say that the speed of light just connects two arbitrarily chosen units, the second and the metre. We can always choose to use units, say, the second and the light second, in which the speed of light is by definition 1. The light second is inconveniently large for ordinary measurements on Earth, but units in which $c = 1$ are routinely used by physicists and astronomers.

Einstein's theory of spacetime has some unexpected consequences, such as the non-existence of absolute time and the "twin paradox". But things moving near the speed of light are commonplace in the laboratory and in the universe, and Einstein's theory is always being used with success.

The picture, as it emerges from Einstein's "special relativity" of 1905, is that spacetime provides the fixed arena within which particles move and electric and magnetic fields "exist". The fields are defined indirectly (as in Sections 3.1 and 3.2). They have energy and momentum associated with them, but they seem to resist being modeled in any "mechanical" way. No aether plays a useful role in the understanding of electromagnetism (or light). There is a picture by M. C. Escher that strikingly brings home to us the aether-less light waves that we seem to be forced to try to imagine. The picture, called *Ripples Surface* (reproduced in Hofstadter's *Gödel, Escher, Bach* and in *Escher on Escher*), shows ripples (with distorted reflection in them), but the water surface that might be rippling is just not there.

# 6

# LEAST ACTION

## 6.1  What This Chapter Is About

The titles of most of the chapters in this book convey, I hope, something to the reader. The title of this chapter is an exception. The idea that there is a quantity called *action*, which takes its least value when the equations of motion are obeyed, is now one of the foundations of classical (as opposed to quantum) physics, and even in quantum physics the action is a basic quantity. This idea is not easily expressed in everyday terms, but I think I would be guilty of some sort of distortion if I omitted it from this book. So what is it all about?

In Section 4.4, I explained Fermat's principle of least time as applied to light rays. The principle says that the path taken by a light ray between two given points is such as to make the time taken by the light a minimum. The principle of least action is an extension of this sort of idea to the motion of particles or to any other time-varying system, like the electromagnetic field. The difference from the case of light is that it is now not just the time that is a minimum, but some less obvious quantity, called *action*. One needs some rule to decide what the action is for a given system. This rule, indirectly, defines the forces operating in the system.

I will use the word *least* throughout, but as I will explain in Section 6.3, this is not quite accurate. In some cases, we need a property a little more general than being least (i.e., minimum), that is, the property of being stationary. But for the moment the important thing is to explain what action is.

## 6.2  Action

*Action* is another of those words chosen for the idea they convey in ordinary usage, but endowed with a special technical meaning by scientists. *Force* and *energy* are earlier examples that we have met in Chapters 1 and 2. The use of the word *action* (the same in French as in English) in its modern sense was begun in 1744 by the French supporter of Newton Maupertuis. Almost immediately the idea was used more clearly by the great Swiss mathematician Euler (who, generously to my mind, insisted on Maupertuis's priority).

The general mathematical formulation of Newtonian mechanics was constructed by French and Swiss mathematicians in the eighteenth century. New refinements followed in the nineteenth century. The most important of these was due to the Irish mathematician William Rowan Hamilton (1805–1865), who was also a virtuoso linguist; a friend of Wordsworth and Coleridge, he himself wrote poetry (though Wordsworth advised him not to).

Hamilton found a more general form of the principle of least action, which is the one I shall explain. Action is a property of a whole motion of a body, a whole history from some initial time to some final time. Hamilton defined it to be

action = [time average of (kinetic energy − potential energy)] × $t$

(where $t$ is the time taken in the motion). To understand this, remember that the total energy is the kinetic energy *plus* the potential energy. In contrast there is a minus sign in the definition of action.

Hamilton's principle is

- From all possible (false) motions, starting at a point $P$ at a time $t_P$ and ending at a point $Q$ at a time $t_Q$, the true motion has least action.

I will not explain how Hamilton's principle is derived from the laws of motion, but will just give a simple example of its working. Take a weight dropped (released from rest) under the Earth's gravity and allowed to fall freely (neglecting friction due to the air) for 10 seconds. It falls approximately 500 metres. Plot the height against

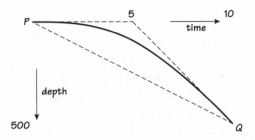

FIGURE 6.1 A graph of height (plotted vertically) against time (plotted horizontally) for a falling weight. The true motion is given by the parabolic curve. Two examples are shown (by dashed lines) of "false" motions, which have the same initial and final times and position.

the time that has elapsed. The graph is part of a parabola, as shown in Figure 6.1. The action calculated for this true motion comes out to be $\frac{1}{3} \times 10^5$ joule seconds. The action principle asserts that any other motion (not obeying the equations of motion) must have a larger action. The figure shows two such "false" motions, chosen quite arbitrarily, which have the same starting and finishing times and heights as the true motion. The lower false motion is a descent with a constant speed of 50 metres per second. The action comes out to be $\frac{3}{8} \times 10^5$. This false motion has less average kinetic energy than the true one, but it also has less potential energy (it falls too quickly), and the latter effect is dominant.

In the upper false motion, the particle remains at rest for 5 seconds and then drops with a constant speed of 100 metres per second. The action is $\frac{3}{8} \times 10^5$ joule seconds. In this case, the average kinetic energy and the average potential energy exceed that for the true motion, but the former effect dominates. The true motion is the one that achieves the optimum balance between kinetic and potential energy.

Of course, these two examples are very artificial. I have chosen them just because they are simple to work out. The action principle asserts that the true motion (that is, the one that satisfies the equation of motion) has less action than *any* other motion one might think of. In a spacetime graph like Figure 6.1, a *motion* means a curve going from $P$ to $Q$.

## 6.3 Minimum or Just Stationary?

(The reader can omit this section with little loss.)

Now I must explain, as promised, why, in general, it is more accurate to say *stationary action* than *least action*. Consider this example. Take two very smooth spheres, one just inside the other, with a small ball sliding about in the space between them (the space being assumed just thick enough to accommodate the ball). Assume that everything is so smooth that friction can be ignored. Take this contraption far out into space, so that gravitation can be neglected. The true motion is very simple: the ball always moves along a great circle (a circle on the sphere of maximum size, so that its centre is at the centre of the sphere). Because there are no forces in the direction of its motion, the ball's kinetic energy remains constant, and there is no potential energy. So the action is just the kinetic energy times the time taken.

Take given initial and final points $P$ and $Q$, as in Figure 6.2, for example. Any true motion on a sphere like this (under no forces) is along a portion of a great circle. The figure shows the great circle through $P$ and $Q$. There are two possible paths from $P$ to $Q$, the short one round the front and the long one going all the way round the back through the point $P'$. On the longer path, the speed has

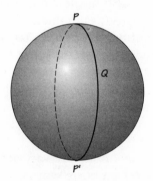

FIGURE 6.2   Paths of a ball confined to roll in the space between two very smooth spheres. The ball is to go from point $P$ to point $Q$ in a fixed time. There are two possible true motions, the short way round the great circle and the long way. The first minimizes the action; the second only gives a stationary value.

to be higher in order to complete the journey in the given time. Therefore the action for the longer path exceeds the action for the shorter path.

The short path does minimize the action. What about the long path? Consider some false paths very close to this long one. One such path is an arc of a circle on the sphere a little bit shorter than the great circle. This false path has *less* action than the true one. But one can also construct paths, say wiggling around a little on either side of the great circle, that have more action than along the great circle. So the longer true (great circle) path is neither a minimum nor a maximum of the action. Nevertheless, it is a *stationary* value of the action. To explain what this means, take a simpler model.

Consider how the height of land above sea level varies with position. There are maxima at the tops of mountains, and minima, often at the bottom of lakes. But take the top of a pass between two valleys. If one walks over the pass from valley to valley, one encounters a maximum of height, but walking, from a mountain on one side to a mountain on the other, one meets a minimum. The top of the pass is therefore neither a true maximum nor a minimum. It is called a *stationary point* (or sometimes an *extremum*). At the exact top of the pass, the ground is level, sloping neither up nor down (just as it is level at the top of a mountain). A boulder can balance there without rolling down.

This is only an analogy. The height in the model corresponds to the action. But we are considering the variation of the height with position, say, with latitude and longitude. On the other hand, we are concerned with the variation of the action with *all* possible (false) motions – a much greater range of variations. In spite of this, the idea of a stationary point, which is a maximum for some variations but a minimum for others, is common to the two cases.

I will continue to use the inaccurate word *least* where I should strictly write *stationary*.

## 6.4 Why Is the Action Least?

Just like Fermat's principle of least time in optics (Section 4.4), the principle of least action raises questions in one's mind. How does a moving particle "know" the path that makes the action least? How does it "sample" other (false) paths to find the right one? A conservative answer to these questions is just to say that mathematics shows that the least path satisfies Newton's equations of motion. But Maupertuis, at least, thought that the principle went deeper. In 1746, he read a paper to the Berlin Academy, "Sur les loix du mouvement et du repos déduites des attributs de Dieu". The final title was toned down a bit: "Les lois du mouvement et du repos déduites d'un principe métaphysique" (see Beeson's book).

But there is now a physical answer to the preceding questions. Remember that, in the case of optics, the principle of least time turns out (see Section 4.6) to be only an approximation: in the wave theory of light, the waves *do* "sample" all paths; the least time path is just the one near which the waves reinforce each other most. In the present case, of particle motion, quantum theory eventually provided a similar resolution. The principle of least action is just the classical (that is, without quantum theory) approximation. In quantum theory, all motions *are* "sampled". Reinforcement takes place near the classical motion of stationary action. I will tell this tale in Chapter 8.

This story shows how dangerous it is to draw theological or philosophical conclusions from science. Few, if any, scientific theories are final: each one is destined to be subsumed into some more complete theory (in this case, classical mechanics was subsumed into quantum theory). Grand deductions may be premature.

## 6.5   The Magnetic Action

Thus far in this chapter I have assumed that a potential energy exists, as it does, for example, in Newtonian gravity and in electrostatics (see Section 2.3). What about an electrically charged particle moving in a (time-independent) magnetic field? If the particle is at rest, it experiences no force. If it is moving, the force (given in Section 3.3) is always at right angles to the direction of motion. Such a force does not alter the speed or the kinetic energy of the particle: it affects only the *direction* of motion. (The force does no work on the particle, according to Figure 2.2.) Thus the kinetic energy by itself is constant: there is no potential energy. So Hamilton's action will not do. There has to be some way for the action to "know" about

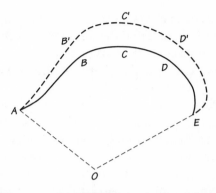

**FIGURE 6.3** One possible definition of "flux" for an open curve. Let *ABCDE* be the open curve. Choose a point *O*. Define the "flux" to be the flux through the closed loop *OABCDEO*. If *ABCDE* is changed to another curve *AB'C'D'E* with the same end points, the change in the "flux" is independent of the choice of *O*: that change is the flux through *ABCDED'C'B'A*.

the magnetic field. In order to explain what the correct action is, I must first make a little detour.

In Section 3.4, we noted that, for any *closed* loop in a region where there is a magnetic field, there is a well-defined quantity called the *magnetic flux* through that loop: it is simply the number of magnetic lines of force threading through the loop. We need to define a "flux" for an *open* curve. We cannot do this by counting lines of force through the curve, because "through" means nothing for an open curve.

One possible way to proceed is shown in Figure 6.3. Here the open curve in question is *ABCDE*. Choose, at will, some point *O*. Define the "flux" for the open curve *ABCDE* to be the flux through the closed loop *OABCDEO*. Not only is the point *O* chosen arbitrarily, but there are other quite different ways in which the "flux" for *ABCDE* might have been defined. There is much ambiguity.

So have we done anything useful? The answer is that there is something that is not ambiguous, not dependent on the choice of the point *O*, for example. That unambiguous thing is the *change* in the "flux" for the open curve, when that curve is deformed to another open curve with the same end points: for example, to *AB'C'D'E* in the figure. Indeed, that change is just the flux through the closed curve $ABCDED'C'B'A$. Thus the ambiguity in the "flux" for an open curve is connected with its two end points; given those end points, the dependence on the shape of the curve in between is unambiguous.

The phrase "flux for an open curve" is clumsy and is not a recognized usage. The technical term used by physicists is *Wilson line integral* and by mathematicians *holonomy*; neither is very descriptive. I shall use the acronym *FFOC* (Flux For an Open Curve), in the hope that this is memorable but not offensive.

There is one important property that a definition of a FFOC should satisfy. In Figure 6.3, the FFOC for *ABCDE* should be the sum of the FFOCs for the two pieces *ABC* and *CDE*. It is easy to see that the definition used in the figure does have this property.

So FFOCs are inherently ambiguous. But if we make use of a FFOC in the course of a physical calculation, the final answer should be unambiguous, provided we do everything in a consistent way.

For a complicated historical reason, the ambiguity in a FFOC is called the *gauge* ambiguity. The choice of a particular FFOC (for example, defined as in Figure 6.3) is called a *choice of gauge*. The independence of the final answer on the gauge choice is called *gauge-invariance*.

(The historical reason for the word *gauge* is that in 1918 Hermann Weyl tried to construct a unification of electromagnetism with gravity. In his theory the choice of gauge had something to do with a choice of scale of distances: hence the word *gauge*, as in the sense of a standard calibration – the German word used by Weyl was *Eich*. As Einstein pointed out, Weyl's theory led to physical contradictions. But the word *gauge* stuck.)

Ambiguities like this do occur in mathematics. Here is a simple model, which has something in common with the gauge ambiguity. Take $\sqrt{4}$. This is ambiguous: both $+2$ and $-2$ are square roots of 4. Nevertheless, it is sometimes sensible to write just $\sqrt{4}$ in the middle of a calculation: the final answer should be independent of which square root you choose, provided you make the choice consistently.

The seemingly perverse step, of going from the unambiguous flux through a loop to the ambiguous FFOC, has had a profound influence upon the development of physics in the twentieth century, as we shall see in later chapters.

After this long digression, we can now return to the problem of finding the action for a charged particle moving in a magnetic field. Any motion of the particle in space, whether it be the true or a false one, defines an open curve connecting the initial and final points *A* and *E*. Take a FFOC associated with *that* open curve. Multiply by the electric charge of the particle. That, then, is the magnetic piece in the action. In brief:

action = kinetic energy

+ [(charge) × (a FFOC along the particle path)].

The *magnetic action* defined like this is gauge-ambiguous, because the FFOC is ambiguous. But the principle of least action compares different motions with the given end points *A* and *E*. The change in the FFOC between these two paths is unambiguous, as illustrated in Figure 6.3. If the true motion makes the action least for one choice of the ambiguous FFOC, it makes it least for any other choice.

One can verify, though I will not do so here, that the principle of least action applied to this magnetic action leads to the correct Lorentz force (Section 3.3) on a particle in a magnetic field.

Note that the FFOC term in the magnetic action is not a potential energy. The notion of action is more general than that of energy.

One more historical aside. Maxwell originally wrote his laws of electromagnetism partly in terms of gauge-ambiguous quantities (actually FFOCs, but for the special case of a very short curve). This caused some confusion, and later workers like Heaviside were pleased to find that the laws could be written entirely in terms of the electric and magnetic fields (which, being measurable, are gauge-unambiguous). It is this latter form of the laws that I have described in Chapter 3. Gauge-ambiguous quantities are not necessary to write the laws of electromagnetism; they become necessary only if one wants to find a principle of least action for an electrically charged particle.

## 6.6 Time-Varying Fields and Relativity

There are two shortcomings to the actions introduced in the Sections 6.2 and 6.5. Firstly, I have assumed that the electric and magnetic fields were not changing with time. Secondly, it is not obvious that the actions are compatible with Einstein's theory of special relativity (see Chapter 5). We can remedy both shortcomings at once.

To make a theory compatible with relativity, we should express everything in terms of spacetime, not space and time separately. There are two things we must put into a spacetime context: the motion of the particle and the definition of a FFOC in terms of the fields.

The motion of the particle is easy. It defines a curve in spacetime, the world-line of the particle. This world-line begins at the event defined by a position $P$ and time $t_P$ and ends at the event defined by a position $Q$ and $t_Q$ (remember that an event, a point in spacetime, is specified by a position and a time). Along this world-line, proper-time is defined (see Section 5.5). The relativistic form for the action of a freely moving particle is the simplest thing one could think of:

free action $= -$(mass) $\times c^2 \times$ (total proper-time along world-line).

(Here $c$ is the speed of light.) We can see roughly that this action is something like that in Section 6.2 if we remember that

$$(\text{mass}) \times c^2$$

is Einstein's famous energy of mass (see Section 5.11). The minus sign here means that this energy is treated as potential energy.

The least value of this action means (because of the minus sign) the greatest value of the proper-time. In Section 5.5 we learned that the longer a (time-like) curve looks in a spacetime diagram, the less the proper-time it represents. Therefore, the maximum proper-time is represented by a straight line in the space-time diagram. This is the free motion, as we know.

In order to introduce the electromagnetic field, the next step is to generalize to spacetime the idea of the flux through a loop. Loops in space have to be generalized to loops in spacetime. This is not difficult. We have just to imagine a closed curve drawn in a spacetime diagram (representing four-dimensional Minkowski space). The direction of the curve may be space-like in some parts and time-like in others. There are many more curves than exist in three-dimensional space (twice as many, in fact). Because of this, the flux through them can be electric as well as magnetic (or generally some combination of electric and magnetic). Thus the whole electromagnetic field is specified by the fluxes.

Having made this generalization, we may make the step to "flux" for an open spacetime curve, just as we did to an open space curve in Section 6.6. I will continue to use the acronym FFOC for this generalization. These FFOCs are again gauge-ambiguous quantities, and the ambiguities are associated with the spacetime points (events) at the ends of the open curve.

Since all this occurs in spacetime, it is no surprise that time-dependent fields can be handled.

It is now easy to guess the action for a charged particle moving in a general electromagnetic field. It is

$$-(\text{mass})c^2 \times (\text{proper-time})$$
$$+ (\text{charge}) \times [\text{FFOC for particle world-line}].$$

The world-line begins at the event defined by the position $P$ and the time $t_P$ and ends at the position $Q$ with time $t_Q$. For a slowly moving particle in time-independent fields (electric and magnetic), this action reduces to the ones appearing earlier in this chapter.

In spacetime language, Hamilton's action principle can be restated:

- From all possible (false) world-lines of the particle, with given initial and final events, the true one is that which makes the total action least.

## 6.7 Action for the Electromagnetic Field

So far this chapter has been about the action principle for the motion of particles, perhaps moving in given electric and magnetic fields. But Maxwell taught us that these fields themselves are dynamical quantities, with their own equations of motion. So is there an action principle for the electromagnetic *field*, which is equivalent to Maxwell's equations? A particle has just six degrees of freedom: the six numbers needed to specify its position and velocity. Fields have an infinity of degrees of freedom: to specify their values at every point of space. Therefore, the action principle for a field must in some sense be more complicated than that for a particle. The action is distributed over space as well as time.

The recipe for finding the action for the electromagnetic field may be stated as follows:

- Take a small region of spacetime. Work out the quantity

$$\frac{1}{2}[(\text{electric field})^2 - c^2(\text{magnetic field})^2],$$

where the fields are evaluated in that small region of spacetime. Multiply this by the "volume" of that region of spacetime. Add the contributions over all the small regions between some initial time $t_1$ and some final time $t_2$.

Here, the "volume" of a small region of spacetime is a small volume of three-dimensional space multiplied by a little time interval. When we add the contributions, we include all of space, or at least all where the fields are present. But we limit the time range by $t_1$ and $t_2$ (in this respect, space and time are treated differently).

If we compare this action with the one in Section 6.2, we see that the square of the electric field appears like kinetic energy and the square of the magnetic field appears like potential energy. This is consistent with the expression for the energy in Section 3.5, where the two terms occur added together.

The least action principle for the electromagnetic field is a bit complicated. (One cannot just take the least value of the preceding action for *any* values of the fields, because it can be made as negative as we like by choosing the magnetic field to be very big.) First the action must be expressed in terms of FFOCs (FFOCs for very short, straight curves are sufficient). Then the principle says that the total electromagnetic action is less for the true values of these FFOCs than for any false values. Finally, the electric and magnetic fields are deduced from the true FFOCs, and they can be shown to satisfy Maxwell's equations.

## 6.8   Momentum, Energy and the Uniformity of Spacetime

Section 2.3 introduced the important ideas of the conservation of total momentum and energy. These principles may by summarized as follows:

- For a *closed* system, one can define total momentum and total energy so that these quantities do not change with time.

An important word here is *closed*. If the system is not closed, that is, if there are forces acting on it from outside, energy and momentum are not in general conserved. To get conserved energy and momentum, one would have to include with the system the things that were producing the outside forces. No system (except perhaps the entire universe) is exactly closed, but certain systems may (for certain purposes) be closed to a good approximation. Thus the momentum of the Earth is certainly not constant, but the momentum of the whole Solar System is constant to a much better approximation.

The principle of least action offers a new insight into these conservation laws of momentum and energy. It shows that they are connected with, respectively, the uniformity of space and the uniformity of time. By *uniformity*, I mean that no point of space or instant of time is special (as, for example, the centre of the universe – at the centre of the Earth – was a special point for Aristotle). More precisely, the action should make no reference to any special point or special time.

Take the motion of the Earth as an example. If we assume the Sun to be fixed at some point, the gravitational potential energy of the Earth will refer to that fixed point, and so therefore will the action. Correspondingly, the momentum of the Earth is certainly not constant (the direction of motion varies throughout the year). All this is because we have not taken a closed system. If we include the Sun as well as the Earth in the system (ignoring the rest of the Solar System for simplicity), the total momentum of the Earth *and* the Sun is to a good approximation conserved. We have now introduced more variables into the action: the position of the Sun as well as the Earth. The gravitational potential energy now depends upon the positions of Earth and Sun relative to each other: it makes no reference to any fixed point in space. The same is true therefore of the action.

In a similar way, conservation of energy holds when the action makes no reference to any special time. I will give a (rather artificial) example.

Suppose that Earth orbited not the single Sun, but a double "sun" consisting of two stars moving round each other in small orbits. Suppose we take the orbits of the suns as given (perhaps because they had been very carefully observed) and try to calculate the motion of Earth. As the suns move in their orbits, the gravitational force they exert on the Earth varies slightly with time. This has the consequence that the total energy of the Earth (kinetic plus potential) is not constant in time. It is as if the motion of the suns can "crank up" or "crank down" the energy of Earth. Corresponding to this, the action for Earth by itself does refer to special times – times as defined relative to the suns' motions.

To recover energy conservation, it would be necessary to treat the suns' positions on the same footing as Earth's. Then the energy would include the suns' energies as well as Earth's, and the total would remain constant. The action for the complete system would depend on the motions of all three bodies and no longer refer to any externally defined times.

This connection, with the uniformity of space and time, provides the deepest understanding we have of the conservation laws of energy and momentum. It is consistent with the fact that, in relativity theory, momentum and energy are treated together as a 4-vector (see Section 5.11) just as are position and time.

## 6.9 Angular Momentum

Momentum and energy are not the only things that satisfy conservation laws. Another important example is *angular momentum*. This has the same relation to rotational motion that momentum has to motion in a straight line. For a particle moving in a circle, the angular momentum (about the centre of the circle) is the mass times the speed multiplied by the radius of the circle. This definition can be generalized to other motions. It is because of conservation of angular momentum that a spinning ice-skater spins faster when he draws in his arms.

Conservation of angular momentum is connected with the *isotropy* of space, that is, the absence of any preferred direction. The gravitational potential energy (see Section 2.3) of the Earth in the Sun's gravitational field depends on distance only and so makes no reference to direction. Correspondingly, the Earth's angular momentum about the Sun is conserved (neglecting the motion of the Sun and the ignoring the other planets). On the other hand, in the preceding imaginary example of a planet orbiting a double sun, the gravitational potential would depend upon the direction of the planet relative to the orientation of the two suns. The angular momentum of the planet *by itself* would not be constant: it could be cranked up or down by the motion of the suns. Once again, conservation is restored if the suns' angular momentum is included.

If we talk about isotropy of space, what about "isotropy" of spacetime? Are there more conservated quantities like angular momentum? To answer these questions, we must remember that the momentum and energy of the system already define a direction in spacetime, and to this extent the "isotropy" is inevitably broken in some sense, leaving only a sort of residual isotropy. This is easiest to understand if the momentum happens to be zero: then what remains is the isotropy of space only. This simple situation, with zero momentum, is always attained if we describe things from the point of view of an observer moving with the centre of mass of the system.

## 6.10   Conclusion

The reader may well wonder where the physics is in all this talk of "least action". We seem to have said a lot, but nothing new about nature. Nevertheless, least action is almost universally used by physicists. It has the following advantages:

- A single quantity, the action, encapsulates all the laws of motion.

- The laws deduced from the least action principle are guaranteed to be self-consistent.
- Conservation laws (like those of energy, momentum and angular momentum) can be deduced from simple properties of the action.
- The action is a key quantity in the formulation of quantum theory.

It will be sufficient if the reader who has persisted through this chapter has some idea of the meaning of action and least action.

# GRAVITATION AND CURVED SPACETIME

*How Einstein discovered that what we think of as gravitational force is nothing but an effect of geometrical curvature of spacetime.*

## 7.1   The Problem

In 1906, Einstein, having the previous year explained Brownian motion (see Appendix C), invented his special theory of relativity (see Chapter 6) and, having begun the quantum theory of light, was promoted to technical expert *second* class in the patent office at Bern. In 1907 he began his great struggle to create a new theory of gravity, which culminated in his "general theory of relativity" in 1915.

Why was Einstein dissatisfied with Newton's theory of gravity? It had agreed with observations for more than two hundred years, predicting, for example, the existence of the planet Neptune (the work of Adams and Leverrier), which was discovered in 1846. There was one exception: a small discrepancy in the orbit of Mercury, announced by Leverrier in 1859. Several attempts had been made to explain this (for example, as due to a planet, Vulcan, between Mercury and the Sun), but none was generally accepted. So far as I know, this problem with Mercury was not important in motivating Einstein.

Einstein saw two problems with Newtonian gravity (Chapter 1). First: it was not consistent with Einstein's special theory of relativity – his account of spacetime. Second: it offered no explanation of the equality of inertial and gravitational mass (see Section 1.7). I will explain these two points in turn.

In Newton's theory, gravity is transmitted instantaneously; indeed the theory is formulated in the context of an absolute time. All this is forbidden in Einstein's analysis of spacetime, in which nothing can be transmitted faster than light, and there is no absolute time.

One might think that there should be a simple solution to this dilemma. Newton's gravity looks very like electrostatics (Section 3.1): they both have inverse-square laws of force. For moving charges, electric forces are described by Maxwell's electromagnetism (Section 3.5) in a manner perfectly consistent with Einstein's spacetime. Why could there not be a sort of "Maxwell theory of gravity"? The answer to this is that in electromagnetism like charges inevitably *repel*, whereas in gravity *all* "charges" (i.e., masses) are "like", and all attract. The reason that like electric charges repel can be seen from Faraday's lines of force, as in Figure 3.1: to get the charges nearer, the lines of force must be crowded together; that means that energy must be put into the electric field. (When unlike charges get nearer, on the other hand, the lines of force that join them shrink, and so energy is released from the field.) So a relativistic theory of gravity has to be something quite different from Maxwell's electromagnetism.

The equality of gravitational and inertial mass was explained in Section 1.7. In Newton's equation of motion under gravity, the inertial mass appears on one side (multiplying the acceleration) and the gravitational mass appears on the other (in the formula for the gravitational attraction). Experiment shows that these two masses are equal, to a very high accuracy, for all forms of matter. They therefore cancel out, and everything accelerates under gravity the same way. According to Newton, this cancellation seemed to be just an accident. Why put the two masses in, just to cancel them out again? Here was the vital clue that Einstein seized upon. Einstein wrote later:

> I was sitting in my chair at the patent office at Bern when all of a sudden a thought occurred to me: "If a person falls freely he will not feel his own weight". I was startled. This simple thought made a deep impression on me. It impelled me towards the theory of gravitation.

(History does not relate how Einstein was getting on with his patents work at the time.)

We now have an example that was not available to Einstein: the weightlessness of astronauts in freely orbiting space vehicles. If the rocket motor is off, the astronauts, their toothbrushes, their razors and so on, all float weightlessly within their vehicle. From within the vehicle, without looking out, there is no way to know they are in the Earth's gravitational field, rather than in deep space. If the gravitational and inertial masses were not exactly equal for all substances, this would not be true. Suppose, for example, the ratio of gravitational to inertial mass were slightly less for a toothbrush than for the material of the spacecraft. Then a toothbrush would move "up" inside the vehicle, away from the Earth. But this does not happen.

Surely, the spacecraft and all its contents must be moving, in some sense, on a natural track. By contrast, we, standing on the Earth, know about gravity because the ground is pressing on our feet and preventing us from following this natural track of free fall.

Does this mean that gravitation is abolished? Certainly not. If the astronaut looks out of the window and sees another spacecraft passing near on another slightly different orbit, she can infer that they are not both moving in deep space, for if they were, one would not accelerate relative to the other.

Even within the spacecraft gravity could in principle be detected. If the vehicle is 10 metres across, objects at the top have, relative to the centre, an acceleration of about a millionth of the acceleration due to gravity on Earth, that is, about $10^{-5}$ metre per second per second. An object starting away from the centre will drift into the side before an hour is up. Gravity is abolished within the craft only to the extent that it is very small compared to the size of its orbit. These effects, due to the variation of the strength of gravity between nearby locations, are called *tidal forces*. The tidal forces at the Earth due to the Moon are the (main) cause of the tides in the oceans (see Section 1.7).

What we see then is that gravity is abolished *locally* (that is, over a very small region), but if we compare different localities we are certainly aware of gravity. Gravity is all about how to fit together the small regions of local weightlessness. By 1912, Einstein (now a professor in Zürich) had realized that the key to this fitting together lay in the mathematical theory of *curvature*. This mathematics I will now explain.

## 7.2  Curvature

The idea of the *intrinsic curvature* of a surface was first clearly developed by Karl Friedrich Gauss (1777–1855), one of the very greatest of mathematicians ever. A spherical surface (say, the skin of an orange) has *intrinsic* curvature; a cylindrical surface does not. This is because we can, after cutting it open, flatten a cylinder without distorting it, but a sphere we cannot. Any map, of, say, North America, on the flat page of a book inevitably distorts something.

When we imagine a curved surface, we picture it *embedded* (as mathematicians say) in our three-dimensional space. Gauss saw that some properties (the intrinsic ones) of a surface can be defined without reference to a space in which it is embedded. The surface could be studied by a race of blind ants who were not able to leave the surface and who could investigate it only with their legs and antennae.

The important properties (for present purposes) of a surface are the following:

- Locally, on a small enough scale, the surface is well approximated by a piece of a flat plane. We do not notice the curvature of the Earth if we confine ourself to a region no bigger than, say, London.
- We can draw curves on the surface.
- We can measure the distance along such a curve, between any two given points. One of the ants mentioned could count its footsteps along the curve.
- We can define a *geodesic* to be a curve on the surface that is "as near straight as possible". More precisely, given two points, we can consider all curves joining them and define a geodesic to be one such curve that makes the distance along it least, or more generally, stationary (stationary in the sense explained in Section 6.2). On the Earth's surface, geodesics are great circles, that is, circles whose centres are at the

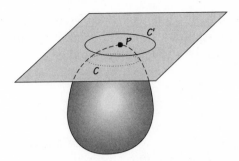

**FIGURE 7.1**  A definition of curvature. $P$ is a point on the surface. The dotted curve is a fixed distance from $P$. The circle $C'$ above it has the same radius but is on a flat plane. The dotted curve is shorter than the circle (in this example) by an amount that determines the curvature.

centre of the Earth. Geodesics are equally well defined, if less easy to visualize, on any surface – the surface of a potato, for example.

- We can define the *curvature* of the surface at any point, in a way I will now explain.

Here is a recipe (see Figure 7.1) for defining curvature at a point, $P$, on a surface (one should imagine a surface more general than a sphere, say, the surface of a pear):

- On the surface, draw a closed curve all of which is at some small fixed distance $d$ from $P$ (so this curve is as near to a circle as one can get on the surface). Measure the length of this small closed curve. Then the (Gaussian) curvature is defined by comparing the length of the small curve with the length (circumference) of a circle with the same radius $d$, but drawn in a plane:

$$\frac{\text{length of curve}}{\text{circumference of flat circle of same radius } d}$$

$$\approx 1 - \frac{d^2}{6} \times (\text{curvature}).$$

(The factor $\frac{1}{6}$ is used just to define curvature in a convenient way.) This approximate relation becomes more and more accurate as $d$ is chosen smaller and smaller. Clearly curvature has the dimensions of 1 over a length squared. For a sphere, the curvature is the same everywhere and can be shown to be equal to

$$\frac{1}{(\text{radius of sphere})^2}.$$

Naturally, the bigger the sphere the less the curvature.

The examples mentioned are all *closed* surfaces, but a surface may equally well be *open*, going on forever.

The curvature can be positive or negative. It is positive for any convex surface. An example of curvature that is negative is afforded by the midpoint of a saddle. A small curve drawn around this point is *longer* than the corresponding flat circle.

A surface is two-dimensional: fixing a position on it requires two numbers, like latitude and longitude. One can also think of three-dimensional spaces that have curvature. These are hard to imagine, because they would have to be embedded in flat spaces of four or more dimensions. But they can be defined mathematically. They have the properties listed, except that curvature is more complicated. The generalized definition of curvature was given by the brilliant mathematician Georg Riemann (1826–1866) in a famous probationary lecture at Göttingen. The definition can be approached as follows.

Near the point $P$ in question, take a very small region of the space (small enough to be very nearly flat). In this region, choose any little plane containing $P$. Draw geodesics starting out from $P$ in any direction in this plane. The set of all these geodesics makes up a two-dimensional surface. Then define the Gaussian curvature of this surface as previously. The result depends upon the orientation of the plane chosen through $P$. In order to specify the curvature of the three-dimensional space, one must have a rule that gives the curvature of the two-dimensional surfaces for any choice of the orientation of the plane. The rule can be expressed in terms of an ellipsoid (like a rugger ball). Given the orientation of the plane, one constructs a radius from the centre of the ellipsoid in the direction at right angles to the plane. The length of this radius can then be used to determine the curvature of the surface.

Thus the curvature, at any point $P$, of a three-dimensional space is specified by an ellipsoid: the lengths of its three axes and its orientation.

There is one particular curvature associated with such an ellipsoid that is especially simple. It may be defined geometrically as follows. Take a surface in the space, such that it is all at a fixed small distance $d$ from the point $P$. We may call this surface a "sphere" – it is the nearest thing to a sphere that can be embedded in the space. Then compare the area of this "sphere" with the area of a sphere of the same radius $d$ in flat space. The curvature in question, which is called the curvature *scalar* (the word *scalar* signifies that no particular direction is associated with it), is defined by

$$\frac{\text{area of "sphere"}}{\text{area of flat sphere of same radius}}$$

$$\approx 1 - \frac{d^2}{3} \times (\text{curvature scalar}).$$

As before, this approximation becomes better and better the smaller we choose $d$.

With more effort of imagination, these ideas can be extended to a four-dimensional curved space, starting off with the curvature of all the geodesic surfaces through the point $P$. But now the information necessary to fix all these curvatures is more complicated. Part of it is again specified by a (generalized) ellipsoid, with four axes. This part of the curvature is called the *Einstein curvature*. But now another 10 numbers are needed in addition. These are called, collectively, the *Weyl curvature*. The reason that four dimensions is qualitatively different from three can be traced to the fact that, in three dimensions but not in four, there is a unique direction at right angles to any flat surface.

In four dimensions, a curvature scalar is defined in a straightforward generalization of the way it is done in three dimensions.

All the curved surfaces and spaces we have mentioned so far in this section are *Riemannian*. That means that any small region of them – small enough to be effectively flat – obeys all the rules of ordinary Euclidean geometry: angles of a triangle add up to 180

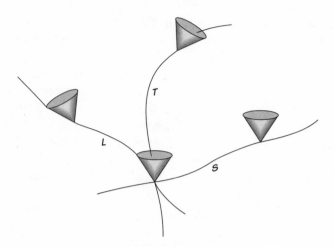

FIGURE 7.2   A representation of part of a curved Minkowski spacetime. Four examples of light cones are shown. The three curves are a time-like geodesic (*T*), a space-like geodesic (*S*) and a light-like one (*L*).

degrees, Pythagoras's theorem is obeyed and so on. For applications to the physics of spacetime, the imagination must be stretched one final time. That is to envisage curved four-dimensional *Minkowskian* spaces. Here any small region is like flat Minkowski space, as defined in Section 5.5. This means in particular that there is a light cone at each point of curved Minkowskian spacetime.

In curved Minkowski spacetime, just as in flat, there are curves representing the world-lines of particles. Along them, proper-times are defined, as the time registered by a clock carried along the curve. Geodesics are defined, and they may be time-like (everywhere inside the light cone), space-like (everywhere outside the light cone) or light-like (everywhere touching the light cone). Figure 7.2 attempts to represent these three possibilities. Looking at this figure and similar ones, we must be very cautious. First, we are trying to represent a curved space on a flat piece of paper, and this involves distortion, like a map of Asia in an atlas. Second, it is a Minkowski space, so that the lengths of the lines drawn in the figure do not necessarily correspond to anything physical (remember, for instance, Figure 5.4). Third (less importantly), the light cones should be drawn small enough so that they lie in a region where spacetime is practically flat.

In Figure 7.2, we can see two indications that the spacetime is curved. The geodesics are not represented by straight lines in the drawing. And the light cones are not drawn parallel to each other.

Curvature can be defined in Minkowski spaces, just as in Euclidean ones, but the definitions have to be modified a bit. The numbers of independent curvatures are the same.

We are now ready to return to the physical application of curvature in the theory of gravity.

## 7.3 Gravity as Curvature of Spacetime

In inventing the geometry of curved spaces, Gauss and Riemann certainly had in mind the possibility that physical space might be curved. This idea was considered by other people in the nineteenth century, for example, Helmholtz and Clifford. Nothing came of this, and by Einstein's time the subject was not well known to physicists.

Let me for a moment follow a dead-end line of thought. In Section 7.1, I remarked that motion under gravity suggests rather that bodies are following some sort of "natural track" than that they are responding to a force. The mathematics of curved spaces has provided such "natural tracks" – geodesics. What is more natural than that gravitation is the curvature of space and the tracks of bodies are just geodesics, that is, paths that minimize distance (or at least make it stationary)?

This does not work, however. Suppose, as an example, that the Earth is following a geodesic in space. Then it should follow that same geodesic whatever its speed. We know that this is wrong. The size of the orbit depends upon the speed of a planet. Outer planets move faster than inner ones.

As often happens in science, an appealing idea needs an extra twist before it works. The extra twist in this case was not available in the nineteenth century. It was Einstein's 1905 discovery of the unity of spacetime. We must identify gravity, not with curvature of space, but with curvature of *spacetime*. In other words, we must use

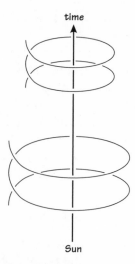

FIGURE 7.3    The helical world-lines, in curved spacetime, of two planets. Time is plotted upwards on the page. The vertical line represents the world-line of the Sun (whose motion is neglected). This diagram does *not* use units in which the speed of light is 1. The vertical axis represents a few years, and the horizontal axis perhaps a hundred million miles.

the geometry of curved, four-dimensional spacetime, as described at the end of the last section. Figure 7.3 indicates the shapes of the world-lines of two planets, moving in circular orbits with the correct speeds. The world-lines are helical. Both world-lines are supposed to be geodesics in some curved spacetime.

In order to make this idea into a quantitative physical theory, two laws are needed: the law of motion of a body, given the spacetime, and the law that determines the curvature of spacetime from the distribution of gravitating matter. For example, to find the Earth's orbit, we must know how the Earth moves in a curved spacetime; and we must know how the Sun determines the curvature of spacetime (for simplicity here I ignore the gravitational force exerted by the Earth on the Sun). The first of these laws we have already stated: bodies (or at least small bodies) move along geodesics. The second law caused Einstein more trouble to get right. I will state it in the form first given by David Hilbert in 1915 (and in fact published just before Einstein produced his final version of his theory).

Before doing this, there is a technical definition that we will need. It is related to describing a curved spacetime.

## 7.4 Maps and Metrics

Curved spacetime is very difficult to visualize. Even when it is possible to embed it in a larger flat space, one might need a space of say six dimensions. The usual way to describe a space (or spacetime) is by means of a *map* with a *metric*. I will explain these two terms in a very simple example: a map of the Earth's surface.

The Earth is (roughly) a sphere and so it has (intrinsic) curvature and cannot be flattened out on a page of an atlas without distorting it in some way. I shall use the word *map* to denote a representation of a curved surface (or, more generally, curved space) on a flat surface (or, more generally, flat space). Take as an example a simple (if not very good) way of representing the Earth's surface on a flat piece of paper. This is called the *cylindrical equal-area projection* and is illustrated in Figure 7.4. The page of paper is rolled up to make a cylinder such that the Earth just fits into it. Any point, say, $Q$, on the Earth's surface, is projected outwards onto the cylinder from the Earth's axis. That is to say, $Q$ in the figure is projected to give

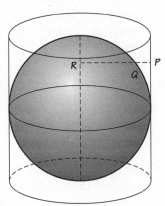

FIGURE 7.4    Projecting the surface of the Earth onto a cylinder. A point $Q$ on the Earth is represented by $P$ on the cylinder, such that the line through $P$ and $Q$ meets the Earth's axis at right angles at $R$. The cylinder can then be unrolled, without distorting it, to make a flat map.

$P$ on the cylinder, so that the line through $P$ and $Q$ meets the axis at right angles (at the point $R$ in the figure). Then the cylinder can be opened up, say, by cutting it along some line parallel to the axis.

This representation introduces some very bad distortions. Lines of longitude come out as equally spaced straight lines. Lines of latitude come out as straight lines also, but they crowd closer and closer together near the two poles. In fact the projection becomes ambiguous at the two poles themselves: there is no unique point on the map to represent the North Pole or the South Pole. Lands near the equator are represented quite well. (There is one good feature of this representation: equal areas on the Earth are represented by equal areas on the map. This is because, on moving away from the equator, distances are compressed in the north-south direction but stretched by a compensating amount in the east-west direction.)

How can we represent the distortions inherent in the map? Take any small region of the Earth's surface (small enough so that it could be represented by a flat map with negligible distortion) and ask how it appears in the map. It is sufficient to take a small circle on the Earth. This will appear (if it is small enough) as an ellipse on the map. We can do this for lots of small regions, sufficient to cover the whole Earth. The totality of all the ellipses thus obtained is called the *metric* (in cartography, it is called the *distortion diagram*). The map, together with the metric, is a way of representing the surface.

Figure 7.5 gives as an example some of these ellipses for the map projection shown in Figure 7.4. The ellipses are actually circles on the equator. Moving south or north they get more and more flattened out, but with their areas all the same. One should imagine that the ellipses are very small and that there are lots of them covering the whole map. In this example, the pattern of ellipses is very simple, but in general (for a map of a potato, for example) the pattern may be complicated and irregular.

Given a map, a knowledge of the metric enables one to work out (a) the (true) distance along any curve, (b) the shape of geodesics and (c) the curvature anywhere.

We can generalize all this to a three-dimensional curved space. The "map" would now be in a flat three-dimensional space, say, a block of perspex. The metric would consist of small ellipsoids (three-dimensional generalizations of ellipses, for example, a rugger

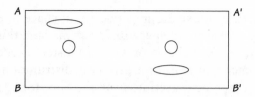

FIGURE 7.5   The metric on the map described in Figure 7.4. The ellipses that make up the metric are circles on the equator and get flattened out approaching the poles. Their areas are all the same in this example. The cylinder was opened up by cutting along a line $AB$, and $A'B'$ is the other side of the cut. Thus these two lines represent a single line of longitude on the Earth. Also, there is no unique point on $AA'$ to represent the North Pole, nor on $BB'$ to represent the South Pole.

ball). We have encountered ellipsoids already in Section 7.2 in connection with curvature. The metric ellipsoids and the curvature ones are of the same mathematical nature, but of course they are not (in general) identical.

For physics, we need a generalization to four-dimensional spacetime. To represent a map on a page, one needs to leave out some dimensions and perhaps resort to some sort of perspective impression.

But the most important thing about spacetime is that the metric is made up of hyperboloids (generalization of hyperbolas) instead of ellipsoids. In the following pages I will not attempt to draw such hyperboloids, for fear of the diagrams' becoming too complicated. But a key feature of Minkowski space is the the *light cone* at each event (see Figure 5.2). In curved spacetime, for each event, there is a light cone lying in a small nearby region. In a map, these light cones will be distorted: stretched, squashed or tipped over. The distortion of the light cones is an important part of the information contained in the metric. I will draw the distorted light cones in some later diagrams.

## 7.5   The Laws of Einstein's Theory of Gravity

After this technical digression about metrics, we can formulate the laws of Einstein's theory. One of these laws we have already met: that particles move along geodesics in spacetime. For example, given the curvature of spacetime near the Sun, this law tells us how the Earth moves. The second thing Einstein needed was a law to fix the geometry of spacetime given the distribution of masses in it, to fix the shape of spacetime near the Sun, for example, given the mass of the Sun.

In Newton's theory, it is mass that is the source of gravitation. From Einstein's special theory of relativity, we know that mass and energy are related. It turns out that Einstein's theory of gravity requires us to think of energy, not mass, as the source of "gravitation" (that is to say, of spacetime curvature). But energy goes along with momentum to make up a 4-vector (Section 5.11), so we must include momentum as well as energy as part of the source of curvature. In general, energy and momentum are distributed over space, for example, over the interior of the Sun, or, much more weakly, over the Sun's magnetic field (Section 3.5). So the distribution of energy and momentum is to be the source of spacetime curvature. This distribution is another of those quantities that may be represented mathematically by a (generalized) ellipsoid. We have already met two of these: the metric and the Einstein curvature.

These remarks are almost enough for us to guess *Einstein's law of gravitation*. It is this: everywhere in spacetime

Einstein curvature $= 8\pi G \times$ (energy-momentum distribution).

Here $G$ is the constant of gravitation, appearing here as it does in Newton's theory. The factor $8\pi$ is just an accident of the way that $G$ is defined. Also, as befits a relativistic theory, I have used units in which the speed of light is 1.

Einstein's law makes sense in virtue of a number of properties of curvature. First, there is a property, which I cannot explain in more detail, that makes the law consistent with (and indeed imply) the conservation of energy and momentum. Second, what happens in empty space, where there is no energy and momentum distributed (for example, the empty parts of the Solar System)? Does Einstein's law imply that there can be no curvature there? It does not, because Einstein curvature is only part of the curvature of four-dimensional spacetime. There are 10 more components, which are not *directly* fixed by Einstein's equations. In fact, Einstein's law gives just the right amount of information from which the metric can be determined (for a given choice of map), and hence all the curvature, both inside and outside the Sun.

In the absence of any matter (or other form of energy and momentum) anywhere, spacetime is flat. Because Newton's constant $G$ is so small, there are significant deviations from flatness only if there are large masses (or more generally large distributions of energy or momentum). This is just another way of saying that gravitation is a very weak force: it becomes so important in the universe only because the combined effects of very many particles all add up.

Gravity, unlike the electric force, is always attractive. This is connected with the fact that energy is always positive. A negative value for the energy of a particle would allow the extraction of an unlimited amount of positive energy by another particle, at the expense of the first's getting more and more negative. It would be like the possession of a bank account which allowed an unlimited overdraft.

In Chapter 6, I described how many of the laws of physics can be stated as least action principles. Einstein's theory of gravity is no exception. The geodesic law of motion is already of this form. A geodesic is *defined* to be a curve of minimum distance between two points. Actually, in Minkowski spacetime, the geodesic world-line of a moving particle makes the lapse of proper-time a *maximum* not a minimum. This is because of the odd geometry of Minkowski spaces: curves that look shorter in diagrams actually represent longer proper-times (Section 5.7). But "least action" should strictly be stationary action (Section 6.3), and a maximum is a special case of a stationary value (the top of a mountain is locally flat).

Thus the action for a particle moving in curved spacetime may be taken to be (in units in which the speed of light $c = 1$):

−(mass) × (total proper-time along a portion of world-line).

The reason for putting in the mass factor will become clear later. Although this definition reads the same when spacetime is curved as when it is flat, it is more complicated to work out for curved spacetime as it requires a knowledge of the metric.

(The alert reader may wonder whether there is any connection between this principle of stationary proper-time and Fermat's principle of least time in optics [see Section 4.4]. The former refers to the motion of a particle, the latter to a light ray, so there is no necessary connection. In fact it can be shown from Einstein's theory that a light flash follows the path of least time (not proper-time) in curved spacetime, provided that the curvature is not changing with time.)

The action for a particle given the metric, then, is just the proper-time. The true motion of the particle is that which makes this action stationary when compared to all possible "false" motions.

We also need an action principle to give Einstein's law of gravitation, connecting the curvature to the distribution of energy and momentum. The thing to be varied here should be the shape of spacetime itself, that is, in practice, the metric. We already have part of the required action: the proper-time for each particle. This depends upon the metric, and so it varies when the metric is varied. It turns out that these proper-times provide the right-hand side of Einstein's law, the energy-momentum distribution (in so far as it is due to particle masses). It is to make this work out right that we inserted the factor of mass in front of the proper-time.

So we only require a new action to give the left-hand side of Einstein's law: the Einstein curvature. This action was discovered by the great mathematician Hilbert at almost the same time as Einstein found his law. It is about the simplest thing one could think of. It uses the scalar curvature, explained in Section 7.2 (or rather this generalized to four-dimensional spacetime). In terms of this, Hilbert's action is given as follows:

- Divide spacetime into a lot of small regions. In each region, take the value of the scalar curvature there and multiply by the spacetime (four-dimensional) "volume" of that region. Add up all these contributions. Multiply by

$$\frac{1}{16\pi G}.$$

As usual, $G$ is the constant of gravitation, and units are used in which the speed of light is 1. (I have put "volume" in quotation-marks because I am extending the word from three to four dimensions. Some people would write *hypervolume*.)

So the complete action principle for spacetime is

- Add together Hilbert's curvature action and the proper-time actions for each particle. Find the metric for which the combined action is stationary. This metric determines the curvature of spacetime.

This is how matter determines the curvature of spacetime, that is, how it acts as a source of gravity. The proper-time actions for the particles have a dual role: they depend upon the particle motions *and* on the metric.

Now that I have stated Einstein's theory of gravity, there are two questions we must answer. Does it give Newton's theory as a good approximation, at least in suitable cases? What does it predict differently from Newton?

## 7.6 Newton and Einstein Compared

At first sight, Einstein's account of gravity seems so utterly different from Newton's that is hard to see how they can be related. Two of the fundamental differences are these.

- Newton's law is linear, Einstein's non-linear. This means that, as an example, doubling the Sun's mass would double the Newtonian gravitational force due to the Sun. But it would not in general double the spacetime curvature in Einstein's theory – the change would be more complicated. The reason for this is that energy (and momentum) is the source of spacetime curvature, but such curvature itself may have energy associated with it, so gravity feeds off itself.
- Newtonian gravity propagates instantaneously by action-at-a distance. In Einstein's theory, gravitational effects (like everything else) cannot be propagated faster than the speed of light.

In spite of these profound differences, Einstein's theory has Newton's as a good approximation, provided that bodies are moving slowly compared to the speed of light and that the curvature is "small". (Here "small" means in comparison with 1 over the square of some relevant distance.) For example, the curvature of spacetime at the Earth's orbit multiplied by the square of the distance from Sun to Earth is about $10^{-8}$. And the square (it is the square that is relevant) of the speed of the Earth round the Sun is also about $10^{-8}$ of the square of the speed of light.

With neglect of this small quantity, Einstein's theory reduces to Newton's. An explanation of this is that Newton's theory is uniquely fixed by the properties of being linear and instantaneous and having the gravitational constant with given dimensions (then the law must be the inverse-square one). As far as the motion of the Earth is concerned, Einstein's theory differs from Newton's by only about 1 part in $10^8$.

However, Einstein's theory gives corrections to the inverse-square law. These involve either the square of Newton's constant or the speed of the planet (compared to the speed of light). For the inner planets, these corrections have relative size about $10^{-8}$. As explained in Appendix A, the inverse-square law has the special property that the orbit "closes up" (that is, gets back to exactly the same point) periodically. Any deviation from an inverse square produces a slight angular shift every time the planet gets back to its closest position to the Sun (the position of *perihelion*).

This effect should be largest for Mercury, partly because it is moving fastest (nearly twice as fast as Earth) and partly because the elliptical shape of its orbit is quite far from being circular. One week after completing his theory, Einstein calculated that it predicted an advance of the perihelion of Mercury by about $3 \times 10^{-5}$ degree per orbit, that is, by about $8 \times 10^{-8}$ of a complete rotation per orbit. This agreed with observations that had been made and that had been unexplained since the work of Leverrier in the nineteenth century.

This success gave Einstein understandable delight:

For a few days, I was beside myself with joyous excitement.

According to Pais's biography, for three days Einstein could do no work.

## 7.7  Weighing Light

Is light affected by gravity? Does light have "weight"? From Newton's theory of gravity one might argue as follows. The orbit of a particle does not depend upon its mass, so why should not light move along the same path as a particle would *if* it had the speed of light? Of course, that speed is very high, so light could not move around the Sun like a planet (in an orbit outside the Sun). But if a light ray passed very near the Sun it should deviate a little from a straight line path. For a ray that just grazes the Sun, the deviation would be $2.4 \times 10^{-4}$ degree.

But this argument is not expected to be reliable. Light, with its constant speed, does not fit into Newtonian mechanics. In order to deal with light, the theory must be consistent with special relativity (in which, in fact, no particle with mass can move with the speed of light). Einstein's gravity theory does provide a consistent way of dealing with light. In flat spacetime, a light flash generates the light cone. In curved spacetime, there is a light cone in any small region of spacetime. It is natural to assume that the history of a beamed light flash is a curve that everywhere touches the local light cone and in addition is a geodesic (slightly extending the definition of geodesic). With this assumption, Einstein showed in 1918 that the bending of a light ray grazing the Sun should be $4.8 \times 10^{-4}$ degree. Loosely speaking, half this deviation comes from the Newtonian-type argument earlier, and the other half comes from the curvature of space (here I mean three-dimensional space, as opposed to four-dimensional spacetime).

This deviation causes a shift in the apparent position of a star when it is in a direction very close to that of the Sun. This shift was observed in 1919. The story of this is interesting, and I will recount some of it as told by Chandrasekhar.

The famous British theoretical astronomer Eddington had been sent copies of Einstein's papers during the Great War by the Dutch astronomer de Sitter, and by 1917 Eddington was well familiar with Einstein's theory of gravitation. Eddington persuaded the astronomer royal, Sir Frank Dyson, of the importance of an observational test. Eddington was a Quaker and therefore a pacifist. He might have claimed deferment from military service as a conscientious objector, with the stigma and unpleasantness that this entailed in the 1914–1918 war. In fact, partly by Dyson's intervention, he was given a special deferment, a condition of which was that he should lead an expedition in 1919 (if the war had ended by then) to check Einstein's prediction.

To allow observation of the bending of light by the Sun, the Sun has to pass close to some bright stars during a total eclipse (to remove the glare of the Sun). By a remarkable chance, these conditions were well fulfilled on 29 May 1919. Two expeditions were arranged, one to Sobral in the north of Brazil and the other (which included Eddington) to the island of Principe in the Gulf of Guinea off the coast of West Africa. In Principe, the day started cloudy, but two photographs were found to contain the necessary images of stars.

The results of the two expeditions were officially announced at a meeting of the Royal Society on 6 November 1919. A. N. Whitehead wrote:

> The whole atmosphere of intense interest was exactly that of the Greek drama: we were the chorus commenting on the decree of destiny as disclosed in the development of a supreme incident. There was a dramatic quality in the very staging – the traditional ceremonial, and in the background the picture of Newton to remind us that the greatest of scientific generalizations was now, after more than two centuries, to receive its first modification. Nor was the personal interest wanting: a great adventure in thought had at length come safe to shore.

(In fact, the measurements made on these expeditions were of poor quality. The bending of electromagnetic radiation has been measured much more accurately since, especially by radio astronomy.)

On this occasion, a British expedition was confirming a German theory (Einstein was in Berlin between 1914 and 1918). This reminds me of the French expedition of 1736 to check Newton's prediction about the shape of the Earth (see Section 1.7). The 1919 expedition was led by Eddington, who was an early expert on Einstein's theory. The 1736 expedition was inspired by Maupertuis, an early champion of Newtonianism in France.

Until the 1960s, observational tests of Einstein's theory were few and not very precise. By now, it is accurately confirmed by measurements within the Solar System. But systems are now known in which the effects of general relativity are more dramatic. A *binary pulsar* consists of two very small and dense neutron stars that orbit each other. One of the neutron stars is a *pulsar*; that means that it emits very regular pulses of radiation. This enables the system to be well monitored. The orbit precesses, as does the orbit of Mercury, but much faster. Also, the orbit has been observed (over many years) to shrink slowly. This is explained as due to loss of energy caused by the emission of "gravitational waves". Such waves are expected on Einstein's theory, being ripples in the curvature of spacetime, propagating with the speed of light. Accelerating masses should emit gravitational waves in much the same way (but much less strongly) that accelerating charges emit electomagnetic waves.

## 7.8 Physics and Geometry

In Chapter 1 we noticed two contrasting strategies for trying to understand the physical world: mechanical model building on the one hand and just plain mathematics on the other. These were typified by, respectively, Descartes's vortex theory of gravity and Newton's approach in the *Principia*. A similar dilemma arose in the nineteenth century in trying to understand electromagnetism (see Chapter 3). Again mechanical models were tried, but Maxwell's theory that finally triumphed was just mathematical.

The problem with a mechanical explanation is that it can never do more than push the questions back to something else that remains to be explained. What was the nature of the medium (aether) in which Descartes's vortices swirled?

How should we place Einstein's theory of gravity? It seems to open up yet another style of explanation, geometrical, which in some ways lies between mechanical and mathematical. Einstein's theory is of course highly mathematical. Einstein had to teach himself the mathematics of curvature, which was not well known at the time. Nevertheless, we do feel that his theory is not *just* equations. We think we can try to picture, however imperfectly, curved spacetime "out there" as some sort of real "thing". For most of us, our mental images contain much that is irrelevant or misleading. We think of space as something like rubber or perspex, not just empty space. We think of it as embedded in more dimensions, which (probably) have no physical significance. We find it very hard to imagine Minkowski spacetime, as opposed to space with a metric as in Figure 7.5. Notwithstanding all this, many people find a geometrical theory a rather satisfying compromise between the mechanical and the mathematical. Such people would be very happy if a geometrical explanation of electromagnetism, say, could be discovered.

Is "spacetime" any less of a slippery concept than the old "aether"? The only answer I can give to this question is that the properties attributed to the aether were mechanical, whereas spacetime is described in geometrical terms.

## 7.9 General "Relativity"?

Einstein called his 1905 theory of spacetime "special relativity" and his 1915 theory of curved spacetime "general relativity". The reason for using the word *relativity* in connection with the 1905 theory is clear. The basic premise is that absolute motion with constant velocity has no physical meaning. Only *relative* motions with constant velocity are observable. Put geometrically, all time-like (or all space-like) directions in spacetime are equivalent. Roughly, spacetime is *isotropic*, like a featureless flat desert in which all directions look the same.

The 1915 theory is quite different. Spacetime is curved, so all directions are not the same, any more than they are in a mountain range. Einstein used the term *general relativity* because one is free to choose any *map* (in the sense of Section 7.4) to describe spacetime. The description that one has is "relative" to the map. But any choice of map describes the same physics. In a similar way, different maps of Africa may make it look a different shape but carry the same geographical information if interpreted correctly. No choice of map is uniquely "right", though some are more convenient than others for a given purpose. But this is a statement about our maps of spacetime, not about spacetime itself. For this reason, I think the term *general relativity* is a misleading one.

## 7.10  Conclusion

Einstein's 1915 theory explained gravity as nothing more than a manifestation of the curvature of spacetime. It naturally explained why all bodies (whatever their composition) move along the same orbits under gravity. It could not have been formulated without the 1905 theory of spacetime. It was consistent with the laws of electromagnetism and unambiguously predicted the bending of light by gravity. It predicted that gravitational waves should exist, and there is now indirect evidence for them, from observations on binary pulsars.

The theory has the consequence that black holes may exist, with strange, but also remarkably simple, properties. As we shall see in Chapter 14, Einstein's theory may be applied to the whole of spacetime, in the subject of cosmology.

# THE QUANTUM
# REVOLUTION

*How, in the first quarter of the twentieth century, the laws of physics were fundamentally altered, in a way with which scientists are still struggling to come to terms.*

## 8.1   The Radiant Heat Crisis

At the beginning of Chapter 5, I mentioned the contradictions lurking in late-nineteenth-century physics. One of these arose from the mismatch between Newton's mechanics (Chapter 1) and Maxwell's electromagnetism (Chapters 3 and 4). This was resolved by Einstein in 1905 with his theory of spacetime. Also, Newton's laws of gravity raised questions that were only answered in 1915 by Einstein's theory of curved spacetime (Chapter 7).

We now come to another contradiction, which first appeared as the mismatch between the statistical theory of heat (Chapter 2) and electromagnetism. This contradiction led, in 1900, to the beginnings of yet another great change in physical thought, the quantum revolution, something with which physicists now, a century later, are still grappling.

Consider an oven (or a furnace) whose walls are raised to some high temperature. Look into it through a small hole. If the temperature is high enough, it will look "white hot". A little less and it will be "red hot". Cooler still and no light will be seen, but heat radiation (infrared radiation) will be felt. In 1860, the physicist Kirchhoff, then in Heidelberg, announced the law that the nature of the radiation in the oven depends *only* upon the temperature,

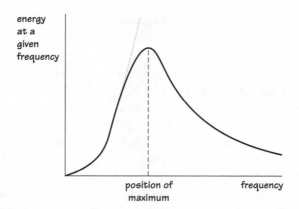

FIGURE 8.1　Radiant energy per unit frequency range plotted against frequency. The shape of this curve depends only upon the temperature, and by suitably adjusting the scales along the two axes, the same curve can be used for any temperature. The curve has a maximum at a definite frequency, which depends upon temperature only. Classical physics, without Planck's quantization rule, predicts the catastrophic rise shown by the thin dotted line.

being independent of, for example, the material of the walls. The radiation actually has a spread of different frequencies (that is, colours), as indicated in Figure 8.1. Kirchoff's law says that this distribution depends upon the temperature only. The problem of finding this distribution remained a challenge for 40 years. The distribution has a maximum at a certain frequency, and this frequency depends only on the temperature. For simplicity, we can focus on the position of this maximum.

What is going on in the oven? The walls are hot, so the molecules in them have random thermal motion. Sometimes, collisions set the electrons in a molecule into motion. The motion of electrically charged particles can emit light or other electromagnetic radiation. The radiation may cross the oven, hit electrons in the material there and perhaps be absorbed again. Eventually a state of equilibrium will be attained, when as much radiation (of any given frequency) is absorbed as emitted. Then the radiation itself has reached the same temperature as the oven. What does this mean?

When a gas has some given temperature, it means that the molecules have a random motion, with an average energy that is determined by the temperature (indeed, can be used to define that temperature). The energy is distributed equally among all the *degrees of freedom* (see Section 2.5), for example, among all the molecules and all the (three) directions in which each can move. So we must ask, What is the meaning of "degrees of freedom" for radiation?

Let us first take a simpler system, a violin string. The complete, unstopped string can vibrate as a whole, giving the fundamental frequency. The wavelength is then twice the length of the string. By touching the string in the middle, the first harmonic can be produced. This is an octave above the fundamental, and the wavelength is equal to the length of the string. Similarly, by touching the string a third, a quarter, and so on, of the way from the bridge, higher harmonics can be made to sound. Each of these forms of motion of the string is called a *mode* of vibration. The mode is defined by the wavelength, and this determines the frequency of vibration.

For a two-dimensional system, like a drum, the modes are more complicated to define. One can characterize a mode by the wavelength together with a direction. The electromagnetic radiation in the oven similarly is made up of modes, which now require a wavelength and a direction in three dimensions to define each of them. The frequency is again given in terms of the wavelength, by dividing by the speed of light.

Each mode corresponds to 4 degrees of freedom. This number 4 is an unimportant detail, but it comes about as follows. First there is a factor of 2 because of the two possible polarizations (see Section 4.8). Second there is another factor of 2 because there are two terms in the electromagnetic energy, electric and magnetic. By the equipartition principle (Section 2.5), the thermal energy should be distributed equally among all the modes:

- *Equipartition law*

energy per mode $= 2k \times$ (temperature).

Here $k$ is just Boltzmann's constant, which converts temperature units to energy ones. Thus the energy per mode in radiation should be similar in size to the energy per molecule in a gas at the same temperature.

But there is no limit to how short the wavelength of a mode can be (it cannot be longer than the size of the oven, but long wavelengths turn out to be no problem anyway). So there seems to be no limit to the number of degrees of freedom. This would imply that radiation could go on soaking up heat energy indefinitely, never attaining equilibrium, in complete contradiction with what is observed.

Actually, there is another way to see that there is a grave problem. This depends upon a "dimensional argument", involving thinking about units (see Appendix D). We want to find the *frequency* that gives the maximum of the energy distribution of the thermal radiation in Figure 8.1. We believe that this depends only upon the temperature. Using Boltzmann's constant $k$, we can convert the units of temperature into units of *energy*. So we want a connection between a frequency and an energy. Is there anything else that might be relevant? If we were talking about a gas, there would be other things like the mass of the molecules. But for pure radiation the only relevant number is the speed of light $c$. It is impossible to match up the units in any relation connecting a frequency to an energy and a speed. The units in which energy are measured include a unit of mass (say the kilogram), but neither speed nor frequency has anything to do with mass. So again we have a contradiction. The only way out is if there is another "constant of nature" that is relevant and that we did not know about.

A decisive step toward the solution of this problem was taken in 1900 by Max Planck (who, interestingly, had been thinking about the units involved the year before – see Appendix D). Planck, one of the great physicists of the twentieth century, was moderate, patriotic and musical: a man of great probity who became a powerful scientific administrator. One of his sons died in the First World War, and another was executed in 1945 for supposed complicity in the attempt to assassinate Hitler.

Between 1895 and 1900, Planck made several unsuccessful attempts to solve the problem of radiant heat. The frequency distribution was being measured more accurately at the time, in Berlin especially. In 1900, Planck made an informed guess of a formula for the frequency distribution, which fitted the new data well. His formula contained a new "constant of nature", which he called "the unit of action" (see Chapter 6 for the definition of action) and denoted by the letter $h$. This has been called *Planck's constant* ever since and is still denoted by $h$.

The modern value of Planck's constant is

$$h = 6.626 \times 10^{-34} \text{ joule second.}$$

The value Planck obtained in 1901 (by fitting his formula to the experimental data) was within 1 or 2 percent of this. He got a similarly accurate value for Boltzmann's constant, which also appears in his formula.

Planck then worked backwards, trying to think up an explanation for his formula. His argument was complicated and only partially logical. Einstein began to work on the problem in March 1905 (before, in the same year, he proposed the theory of Brownian motion [Section 2.5] and invented special relativity [Chapter 5]), but even then it was another 20 years before a proper understanding was attained. I will describe one of the modern derivations of Planck's formula (which he or Einstein might well have given, but did not).

Planck's new and crucial assumption might have been stated

- *Planck's radiation rule:*
  The energy in any particular mode of radiation is a whole number multiple of the *quantum*

$$h \times \text{(frequency)}.$$

This assumption was completely unexpected and unexplained. Nothing like it had been seen before. It does not contradict everyday experience because the *quantum* of energy $h \times$ (frequency) is very small, so that energy can change *almost* continuously. But, in the context of thermal radiation, the energy of a radiation mode is being compared to the energy of a *single* molecule, so that Planck's rule can have dramatic effects.

Planck's assumption overcomes both of the problems mentioned. It introduces a new dimensional quantity, $h$, and so there is no longer any problem with units. In fact the frequency at which the energy distribution has its maximum (see Figure 8.1) is predicted in terms of the temperature and Planck's constant $h$. Thus, for example, a temperature of 10,000 K gives a frequency at maximum of about $6 \times 10^{14}$ (second)$^{-1}$ and a wavelength of about $5 \times 10^{-7}$ metre, in the region of visible light.

As to the problem of the ever increasing number of degrees of freedom, this is solved as follows. For frequencies such that the quantum in Planck's rule is much less than the average thermal energy, the rule has little effect. The equipartition rule is approximately obeyed. This accounts for the left-hand edge of the graph in Figure 8.1. The rise is due to the increase in the number of modes as the frequency increases. But, for high frequencies, the minimum energy allowed by Planck's rule is bigger than the temperature, and such frequencies are seldom excited. This accounts for the fall on the right edge of the graph in Figure 8.1.

## 8.2 Why Are Atoms Simple?

In 1911, the great New Zealand physicist Rutherford, working in Manchester with Geiger and Marsden, discovered that the positive charge in an atom is concentrated in a tiny nucleus, with a diameter about $\frac{1}{10,000}$ of that of the atom. Nuclei of helium atoms (emitted from radium in radioactivity) were directed through gold foil. Mostly they passed through with little deviation, but occasionally large deflections were observed. The helium nuclei had high speeds, 10 percent or so of the speed of light, so, in order to be substantially deflected, they must have experienced a large electrical repulsion by the nucleus of a gold atom. Since the electrical force goes down as the inverse square of the distance, this requires that the helium and gold nuclei must get very close together, and that is possible only if they are both very small. Rutherford is famously quoted

> Quite the most incredible event that ever happened in my life ... it was almost as if you fired a 15-inch shell at a piece of tissue paper and it came back and hit you.

The discovery of the nucleus, along with the discovery of the electron in 1897 (see Section 5.1), completed a rough picture of the atom as consisting of a number of electrons orbiting round the nucleus, with empty space in between. Such a system is analogous to the Solar System, but with electrical instead of gravitational attraction.

This simple picture, however, raises several questions, which might have already have been asked before the discovery of the nucleus, although they would then have been less sharply posed.

First there is the question of atomic stability. An orbiting electron is an accelerating electric charge and as such should emit electromagnetic radiation. The outflow of energy should make the orbit contract and the atom collapse. Why does this not happen? (There is indeed a similar radiation of gravitational waves from bodies in gravitational orbits. It is very slow for the Solar System, but observable in some binary pulsars – see Section 7.7.)

Second there is the question of the size of atoms. Newtonian orbits can be of any size, yet (unexcited) atoms of a given element all have the same size.

The third point concerns *spectral lines*. In 1859, Kirchhoff (mentioned in the last section for his work in the same year on thermal radiation), working with Bunsen, discovered that, when some substances were heated in a flame, the spectrum of the light emitted did not vary smoothly with frequency (that is, with colour) but had narrow regions ("lines") of greater intensity. (The "Bunsen burner" gas flame was useful in these experiments, because the flame itself did not emit much light to be confused with the light from the heated element.) For example, the element sodium has a strong yellow spectral line, causing the yellow light from "sodium" street lights and from table salt dropped onto a gas burner.

Actually, characteristic spectra had been noted already in 1802 by Wollaston, and in 1814 Fraunhofer had observed *dark* spectral lines in the spectrum of sunlight. These dark lines are interpreted as due to the *absorption* in the outer layers of the sun of light of definite frequencies. It later became clear that dark "absorption lines" were at the same positions in the spectrum as bright "emission" lines that could be produced in different circumstances. That is, an atom that can absorb light at some special frequency can also emit light at that same frequency.

The spectral lines are characteristic of different elements and can be used to test for the presence of given elements. Indeed, several elements were discovered indirectly from their spectral lines, like helium in the Sun (the name *helium* from Greek for Sun).

The existence of spectral lines suggests that modes of vibration of atoms exist with characteristic frequencies, like the vibrations of piano strings. But orbits under electric attraction can be of any size, and in ellipses of any elongation, just like the medley of planetary orbits in the Solar System. There is no way to come up with any special periods.

In 1885, a Swiss girls' school teacher aged 60, Balmer, noted that the spectral frequencies of hydrogen could be fitted by the simple formula

- *Balmer's formula:*

$$\text{frequency} = (\text{a constant, } R) \times \left[ \frac{1}{n^2} - \frac{1}{m^2} \right].$$

Here $n$ and $m$ are any whole numbers 1, 2, 3, 4, ... (with $m$ bigger than $n$). The "constant" $R$ (after the Swedish physicist Rydberg) has the value

$$R = 3.28984186 \times 10^{15} \text{ (second)}^{-1}.$$

I quote this number to more decimal places than usual in this book to give some idea of the precision of spectral line studies. Balmer's value was within 1 percent of the true one.

Balmer's formula in some ways reinforces the idea of characteristic periods of vibrations within the atom. The occurrence of whole numbers is slightly reminiscent of the whole number 2, 3, and so on, ratios of frequencies of the overtones of a piano string to its fundamental frequency. But on the other hand the feature of Balmer's formula, that it is the *difference* between something depending on one whole number $m$ and something depending on another $n$, is hard to understand in this way.

## 8.3   Niels Bohr Models the Atom

These problems, and their resolution, were to lead, in the 1920s, to a restructuring of physics, quite as profound as the Scientific Revolution with which I began this book. One man had a specially deep influence, Niels Bohr. Of the physicists of the twentieth century,

Einstein has captured the public imagination most. He often worked on his own, and he had some mysterious way of smelling out the truth about the physical world that can only leave us marvelling. Bohr, on the other hand, although perhaps a larger than life figure, seems not to have been quite so superhuman.

Bohr was born in Copenhagen in 1885. In 1911, he went to Cambridge to work with J. J. Thomson (the discoverer of the electron), but this visit was not a success. In 1912, Bohr left for Manchester, where Rutherford and his assistants had recently discovered the atomic nucleus. Bohr and Rutherford got on very well. In 1913, Bohr arrived at the first understanding of the hydrogen atom, its stability, its size and its spectral lines.

I will give the simplest example of the sort of argument Bohr made at this time. He assumed that in an atom electrons orbit round the nucleus rather like planets round the Sun. The attraction is electrostatic for electrons, instead of gravitational for planets, but obeys an inverse-square law in both cases. Take the simplest atom, hydrogen, with just one electron. Consider the simplest possible orbit, a circular one. A planet could move in such a circular orbit at any radius. Given that radius, the speed is determined by the condition that it must be enough to balance the attraction. Then the total energy (kinetic plus potential) is also fixed.

For the atom, Bohr added on to all this one additional assumption, which was a bit like Planck's assumption about the quantization of the energy of electromagnetic radiation. This assumption was

- *Bohr's rule:*

$$(\text{momentum}) \times (\text{circumference}) = n \times h.$$

Here $h$ is Planck's constant again, and $n$ is any whole number, $1, 2, 3, \ldots$ This rule, like Planck's 13 years before, is ad hoc and unexplained, and just grafted on to the ordinary laws of motion.

Given Bohr's quantization rule together with the Newtonian relation between momentum and radius, the radius is related to $nh$. In fact one easily finds that the radius goes as $n^2$, so that

$$\text{radius} = n^2 \times (\text{Bohr radius}),$$

where *Bohr radius* is the name traditionally given to the radius of the smallest possible orbit, the one for $n = 1$. In fact

$$\text{Bohr radius} = \frac{h^2}{4\pi^2 \times (\text{mass}) \times (\text{charge})^2},$$

where the "mass" and "charge" are those of the electron. Since all these quantities are known, we can work out that

$$\text{Bohr radius} = 0.55 \times 10^{-10} \text{ metre.}$$

Also, the energy is given by

$$\text{energy} = -\left[\frac{(\text{charge})^2}{2(\text{Bohr radius})}\right] \times \frac{1}{n^2}.$$

(The energy is negative because the negative potential energy outbalances the positive kinetic energy.) The details of these relations are not very important. The important points are the dependence on $n$ and the fact that everything else is made up out of the three known quantities, the electron mass and charge and Planck's constant $h$.

We can now see how Bohr's quantization rule solves all the problems mentioned earlier. There is a *minimum* energy allowed for the atom, that corresponding to the value $n = 1$. So the atom must be stable: there is no way it can decay further by emitting energy. Also, there is a natural size for atoms, determined by the *Bohr radius*. It is dimensionally possible to construct this radius because Planck's constant $h$ has been brought into the picture, as well as the electron's mass and charge.

Finally, there is the Balmer formula for the frequencies of spectral lines. This is very naturally explained by combining Bohr's rule for the energies of atomic states with Planck's hypothesis about the quantization of the energy of electromagnetic radiation. Suppose an atom goes from a state with a higher value of $n$ to one with a lower value. There is a change of energy. Suppose this energy is carried away by the emission of light (or other electromagnetic radiation). According to Planck, we find the frequency of this light by dividing the energy by $h$. Thus we get the difference between two Bohr energies, divided by $h$, a form that may be identified with the Balmer formula in Section 8.2. What is more, the Rydberg constant is predicted in terms of the electron mass and charge and $h$. This predicted value agrees well with experimental measurements of spectral frequencies, and this agreement left no doubt that Bohr was on to the truth.

However, Bohr's work raised as many questions as it answered. What if the electron's orbit were not a circle? How to deal with atoms containing more than one electron? Where *was* the electron as it made the *transition* between states with two different values of $n$? How to deal with an electron free outside any atom?

Bohr's work began a decade of struggle to answer these questions. The leading figures in this work were Bohr in Copenhagen, Sommerfeld in Munich, Born in Göttingen and (of course) Einstein in Berlin. It is remarkable that quantum theory, a scientific revolution comparable with that of the sixteenth and seventeenth centuries, should have emerged largely from the economically crippled Germany (and, as we shall see, Austria) of the 1920s.

But in some ways Bohr was at the centre of all this work. It is difficult for those of us who never met him to understand the extraordinary influence that Bohr had. As a lecturer, he was poorly audible and unclear. He was very dependent on others to bounce his ideas off. He did not like "writing up" papers. He talked a lot. He must have been very persuasive, at least with the Danish authorities, and as a fund-raiser.

For my own benefit, as well as, I hope, the reader's, I give a few comments by other people on Bohr. It may be that (in some cases) they tell us more about their authors than about Bohr, but I hope that taken together they give some sort of a glimpse of Bohr himself. Most of my quotations are taken from the authoritative biography by Pais.

> Father figure extraordinary to physicists belonging to several generations. (*Pais*)

> Probably Bohr's most characteristic property was the slowness of his thinking and comprehension. (*Gamow*)

> Einstein appeared forever as his leading spiritual sparring partner: even after Einstein's death he would argue with him as if he were still alive. (*Pais*)

To get into the spirit of the quantum theory was, I would say, only possible in Copenhagen at that time [1924]. (*Heisenberg*)

I have really, in this whole period, been in real disagreement with Bohr ... I [could get] really very angry about Bohr. (*Heisenberg*)

Now I do hope that Bohr will save us with a new idea. I urgently request that he do so. (*Pauli*)

People were pretty well spellbound by what Bohr said.... While I was very much impressed by [him], his arguments were mainly of a qualitative nature, and I was not able to really pinpoint the facts behind them. What I wanted was statements which could be expressed in terms of equations, and Bohr's work very seldom provided such statements. (*Dirac*)

We had long talks, very long talks, in which Bohr did practically all the talking. (*Dirac*)

The discussions between Bohr and Schrödinger began already at the railway station in Copenhagen and were continued each day from early morning until late at night.... And although Bohr was otherwise most considerate and amiable in his dealings with people, he now appeared to me almost as an unrelenting fanatic, who was not prepared to make a single concession to his discussion partner or to tolerate the slightest obscurity. It will hardly be possible to convey the intensity of passion with which the discussions were conducted on both sides .... After a couple of days, Schrödinger fell ill, perhaps as a result of the enormous strain. (*Heisenberg*)

There will hardly again be a man who will achieve such enormous external and internal success, who in his sphere of work is honoured almost like a demigod by the whole world, and who yet remains – I would like to say modest and free of conceit – but rather shy and diffident like a theology student. (*Schrödinger*)

BOHR towering completely over everybody. At first not understood at all ... then step by step defeating everybody. (*Ehrenfest*)

Bohr's principle of complementarity, the sharp formulation of which I have been unable to achieve despite much effort I have expended on it. (*Einstein*)

Recently in London spent a few hours with Niels Bohr, who in his kind, courteous way repeatedly said that he found it "appalling", even found it "high treason" that people like Laue and I, but in particular someone like you, should want to strike a blow against quantum mechanics. . . . It is as if we are trying to force nature to accept our preconceived conception of "reality". He speaks with the deep inner conviction of an extraordinarily intelligent man, so that it is difficult for one to remain unmoved in one's position. (*Schrödinger*, to Einstein)

The President and I are much worried about Professor Bohr. How did he get into the business. . . . It seems to me that Bohr ought to be confined or at any rate made to see that he is very near the edge of mortal crimes. . . . I did not like the man when you showed him to me, with his hair all over the place, at Downing Street. (*Winston Churchill*)

## 8.4   Heisenberg and the Quantum World

The discovery of quantum mechanics involved two different achievements: finding the mathematical equations and understanding (partially at least) their physical interpretation. This was something new in physics. Given Newton's equations of motion and of gravitation, one is not in much doubt about their physical meaning. But in quantum theory, three-quarters of a century after its invention, the physical meaning is still a subject of debate.

Born in 1901, Werner Heisenberg was educated at Munich and Göttingen – two of the centres of the study of the new atomic physics. He met Bohr in Göttingen in 1922. It is difficult not to think of Heisenberg in this period as a golden youth. Success in many fields came easily to him. He was optimistic and a natural leader (in, for example, German youth movements). Bohr was so impressed that he is reported to have said "now it is up to Heisenberg" – up to him, that is, to straighten out atomic theory. Indeed, it was after a visit to Copenhagen in 1924 that Heisenberg had his first great idea.

Heisenberg worried how it could make sense to talk about the *orbit* of an electron when it was making a transition from one allowed state to another. He decided to sweep away reference to anything that was not clearly observable, and not to talk in general about the orbit or the position or the momentum of an electron. He allowed himself to refer to the position only in as far as a particular transition between two allowed states was concerned. For example, in a transition from the state with $n = 2$ to one with $n = 1$, he considered a quantity, related to the position, $x_{21}$. In this way, Heisenberg studied a whole array of variables

$$\begin{pmatrix} x_{11} & x_{12} & x_{13} \dots \\ x_{21} & x_{22} & x_{23} \dots \\ \cdot & \cdot & \cdot \\ \cdot & \cdot & \cdot \end{pmatrix}$$

He wrote down equations connecting these arrays with the energies, and with Planck's constant $h$. (Here, and in most of this chapter, I write things as if space only had one dimension, with just one position variable $x$. It is an easy matter to generalize to three dimensions.)

Max Born in Göttingen, with its great mathematical tradition (going back to Gauss), knew more mathematics than the young Heisenberg. He recognized these arrays as being things, called *matrices*, well known to mathematicians since their invention by Cayley in England in 1858. Heisenberg and Born and his assistant Jordan rewrote everything using the compact notation of matrix theory. I will write one of their equations. Let the symbol $x$ stand for the *matrix* (a whole array, like the preceding one) connected with the position of an electron, and let the symbol $p$ denote a similar matrix connected with the momentum. Then the equation is

• *Born-Heisenberg commutation relation:*

$$x \times p - p \times x = i \frac{h}{2\pi} \equiv i\hbar.$$

Here $h$ is Planck's constant again, and $i$ is the square root of $-1$, $i = \sqrt{-1}$. (The abbreviation of $\frac{h}{2\pi}$ as $\hbar$ has become standard.) This equation is called a *commutation* relation because commutation means changing the order of factors multiplied together. It is perhaps the most revolutionary equation in the whole of physics. It is inscribed on Born's gravestone in Göttingen. Of course, if $x$ and $p$ were ordinary numbers, it would make no sense, because for any ordinary numbers $x \times p = p \times x$. But these symbols stand for matrices, and for these $x \times p$ is not necessarily the same as $p \times x$.

This commutation relation is the real place where Planck's constant $h$ makes its appearance in physics. It is not now in an arbitrary rule, grafted onto the classical equations, as in Planck's and Bohr's quantization rules. Nor is it in some modification of the classical laws of motion. It is at an even deeper level. It tells us something about the mathematical properties of the mathematical objects in terms of which everything is formulated. No equation like this had appeared in science before.

Another remarkable thing about the Heisenberg-Born formula is that it mentions the square root of minus one, $i$. All direct measurements, of lengths or times or masses or whatever, are made in terms of ordinary real numbers. $i$ was an invention of mathematicians, but here it is appearing in one of nature's most basic laws.

It is ironic that Heisenberg's resolution to mention only observable things should have ended up with such an abstract mathematical equation, involving matrices and imaginary numbers. Indeed Heisenberg was unhappy about the direction in which his collaborators Born and Jordan were leading him.

There is another quirk to this story. Trying to convince the sceptical Einstein of the correctness of his new quantum theory, Heisenberg pointed out that he had tried to remove unobservables, just as Einstein had expunged the unobservable notion of absolute time in 1905. Einstein, irritatingly, replied, "A good trick should not be tried twice".

I have written down the Heisenberg-Born law connecting the matrices with Planck's constant, but I have not said a word about how atomic energies or anything else in physics can be deduced. I will postpone this until after describing Schrödinger's alternative form of quantum theory.

## 8.5   Schrödinger Takes Another Tack

The Austrian Erwin Schrödinger, having seen active service in the 1914–18 war, became in 1922 a professor at the Eidgenossische Technische Hochschule (E.T.H.) in Zürich (where Einstein had previously been). In 1926 (aged 39), he invented his own version of quantum theory, very different at first sight from Heisenberg's.

There are many examples in physics of quantities that vary with position in a continuous way. Electric and magnetic fields are of this nature, but let us think of the density and pressure of a gas as specific examples. These obey certain equations that connect their variation in space with their variation with time. These are the sorts of equation that meteorologists try to solve to predict the weather. A particularly simple type of solution (valid when the pressure variations are small) describes a sound wave.

Schrödinger's approach to the study of electrons in atoms was to assume the existence of a quantity that varied in a like manner, with position and with time. This is always denoted by the Greek letter $\psi$ (psi) and called (for want of a better word) the *wave function*. Schrödinger postulated an equation that $\psi$ should obey, a little like the equation for the density of a gas.

The difficult question to answer is what $\psi$ has got to do with a small particle like an electron. Schrödinger himself believed that the electron was somehow smeared out and that $\psi$ determined how densely it was smeared over different points of space. He thus believed that he had replaced all the perplexities of atomic physics by a sort of equation that had been very familiar throughout the nineteenth century. He was wrong in this interpretation, as Bohr and Heisenberg were quick to tell him.

However, leaving aside the physical interpretation for the moment, we can understand one thing about Schrödinger's approach: that is, how it predicted a discrete set of energy states for electrons bound in atoms. We are familiar with wave motions having discrete sets of possible frequencies. A violin string, for example, can vibrate at its fundamental frequency, or its first harmonic (twice the frequency) or its second harmonic (three times the frequency) and its higher harmonics (higher multiples of the fundamental frequency). By connecting energy to frequency by Planck's rule (Section 8.1), Schrödinger could obtain a discrete set of energies.

One may object to this analogy: a violin string is fixed at both ends, but the same is not true for $\psi$ in an atom. What keeps the electron in the atom is the electric attraction to the nucleus, which dies off continuously and slowly, not sharply like the ends of a violin string. But in fact if the violin string were not fixed rigidly at each end, but just embedded in some sort of elastic jelly, it would still vibrate with a discrete set of possible frequencies. All that would happen is that the ratios of the frequencies would not be the simple 1:2:3 ... but obey some more complicated rule.

Thus Schrödinger's equation could be used to find the allowed energies of atoms, using a style of mathematics with which people were relatively familiar. In fact this equation is one of the most used in all of physics. It has been said to "contain most of physics and all of chemistry".

Schrödinger's and Heisenberg's quantum theories look as different as chalk and cheese. But it was soon shown that they are mathematically equivalent. How, then, does Scrödinger's approach lead to the Heisenberg-Born commutation relation in Section 8.4? To answer this, we must find how the objects called $x$ and $p$ in the commutation relation are realized in Schrödinger's formalism. The wave function $\psi$ depends upon the position, $x$, and we can always multiply $\psi$ by $x$ to get another function $x \times \psi$. This much is trivial. But the wave function does not depend upon any variable like momentum; it depends only upon the position $x$ (and upon time). In fact $p$ is related to the *operation* of finding the rate of change of $\psi$ with $x$, that is, finding the slope of the graph of $\psi$ plotted against $x$ (for any fixed time). Having found this, one multiplies by

$$-i\hbar,$$

and this is the effect of $p$ acting on $\psi$. Thus $p$ is an *operator*, operating on the wave function $\psi$.

Because $i$, the square root of $-1$, appears here, the wave function $\psi$ must in general be a complex number. In this respect, it is different from quantities in classical physics, like pressure, temperature and electric field.

It is legitimate to use a mathematical symbol, like $p$, to stand for an operation, provided we have a consistent set of rules governing the use of the symbol. Thus "multiplication" of two operations just means performing one operation after the other. In general, the order in which this is done matters: in multiplication, operators do not *commute*. It is not difficult to see that the successive operations

> (multiply by $x$) *then* (find the rate of change with $x$)

and

> (find the rate of change with $x$) *then* (multiply by $x$)

give different results: the operation represented by $p$ does not *commute* with $x$. In fact the Heisenberg-Born commutation relation is recovered.

Although the mathematical equivalence of the two forms of quantum theory was soon accepted, Schrödinger had quite different ideas from Heisenberg, Bohr and Born about the physical interpretation. Schrödinger thought that he had got rid of the puzzles about visualizing orbits of electrons: there was no point electron at all, just some sort of a smear of electric charge. The Copenhagen/Göttingen physicists vehemently rejected this view. They are generally believed to have been right.

The contrast between Heisenberg's and Schrödinger's outlooks was as pronounced in politics as in physics. Although not (like Einstein and Born, for example) a Jew, Schrödinger left the German-speaking world in 1933 for the British Isles, not to return finally until 1956. The staunchly patriotic Heisenberg remained in Germany throughout the 1939–45 war and played a leading part in the (unsuccessful) German programme on neutron chain reactions. He may have believed that he could best mitigate the evils of the Hitler regime from within.

## 8.6 Probability and Uncertainty

The first key to the interpretation of quantum theory was found by Born in 1926. He stated that Schrödinger's wave function $\psi$ gives the *probability* of finding the electron (or other particle) at a given position at a given time. I must be a bit more precise about this. The wave function $\psi$ is in general a complex number (involving $i$, the square root of $-1$). Complex numbers (see Appendix B) can be represented by positions in a plane, relative to some origin. Such a position may be characterized by two things: its *length*, or distance from the origin, and its *phase angle*, or direction. The probability is in fact proportional to the square of the length of $\psi$. This is a positive number, as a probability must be.

So the complete wave function itself does not have any direct physical significance. It cannot be measured as an electric field is measured, for example. Its phase angle is not directly observable. But this does not mean that phase angles are always irrelevant. Far from it. Wave functions have a *superposition* property (compare Section 4.1): two wave functions may often be added together to give another possible wave function. If the two individually satisfy Schrödinger's equation, then so does the sum. So far as is known, this property of the equation is exactly reflected in nature. When two wave functions are superposed in this way, the *relative* phase angle is of the utmost importance. For example, if the relative phase angle is zero (or small), the two reinforce each other. But if the relative phase angle is 180 degrees (or nearly so), the two tend to cancel each other out.

Why talk of probability? Why does the theory not make definite predictions, as classical physics has always done? I will again postpone consideration of this difficult question, but for the following point.

This is a key physical insight due to Heisenberg in 1927. It concerns the classical notion of the orbit of a body, along which at any time its position and velocity (and therefore momentum) are specified. What happens to this in quantum mechanics? Heisenberg realized that, as a mathematical consequence of the Born-Heisenberg commutation relation (Section 8.4), it is impossible to know, that is, to measure, the position *and* the momentum simultaneously. Indeed, if these two quantities were known they would each be ordinary numbers, and their multiplication would have to commute ($x \times p = p \times x$), contradicting the Born-Heisenberg equation.

Heisenberg went further, showing that the more accurately the position was known the less accurately the momentum must be known, and vice versa. More specifically:

- *Heisenberg's uncertainty principle:*
  Uncertainty in position times uncertainty in momentum is at least of the order of Planck's constant $h$.

I have used the vague phrase "of the order of". One can be more precise, but this requires a more careful definition of "uncertainty in".

Planck's constant $h$ is very small in everyday units, and so the uncertainty relation is practically irrelevant to macroscopic things. It is usually only for molecules or smaller things that the principle is important.

If there is a complete, consistent quantum theory, incorporating Born's commutation relation, then any deduction from it should be consistent with the uncertainty principle. In particular, any attempt to measure position and momentum simultaneously, with more accuracy than the principle allows, should necessarily fail. Heisenberg and Bohr thought out how this would come about in some conceivable measurement devices. For example, to measure position precisely, one might use a microscope. The accuracy of any microscope is limited by diffraction effects (see Section 4.6) connected with the wave nature of the light. These effects are reduced by using short wavelength electromagnetic radiation, like X-rays. But according to Planck (Section 8.1), such radiation has a minimum energy, and some of this energy may be imparted to the particle under study, thereby giving it an undetermined uncertainty in momentum. It can be shown that the uncertainties of this kind are necessarily consistent with Heisenberg's principle.

Thus, in quantum theory, we are forbidden to think about an *orbit*, as we do in classical physics. All we have are the wave function and whatever information we can extract from that.

In Heisenberg's quantum theory, position and momentum appear on the same footing, but Schrödinger's wave function seems to make a special reference to position. The wave function $\psi$ that represents a particle with a definite momentum is as follows. The length of the complex number $\psi$ is the same at all positions, but its phase angle increases steadily in one direction. The rate of increase (and the direction) determines the momentum. Since the length of $\psi$ is constant, the probability (given by the square of that length) of finding the particle is the same everywhere. This conforms to Heisenberg's uncertainty principle: since the momentum is fixed precisely, the position is totally uncertain.

The behaviour of this wave function is indicated in Figure 8.2.

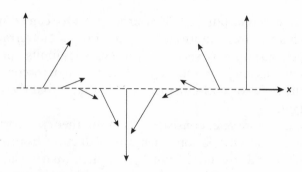

**FIGURE 8.2**   The behaviour of the wave function $\psi$ for a particle with definite momentum, $p$. The lines representing the wave function as a complex number rotate steadily as $x$ increases, making one complete rotation as $x$ increases by an amount $\frac{h}{p}$. $x$ is plotted horizontally and the wave function for any value of $x$ is represented by a line (with an arrow) at right angles to the $x$-axis. In the figure, a segment of the $x$-axis of length $\frac{h}{p}$ is drawn. The figure should be imagined repeating indefinitely in both direction.

## 8.7   Spin

In quantum theory, angular momentum is quantized: that is, only a discrete set of values is allowed. The explanation is as follows.

Angular momentum is to ordinary (linear) momentum as motion in a circle is to motion in a straight line. Let us start with motion in a circle about some single axis; we can come to other axes later. To find the form of a wave function with a definite angular momentum, we can take the wave function with a definite momentum, Figure 8.2, and replace the distance $x$ along a line by a distance round a circle (of radius 1, say). The momentum $p$ then becomes angular momentum. The important new feature is that going right round the circle (a distance of $2\pi$) gets back to the same point. So, we might expect, the wave function ought to be unchanged: that is, the phase angle of the wave function should change by some multiple of 360 degrees: 0, 360, 720, 1,080, ... degrees. This happens if

$$\frac{p}{\hbar} = 0, 1, 2, 3, \dots,$$

or

$$\text{angular momentum} = \hbar \times (0, 1, 2, 3, \dots)$$

This argument is not quite right. The phase angle of a wave function is not always physically relevant, so two wave functions with different phase angles might represent the same physical state. Therefore, going right round the circle might alter the phase angle. It cannot do this in an arbitrary way, however, because we have to make sure that several rotations done in succession affect the wave function in a consistent way. This is not a simple matter to investigate, but the key to the conclusion lies in a fairly simple property of rotations in three-dimensional space. This property is that a rotation through 360 degrees can entangle things in a way that is impossible for a rotation through 720 degrees (a double complete rotation).

Consider a strap that contains a 720-degree twist in it, as in (a) of Figure 8.3. Keep the orientation of each end fixed, but allow yourself to move the bottom around relative to the top, sideways and up and down and so on. Can you get rid of the twist? You can,

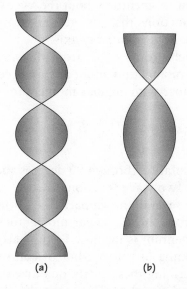

(a)             (b)

FIGURE 8.3   Strap (a) has a twist of 720 degrees in it; strap (b) has 360 degrees. The twist in (a) can be removed by moving the top round a circle about the bottom, always keeping the orientations of the top and bottom fixed. This manoeuvre does not work on case (b).

by taking the bottom around in a circle relative to the top. (Warning: checking this, one needs to hold the top of the strap in one hand and the bottom in the other hand, in order to keep their orientations fixed. But at some point one needs carefully to interchange the two hands in order to be able to finish the manoeuvre.) Now take a strap with only a 360-degree twist in it. The same manoeuvre fails to remove the twist.

This argument is meant to suggest that, although a 360 rotation may change the phase angle of a wave function, a 720 rotation may not. Then the preceding quantization rule for angular momentum is relaxed to

$$\text{angular momentum} = \hbar \times 0, \frac{1}{2}, 1, \frac{3}{2}, 2, \ldots$$

All this refers to the angular momentum connected with rotations about one fixed axis, let us say the $x$-axis. What happens if one does a rotation about another axis, say the $y$-axis. Not surprisingly, this changes the angular momentum about the $x$-axis. But one can find groups of wave functions that just get mixed up with each other when rotations are done about *any* axis.

The simplest such group (apart from the single wave function with zero angular momentum about any axis) consists of just two wave functions, those with angular momentum (about the $x$-axis, say)

$$\frac{\hbar}{2}, -\frac{\hbar}{2}.$$

As an example, a rotation through 90 degrees about the $y$-axis just interchanges these two wave functions.

It turns out that several known particles, for instance, electrons, protons and neutrons, have angular momentum of just this sort. This angular momentum is intrinsic to the particle. It is as if the particle were spinning like a top, but of course quantum theory is essential in describing these small, quantized amounts of angular momentum, and one should not take the classical, "top-like" picture too literally. Such a particle is said to have *spin* $\frac{1}{2}\hbar$.

So an electron can be in the state with angular momentum (in some given direction, say, the $x$-axis) $\frac{1}{2}\hbar$, or it can be in the state with the opposite value. People sometimes call these "spin up" and "spin down" states. The wave function for an electron, as well as depending upon position and time, depends upon the spin. To show this, we may label the wave function with "up" or "down":

$$\psi(up), \psi(down)$$

(where for simplicity I have ignored the position and time dependence). According to the *superposition* property, wave functions like

$$\psi(up) + \psi(down)$$

should also exist. In fact this particular example corresponds to the spin being aligned along a direction at 90 degrees to the $x$-axis.

The next simplest possibility is the group of three wave functions with angular momenta about the $x$-axis

$$+\hbar, 0, -\hbar.$$

A particle with these intrinsic angular momentum states is said to have spin $\hbar$. Examples of such particles are known. They are the $W^-$, $W^+$ and $Z^0$ particles, which have, respectively, the same electric charge as an electron, the opposite electric charge and no electric charge. The $W$ particles have the mass of about 86 protons and the $Z$ particle about 97 protons: that is, to say they are similar in mass to the atoms of medium heavy elements like silver.

Most particles with spin behave as tiny magnets, with the direction of each magnet aligned along the direction of the spin. This is just as expected if there is electric charge within the particle carried round in circles by the rotation. We know that an electric current circulating in a loop produces a magnetic field (indeed the Earth's magnetism is generated in this way). In the case of a particle like an electron, of course, we should not take this classical picture too literally. Even the neutron, with zero total electric charge, is magnetic.

This magnetism allows the spin directions of particles to be manipulated by using magnetic fields, including the magnetic fields in electromagnetic radiation. The technique of nuclear magnetic resonance (NMR) is based upon detecting the radiation emitted when the spin directions of atomic nuclei are changed by magnetic fields.

We can illustrate the quantum behaviour of spin as indicated in Figure 8.4. A beam of neutron (or it could be atoms with spin) comes in from the left and enters a region where there is a magnetic

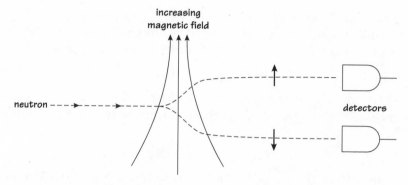

increasing
magnetic field

neutron

detectors

FIGURE 8.4    A beam of neutrons, for example, comes in from the left. They all have spin along the $x$-axis. They enter a region where there is a magnetic field along the $y$-axis and increasing in strength as $y$ increases (the lines of force are shown as continuous curves). The wave function of each neutron is a superposition of wave functions for spin "up" and spin "down" along the $y$-axis. Each neutron therefore has an equal probability of being deflected up or down. If a neutron enters the top detector its spin is "up".

field pointing upwards (along the $y$-axis) and also increasing in strength upwards. Such a magnetic field exerts a force on the magnet that is associated with the neutron's spin. If the spin is "up" (along the $y$-axis, that is), the force is also up. If the spin is "down", the force is down.

Now suppose that the beam of neutrons is prepared so that the spins are all aligned along the $x$ direction (which happens to be the direction of motion). In order to work out how they behave, we have to write their spin wave function in terms of wave functions for spin "up" and "down" along the $y$ direction. The result was stated previously: it is just the sum of these two latter wave functions. The interpretation of this is that there are equal probabilities for the neutron to have spin "up" and to be deflected upwards or to have spin "down" and be deflected downwards. The magnetic field therefore separates the beam (which had spins along the $x$) into two diverging beams that have spins "up" and "down" along the $y$ direction. Any individual neutron has an equal probability of being deflected up or down and being detected in one or the other of the detectors in the figure.

## 8.8 Feynman's All Histories Version of Quantum Theory

I will now jump forward to 1948, when yet a third, equivalent formulation of quantum theory was developed by that charismatic, even notorious physicist Richard Feynman. Schrödinger's equation specified the way in which the wave function changed in a very small interval of time. Feynman's principle gives the time development over any interval of time, albeit in rather abstract form.

Feynman's principle goes like this. Suppose we know the wave function $\psi$ at some initial time and we want to deduce the wave function at any later time. I will set out Feynman's procedure in steps:

(a) Take any value of the position $x$ at the initial time, and any value of the position, call it $x'$, at the later time.

(b) Choose *any* conceivable history (that is to say, motion), starting at $x$ at the initial time and ending at $x'$ at the final time.

(c) Work out the *action* for this history, just as in Section 6.2.

(d) Divide this action by $h$, and define the complex number that has length 1 and phase angle given by the action divided by $h$.

(e) Add up these complex numbers for *all* possible histories! Feynman called the result of doing this "the amplitude for going from $x$ at the initial time to $x'$ at the final time". This amplitude is again a complex number (in general), but its length need not be equal to 1.

(f) Finally, take this amplitude, multiply it by the wave function at the initial time and add up the results over all possible values of $x$. The result is the wave function at the final time (expressed in terms of the position $x'$).

The procedure is illustrated in Figure 8.5.

Feynman's principle deserves several remarks in explanation.

When the complex numbers in step (e) are added up, these complex numbers will not usually have the same directions, so the result is by no means the same as adding the lengths of the complex

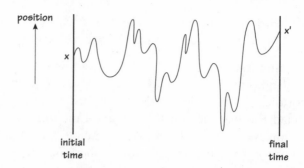

FIGURE 8.5 Feynman's sum over histories. A "typical" history is shown. A sum over "all" histories is required, then a sum over the initial position, $x$, after multiplication by the wave function $\psi$ at the initial time.

numbers. In fact, the complex numbers might have exactly opposite directions; then they would just cancel out.

In adding over "all" histories, very sharply wiggly motions must be included, as the figure illustrates. It is a difficult mathematical problem to give a careful definition of the notion of "all histories". What Feynman did was to replace continuous space and time by a fine mesh of a large but finite number of points at a large but finite number of times. Then the notion of "all histories" is perfectly clear: it just means hopping from point to point at successive times in all possible ways. This procedure gives some approximation to what is required. It is assumed that this approximation can be made better and better by making the mesh of points and times finer and finer (that is, having more and more points and times).

Schrödinger's equation can be deduced from Feynman's principle by taking the special case in which the lapse of time from the initial to the final time is very small. Born's commutation relation can also be deduced. But note that, in spite of this, Feynman's principle does not explicitly mention any matrices or operators, but only ordinary numbers.

To describe the behaviour of most macroscopic objects, classical mechanics is usually sufficient. This means that we expect that classical mechanics ought to be a very good approximation to quantum theory for macroscopic systems. Feynman's principle allows us to see how this happens, at least in a rough way. Suppose that, in step (d), the action is always very large compared to Planck's constant $h$. (Remembering that $h = 6.6 \times 10^{-34}$ joule seconds, we would expect this condition to hold for "ordinary" macroscopic things.) Then the phase angle of the complex number in step (d) is very large, let us say many millions of degrees. Of course 367 degrees, for example, is the same as 7 degrees, and any phase angle is equivalent to some angle between 0 and 360 degrees (in the present context).

Now imagine the contributions to step (e) from a group of several close-together histories. The corresponding phase angles will probably differ from each other by small percentages, but small percentages of many millions of degrees are still substantial numbers of degrees. Thus the complex numbers in step (d) corresponding to this group of histories will have all sorts of different directions and so tend to cancel out on average. (Actually, the length of the sum will go as the square root of the number of orbits, not as this number itself.) This behaviour is illustrated in Figure 8.6.

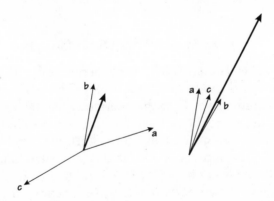

FIGURE 8.6 Representative contributions to Feynman's amplitude sum in the classical limit. The figure on the left shows how such contributions in general do not add up in a constructive way. a, b, c represent the complex numbers corresponding to three nearby histories, and the thick arrow represents the sum of these three. The figure on the right shows how the contribution is enhanced for orbits very near to the classical orbit. In this case, b corresponds to the classical motion and a, c to two nearby histories, giving a larger amplitude (the thick arrow).

Now take the particular motion given by the solution of the classical equations of motion. According to the principle of stationary action (see Chapter 6), this motion makes the action have a stationary value (usually the minimum value). This means that for histories close to the classical one the value of the action is varying slowly (just as the top of a hill is locally flat). Therefore histories near the classical one add up constructively and give the dominant contribution to Feynman's sum. This is how classical dynamics arises from quantum theory as a very good approximation in suitable circumstances.

We are now able to answer the question posed in Chapter 6. How, in classical mechanics, does a particle "know" how to move so as to minimize the action? According to quantum theory, it really "smells out" *all* histories, but the contributions from near the classical motion dominate Feynman's amplitude sum. This is analogous to the way that Fermat's least time principle (see Section 4.4) comes out of the wave theory of light as an approximation (see the end of Section 4.6).

## 8.9   Which Way Did It Go?

In this section I will describe a type of experiment that shows clearly some of the weirdness of quantum theory. This is an *interference* experiment, in some ways like the interference of light explained in Section 4.6. But there the interference was in waves in the electric and magnetic fields (which constitute light): measurable physical things. Here, the interference is related to the wave function for a particle, and this is not a directly measurable quantity. The interference is between two parts of the wave function corresponding to different motions of the particle.

Diffraction (which is intimately related to interference) of the wave function associated with particles was first observed in 1927 by G. P. Thompson in Aberdeen and by C. J. Davisson and L. H. Germer working at (what later became) the Bell Telephone Laboratories. The particles used then were electrons (appropriately enough, since Thompson was the son of the electron's discoverer, J. J. Thompson). I will describe modern experiments done with neutrons (obtained from nuclear reactors).

The experiments make use of perfect crystals of silicon, cut into appropriate shapes. A neutron passing through such a crystal experiences nuclear forces when it is near the nuclei of the silicon atoms, and classically these forces would change its direction of motion. In quantum theory, according to Feynman's formulation (Section 8.8), we must consider all motions, find an amplitude for each and add them up. For some groups of motions the combined amplitude is big, for others small. The process of finding out whether amplitudes add up constructively or destructively (cancelling out) is analogous to finding out in wave optics whether Huygens's wavelets add up or cancel out (see Sections 4.5 and 4.6).

In the silicon crystal, the atoms are arranged in regular arrays, and a plane of atoms acts as a half-silvered mirror: the Feynman amplitudes add up constructively both in the forward direction of the neutrons and in the direction corresponding to reflection. There is a 50 percent probability that a neutron should pass through and a 50 percent probability that it should be "reflected" back.

The experiment makes use of four such "semireflections" in turn. For each of these separately, there is a 50 percent probability of the neutron's passing through and a 50 percent probability of its being "reflected" (that is, bouncing back). However, the complete experiment requires us to consider all possible Feynman paths all the way through the apparatus.

The experimental arrangement is sketched in Figure 8.7. A single crystal of silicon is carved so that the thin slices shown rise above the body of the crystal, and the sides of these slices are parallel to regular planes of atoms within the crystal. Each slice therefore acts on a neutron as a partially reflecting mirror. The neutrons enter from a source $S$ on the left.

The dashed lines represent possible paths through the system. (Actually, each dashed line represents a bundle of paths through the crystal whose amplitudes add up constructively.) The paths either leave the experiment after going through $A$ or $B$ or end up at one of the detectors $D_1$, $D_2$. In fact, the amplitudes for the two paths $XAYD_2$ and $XBYD_2$ cancel out, whereas those for $XAYD_1$, $XBYD_1$ add up. Thus the neutron is detected by $D_1$ but not by $D_2$.

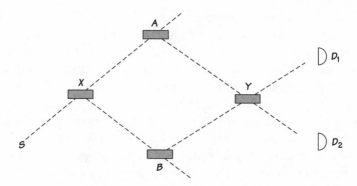

**FIGURE 8.7**    A neutron interference experiment. The thin slabs at $X$, $A$, $Y$, $B$ represent slices formed by cutting away a silicon crystal (about 1 cm long), with sides parallel to the planes of atoms within. There is a source of neutrons at $S$ and neutron detectors at $D_1$ and $D_2$. Taking account of the possibility of transmission or reflection at each part of the crystal, there are four relevant Feynman paths for the neutron: $SXBYD_1$, $SXAYD_1$, $SXBYD_2$ and $SXAYD_2$. The amplitudes for the first two add up constructively; those for the second two cancel. So there is 0 percent probability for $D_2$ to detect the neutron, but 50 percent probability for $D_1$ to detect it. (The remaining 50 percent probability is for the neutron to be transmitted through $A$ or $B$, out of the experiment.)

It is impossible to say which path, through $A$ or $B$, a neutron takes. If, in an attempt to find out, a detector is introduced into one of the paths, say, between $X$ and $A$, the interference is destroyed and the neutron has an equal chance of reaching $D_1$ or $D_2$. If the spin of the neutron is flipped over in a direction along one of the paths (by suitable electromagnetic fields), the amplitudes cannot interfere, and again there is an equal probability of the neutron's being detected by $D_1$ and $D_2$.

The rate at which neutrons enter the system can be made so small that there is never (or very rarely) more than one neutron in the system at any one time. Thus there is no question that the interference is between different neutrons. The statements made earlier are all about a *single* neutron. On the other hand, the predictions are about probabilities, and like all such predictions they are only to be tested by recording the results of a sequence of experiments, that is, the passage of many neutrons one after another. (Actually, the detection of a single neutron by $D_2$ would be sufficient, ideally, to contradict the prediction of zero probability.)

The lesson of these experiments (and lots of similar ones) is that in quantum theory we are forbidden to visualize any one path for the particle. Rather, it somehow pervades all possible paths. This does not mean that half a neutron is in one path and half in another. If a detector is introduced between, say, $X$ and $A$, it detects either one neutron or none.

## 8.10  Einstein's Revenge: Quantum Entanglement

Einstein was never happy with quantum theory in its mature state in the mid-twenties, even though he had played a big part in its development. At a conference held in Brussels (the last at which the grand old man Lorentz presided, shortly before his death), there were celebrated arguments between Einstein and Bohr. According to Otto Stern (quoted in Pais's book on Einstein):

> Einstein came down to breakfast and expressed his misgivings about the new quantum theory, every time [he] had invented some beautiful experiment from which one saw that [the theory] did not work.... Pauli and Heisenberg, who were there, did not pay much attention, "ach was, das stimmt schon, das stimmt schon" [ah, well, it will be all right, it will be all right]. Bohr, on the other hand, reflected on it with care and in the evening, at dinner, we were all together and he cleared up the matter in detail.

There is little doubt that quantum theory is consistent within its own terms. But it is very strange, and Einstein (together with Boris Podolsky and Nathan Rosen) eventually, in 1935, put his finger on one of its weirdest features. The consequence of quantum theory that Einstein and his collaborators described, which I will explain, is often called the EPR "paradox". I do not think that the EPR paper sparked off anything very constructive until the 1950s. Then, two men especially, David Bohm and John Bell, had the independence and courage seriously to question the Copenhagen orthodoxy. John Bell in particular chipped away, with wit and eloquence, gradually persuading people to worry more about quantum theory.

Now to describe a form of the EPR "paradox". Take two protons. They each have spin $\frac{\hbar}{2}$, so they both can have their spins in either of two settings, "up" and "down" (along some arbitrarily chosen axis). For the two protons together there are four possible arrangements of spins: "up up", "down down", "up down" and "down up" (where the first word refers to the proton at one position, and the second to that at another). If we change the axis used to define "up" and "down", these four spin states in general get rearranged. But there is one particular superposition that is independent of the choice of axis. This is the wave function

$$\psi(\text{"up" "down"}) - \psi(\text{"down" "up"}).$$

Remember that in quantum theory we are allowed to superpose wave functions in this sort of way, with coefficients that are in general complex numbers (but just $+1$ and $-1$ in this case). This particular spin configuration is called *singlet*. It is the unique wave function for which the total angular momentum of the two protons is zero in *any* direction.

Suppose we prepare the two protons in the this singlet spin state and allow them to move apart, say, some metres away from each other, being careful that nothing disturbs the spins in the process. Then there is a very strange quantum correlation between the spins. Let us say that one proton has moved to the "right" and the other to the "left", as indicated in Figure 8.8. Suppose we measure the spin of the "right" proton along an arbitrary direction (call that direction $d$). We get the results "up" and "down" with equal probability. But if the spin of the "left" proton is also measured, along some direction $d'$ the two sets of results are correlated. For example, if the directions $d'$ and $d$ are the same, then we

left             right

**FIGURE 8.8** The idea of the EPR "paradox". Two protons, in the singlet spin configuration, are moved apart, one to the right and one to the left. Measurements of their spins are made along directions $d$ and $d'$, respectively.

get "down" on the left when there is "up" on the right, and vice versa.

Is this a paradox? Not really. Suppose there are two closed envelopes, one containing a piece of paper with the word *up* written on it and the other one with *down* on it. If one envelope is given to a woman on the right and the other to a man on the left, without anyone's knowing which is which, then the results of opening the envelopes are correlated. If she finds "up" he must find "down", and vice versa. There is no mystery about this. There are real physical pieces of paper situated at the two positions (right and left). It is just that the two people are initially missing information about them.

Could something like this be happening with the proton spins (as Einstein believed must be the case)? Such an explanation is called a *local hidden variable theory*. It assumes the existence of some real physical thing on the right and another on the left, which we do not know, but to which we can assign correlated classical probabilities. Bell proved that the correlations predicted by quantum theory *cannot* in general be reproduced by any local hidden variable theory.

Bell's proof requires us to look at something more complicated than the simple case $(d = d')$ mentioned. Let us record the results of the spin measurements by a $+1$ for "up" and $-1$ for "down". Let $S(d)$ be the result for a particular measurement in direction $d$ on the right, and $S(d')$ the result in direction $d'$ on the left, so that $S(d)$ and $S(d')$ are each $\pm 1$. The prediction of quantum theory is very simple:

average of $[S(d) \times S(d')] = -$(cosine of angle between $d$ and $d'$).

(Here *average* means an average over many repetitions of the experiment.)

Suppose there were some "hidden variable", some quantity $u$ ($u$ for unknown) that determines the spins, but that we do not know. Suppose further that the spin on the right in direction $d$ is given by some dependence $f(u, d)$ and likewise on the left by $f(u, d')$. Each of these can only have the values $+1$ or $-1$. The important assumption is that there *are* definite spins, on the right and on the left: we just do not know them because we do not know the value of $u$. Then we would have that

average of $[S(d) \times S(d')] = $ average over $u$ of $[f(u, d) \times f(u, d')]$.

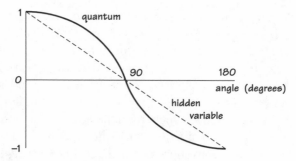

FIGURE 8.9   Quantum theory and a particular hidden variable model compared. The angle between $d$ and $d'$ is plotted horizontally, and the average of $-S(d)S(d')$ vertically. The quantum prediction is the black curve; the hidden variable prediction is the dashed curve.

But Bell was able to prove that the last two equations are inconsistent: it is impossible to reproduce the quantum result for *any* choice of $u$ or $f$.

Here is a particular *example* of a plausible choice of $u$ and $f$, which fails to reproduce the quantum result (as, according to Bell's theorem, it must). Suppose $u$ to be a direction, like $d$ and $d'$. Let

$$f(u, d) = +1 \text{ when the angle between } d \text{ and } u \text{ is acute}$$

$$= -1 \text{ when the angle between } d \text{ and } u \text{ is obtuse.}$$

Figure 8.9 shows the value of the average of $S(d)S(d')$ as given by this "hidden variable" example compared to the average predicted by quantum mechanics. The quantum correlation is greater than the "hidden variable" one (except at 0, 90 and 180 degrees). This single example does not prove anything, but it does perhaps suggest the way in which hidden variable models fail.

Thus, *if* quantum theory is right, Bell proved that the correlations in the EPR situation involve something more subtle than missing classical information about real things situated separately at the right and at the left.

But *is* quantum theory right? Maybe experiment would show that quantum theory (although it has worked so well before) fails in the particular EPR situation. Experiments have been carried out since 1972, especially at Orsay near Paris and more recently at

Innsbruck. The experiments are difficult, but they have become more and more accurate and possible loopholes have gradually been closed. It is now almost certain that quantum theory works and that local hidden variable models are ruled out. Einstein was right to say that quantum theory is weird; he was wrong to think that nature might not appear to be weird too.

The strange quantum correlation between spins (or anything else) at distant regions of space was termed *entanglement* by Schrödinger in 1935. Somehow, the two separated protons (in our example) behave as if they are still in some subtle non-local communication with each other. It is a feature of quantum theory that we cannot visualize but that nature seems to demand we accept.

## 8.11  What Has Happened to Determinism?

We must now try to face the most difficult question posed by quantum theory: how and why is determinism lost?

Classical physics is thought to be deterministic: given full initial conditions for an isolated system, the future of that system is uniquely determined by the laws of nature. Of course, there are some weasel words in this assertion. It may be very difficult to get the "full" initial conditions. No system can ever be exactly "isolated". Nevertheless, in simple cases, there seems to be nothing to stop one from getting closer and closer to the ideal situation envisaged in the preceding statement.

We must also distinguish between determinism and predictability. Many classical systems are chaotic (see Section 2.8), which means that there are severe limitations upon how far we (or our computers) can predict their future. But this does not mean that the future is not, in principle, determined.

Now I will describe the situation in quantum theory, according to the "Copenhagen" view, as developed preeminently by Bohr. According to this, it is essential to the interpretation of quantum theory that there should be a *classical* world of macroscopic instruments: detectors, laboratories, recording devices, computers, perhaps people. These are well described by classical theory, or at least the important relevant features of them are. We use these instruments to study microscopic quantum systems. The business of quantum theory is to predict the results of experiments, done with macroscopic apparatus on microscopic systems.

Any measurement involves arranging for some sort of correlation between a measuring device and the object being measured, for example, by lining up a measuring tape against the side of a carpet. In classical physics, we do not normally have to worry about the effect this process has on the object. In quantum theory, as we shall see, things are different.

I will use the experiment illustrated in Figure 8.4 as an example. Here a neutron enters from the left, and we are trying to measure its spin in the $x$-direction (whether it is "up" or "down"), using the magnetic field and the detectors on the right. In Section 8.7, I assumed that the neutron entered with its spin aligned along the $y$-axis, so that its wave function was the superposition

$$\psi(\text{up}) + \psi(\text{down}).$$

It will now be useful to consider a more general case in which the wave function is

$$a \times \psi(\text{up}) + b \times \psi(\text{down}),$$

where $a$ and $b$ are any complex numbers. What matters is the ratio $b/a$, which is also a complex number. This complex number has a magnitude and a phase angle: to specify the state we must know both of these.

Now consider the wave function describing the neutron after it has been through the magnetic field but before it has entered a detector. The position (deflected upwards or deflected downwards) is relevant now, as well as the spin. The wave function is now a superposition of the form (using the capital Greek letter $\Psi$ for the position part of the wave function)

$$a \times \psi(\text{spin up}) \times \Psi(\text{deflected up})$$
$$+ b \times \psi(\text{spin down}) \times \Psi(\text{deflected down}).$$

The simple spin wave function in the preceding paragraph is no longer present in its original form. Instead, the spin and position parts of the wave function have become *entangled*. The information is still present, in the complex numbers $a$ and $b$, but the wave function of the measured object (the neutron) has been modified.

At this point, the measurement of the neutron spin has reached its first stage: it has been correlated with the neutron's position, and we may think that position is more accessible than spin. The next stage is to measure the neutron's position by using the detectors to see whether it has moved upwards or downwards. We might try to treat the detectors by quantum theory and extend the preceding sort of analysis to include wave functions of the detectors. The result would be an entanglement of the detectors' wave functions with the neutron's. The coefficients $a$ and $b$ would still appear in this entangled superposition.

Actually, this talk of treating the detectors by quantum theory is hopelessly oversimplified. A detector is a macroscopic system, with very many degrees of freedom. It makes sense for an atom to say it is in one or the other of a few possible states, because an atom is a simple system and the states have separated energies. If an atom is in its state of least energy, for example, it will not normally be able to receive enough energy to get into any other state.

A macroscopic thing, on the other hand, has an enormous number of quantum states with almost the same energy. For example, suppose the system is a smallish crystal, say a millimetre long. The crystal can vibrate, like a violin string. In quantum theory, the longest-wavelength vibration will have the least energy, and this comes out to be so small that these vibrations are excited at a temperature of about $10^{-4}$ K. So, in reality, a detector is not in any single quantum state, but in some mixture of very many states. There will be frequent transitions among all these states, and when this happens the phase angle changes. Thus the *coherence* (the simple relation between the phases of the complex coefficients in a superposition) gets lost.

Quantum systems are much more vulnerable than classical ones to extraneous interactions (with the environment). This is because of the importance of coherence, and the phase angles are exquisitely sensitive to small, seemingly irrelevant effects.

A situation in which the relative phase of two parts of a wave function is known is called *coherent*. The loss of this information in the macroscopic detector is called *decoherence*. Thus decoherence is characteristic of macroscopic systems used to measure quantum systems. As a result of all this, the information contained in the *phases* of the numbers *a* and *b* (earlier) is lost. What remains is the information contained in their *lengths*. How does this translate into the results of the experiment, as obtained from what the detectors register? The "Copenhagen" answer to this is that the ratio of the squares of the lengths of *a* and *b* determines the relative *probability* for the two detectors to register. Thus in the particular case of Section 8.7, where $a = b = 1$, the two detectors are equally likely to register.

The only way to verify a prediction about probability is to repeat the experiment many times and find how many times each detector registers.

It is not so surprising that the result of a single experiment is not determined. Because of decoherence, we have lost the information contained in the phases of *a* and *b*. Before the neutron enters a detector, the phase information is still present. We could have made use of it by deflecting the two paths together again and observing interference, somewhat as in Figure 8.7. But this would not be what we set out to do: to measure the neutron spin along the *x*-axis. If we want to do this, we must use detectors as in Figure 8.4 or something similar; then only probabilities are predicted.

Macroscopic, classical objects play a dual role in the interpretation of quantum theory. On the one hand, they destroy coherence and information. On the other hand, they seem to be necessary for us to know anything about what happens in the world.

Where does the information about quantum phase angles go when decoherence takes place? Take a detector that has detected a particle that was in an entangled state. Suppose it is struck by an air molecule or a photon from the cosmic microwave background. The molecule (or photon) may bounce off, leaving the detector in a different quantum state. Then the detector and the receding molecule (or photon) are in an even more complicated entangled state. Very quickly, a sort of web of entanglement is spreading out into space. In principle the information may still be somewhere, but it is hopelessly irrecoverable.

Many people, from Einstein on, have been unsatisfied with the Copenhagen interpretation of quantum theory. However that may be, there are two trends in contemporary physics that are forcing people to think hard about the meaning of quantum theory. One such trend is the development of larger devices in which decoherence is *avoided*, thus breaking down the divide into microscopic (quantum) *or* macroscopic (classical). Quite apart from the theoretical interest in such work, possible future applications include quantum computing and secure long-distance communication using quantum cryptography. In both cases, quantum entanglement is expected to make fundamental advantages over classical systems possible.

The second trend is the attempt to think about quantum cosmology, to apply quantum theory to the history of the *whole* universe. At very early times, there may be nothing that behaves classically, so the Copenhagen interpretation cannot be used. Suppose we knew the "wave function of the universe"; what would it *mean*? Can we understand why *some* properties of *some* things emerge that behave classically to good approximation?

These are not idle questions. Some people, for example, hope to attribute the formation of galaxies ultimately to the *quantum* uncertainty principle, operating in the very early universe.

## 8.12  What an Electron Knows About Magnetic Fields

This section is about the quantum theory of electrically charged particles in the presence of magnetic fields. We shall see that particles may be affected even when they move *near* but not actually *through* the magnetic field. The section will also lead up to one of the most important ideas to appear in physics in the second half of the twentieth century – *gauge invariance*.

In order to apply Feynman's sum over histories principle (Section 8.8) to a charged particle in a magnetic field, we must know the action for such a particle. This action was constructed in Section 6.5. It made use of the idea of a magnetic "flux" (which I termed FFOC) associated with an open curve, namely, the curve given by the particle's path. As explained in Section 6.5, such a notion of "flux" is inherently ambiguous, its ambiguities being associated with the end points of the curve.

In Feynman's principle, the action gives the change of phase angle for each of the histories in the sum. As the action is ambiguous, it follows that the phase angles of the wave functions must share this ambiguity. The phase angle of the wave function must be ambiguous at each point of space and each time, capable of being adjusted as we please. Any choice of the wave function's phase angle implies a choice of the FFOCs associated with open curves: the two things must be chosen consistently.

The principle just stated is called *gauge-invariance*. The historical accident that led to the odd word *gauge* was mentioned in Section 6.5. People speak of "invariance" because the physics is *not* changed when the wave function phase angles and the FFOCs are changed together consistently (such changes are called *gauge transformations*). The principle of gauge-invariance in quantum theory was first recognized in 1927 by Vladimir Fock and Fritz London.

There is no immediate contradiction in having an ambiguity in wave function phase angles. The phase angle of a wave function is not directly measurable: it is the length of the wave function that determines physical probabilities.

We can now draw a very surprising conclusion. Suppose electrons pass near and at right angles to a very thin "whisker" of magnetized iron, as illustrated in Figure 8.10. There is a magnetic field within the iron, but none in the space outside. According to Feynman's

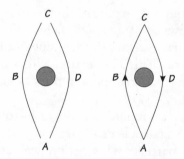

FIGURE 8.10   The Bohm-Aharonov effect. The shaded region is the cross section of a thin whisker of magnetized iron. An electron moves from *A* to *C*. The paths *ABC* and *ADC* are two motions to be included in Feynman's sum over histories. The right-hand figure illustrates how the difference between the FFOCs associated with the two paths is the same as the flux through the closed curve *ABCD*. (Compare Figure 6.3.)

principle, we must, in quantum theory, include all possible motions from, say, $A$ to $C$. In particular, some paths go to the left of the iron whisker, like $ABC$, and some to the right, like $ADC$. In either case, the change of phase angle (between $A$ and $C$) is given by the FFOC associated with the path in question. What is important is how these two changes of phase angle compare with each other. If they are the same, the two contributions at $C$ would just add up. If they differ by 180 degrees, the two contributions would just cancel. Thus the probability of finding the electron at $C$ depends upon the difference of the two FFOCs.

But the difference between the FFOC associated with $ABC$ and that with $ADC$ is just the flux through the *closed* curve $ABCD$. This flux is a physical, unambiguous thing: it is determined just by the number of magnetic lines of force passing through $ABCD$ – in other words, by the lines of force in the iron whisker. Thus the probability of finding an electron at $C$ is affected by the magnetism in the iron. But, for this conclusion, there is no need for the electron to pass *through* the iron (where the magnetic field is); it is sufficient to pass *near*.

Nothing like this could occur in classical physics. That is because classically the electron must pass either to the right or to the left. It cannot do both at the same time.

The unexpected behaviour of charged particles near magnetic fields was predicted by Aharonov and Bohm in 1959 (although this could have been done almost 30 years before) and has been experimentally verified since. It shows very clearly that we have done the quantum theory in a magnetic field correctly: that the phase angles are connected directly (by the gauge-invariance principle) with the FFOC.

We shall see, in Chapter 12, how the idea of gauge-invariance has been generalized in a most fruitful way.

## 8.13   Which Electron Is Which?

There is a question that comes up in quantum theory: can one tell one electron from another? or one proton from another? Are similar particles *distinguishable*? In classical physics, this question does not arise. If you have a number of electrons, you could note the position of each at some initial time, then follow their subsequent motion. Each one would be effectively "labelled" by its initial position. In quantum theory, the uncertainty principle forbids this. One cannot in general determine which electron started where.

Let us assume, then, that in quantum theory electrons are indistinguishable from one another. Interchanging two of them should then make no difference. The obvious way to implement this is to assume that the wave function is unchanged when two electrons are swapped over. For example, the wave function of a helium atom (which contains just two electrons) depends upon the positions of each of them (and also upon their spin directions, but let us ignore this complication for the moment). From the preceding argument, we would expect the wave function to have the property that it was unchanged when the two positions were interchanged.

However, there is another less obvious possibility. In quantum theory, wave functions are not directly observable. The physical things are probabilities that are calculated from the lengths of the wave functions. It would be sufficient if these probabilities were unchanged when two electron were swapped over. One particularly simple way to arrange this is for the wave function to be multiplied by $-1$ each time two electrons are swapped. Particles that have this $-1$ are called *fermions*, after the Italian nuclear physicist Enrico Fermi. Particles for which the wave function is unchanged (or, of you like, multiplied by $+1$) are called *bosons* after the Indian theoretical physicist Satyendra Bose.

So which are electrons, bosons or fermions? One way to answer this is to look to experiment. It turns out that electrons are fermions, and all of chemistry depends upon this fact. If electrons were bosons, chemistry would be a very dull subject (if it deserved to be called a subject at all), and the world would be dull and without life.

Take the wave function of the helium atom again and consider its value when the two electrons are at the same point. If they are fermions, this value must be zero (for it is equal to $-1$ times itself). In general, the wave function for a number of fermions must be zero whenever any two coincide in position. This means that the wave function is small whenever two of them are near each other. Fermions "want to keep apart". This (roughly) is the *exclusion principle*, formulated by Wolfgang Pauli in 1924.

I have oversimplified. Electrons have spin $\frac{1}{2}$ and can be in one of two spin states, "up" and "down". So in fact *two* electrons can be at the same point provided they are in opposite spin states. But a third electron could not be put at the same point, since its spin state would be the same as one or the other of the first two.

Consider how these facts affect chemistry. The two electrons in a helium atom can be close together near the nucleus, forming a compact and unreactive system. A lithium atom has three electrons. Two of these can fall near to the nucleus, as in helium, but the third must find somewhere else to go, farther from the nucleus. This electron is easily detached from the atom in chemical reactions; that is why lithium is chemically reactive (forming compounds like lithium chloride).

It turns out the next important number of electrons is 10 – the element neon. Two of these are near the nucleus, as in helium, and eight can just fit in farther away. (This number 8 is $2 \times [1 + 3]$, where the 2 is the two spin directions, and the 1 and the 3 are from possible angular momentum states of the motion of the electron round the nucleus.) Neon, like helium, is an inert gas, very reluctant to combine chemically. One more electron (making 11) gives the reactive light metal sodium (it reacts with cold water). One less electron gives the reactive gas fluorine, reactive because it readily takes up an extra electron to get the same stable structure as neon.

Proceeding in this way, the properties of all the chemical elements are explained, as they had been systematized in the periodic table of Mendelayev, in the late nineteenth century.

If electrons were bosons, the world would have been utterly different. Atoms with more and more electrons would have them just packed closer and closer round the nucleus (closer because of the greater attraction of the higher positive charge on the nucleus) in uninteresting blobs.

In some solids, electrons can "hop" from atom to atom. In electrical insulators, the exclusion principle forbids this "hopping": all the relevant electron states are full. In metals on the other hand there are more relevant states than electrons. The electrons can "hop" without hindrance, and thus electric currents flow. Semi-conductors, like silicon, occupy an intermediate position. A few electrons are available to "hop", their number dependent upon the temperature. The current that can flow can be delicately controlled by applying electric fields, thus allowing the use of semi-conductors in transistors.

Protons and neutrons, the constituents of atomic nuclei, are also fermions. Neutron stars are like giant nuclei, made predominantly of neutrons. They have masses similar to the Sun's but sizes of a few kilometres, so their density is some $10^{15}$ times that of ordinary matter on Earth. It is the exclusion principle, keeping the neutrons apart, that prevents such stars from collapsing under their own intense gravitational attraction.

Neutron stars are also called *pulsars* because of the pulses of radio waves they emit (as they rotate), with extraordinary regularity; the interval between the pulses is of the order of a second.

What about bosons then? The basic constituents of ordinary matter, electrons, protons and neutrons, are all fermions. But an atom of helium, for example (containing 2 electrons, 2 protons and 2 neutrons), may itself be considered as a boson (provided it remains in its state of lowest energy). This fact affects the properties of helium gas and liquid helium.

Since 1995, several experiments have been done in which very-low-density gases of atoms have been cajoled into a single quantum state. The atoms are trapped and slowed down by ingenious arrangements of magnetic fields and laser beams and are thus cooled to temperatures of some $10^{-6}$ K.

In the next chapter, we shall see that *relativistic* quantum theory shows the properties of bosons and fermions in quite a new light.

## 8.14 Conclusion

Quantum theory is necessary to explain many features of the physical world, particularly as regards microscopic systems. According to quantum theory, a particle has no definite orbit, a system no definite history. Instead, all histories have to be included in order to work out probabilities. These probabilities are interpreted to refer to the results of measurements carried out on the quantum system using a macroscopic (nearly classical) apparatus. A more general interpretation is not yet known.

Two by-products of quantum theory, which will figure prominently in later chapters, are the notions of gauge-invariance and of bosons and fermions.

# QUANTUM THEORY WITH
# SPECIAL RELATIVITY

*How the marriage of quantum theory with special relativity gives
an account of the creation and annihilation of particles.*

## 9.1 Einstein Plus Heisenberg

The previous chapter began with radiation (as indeed did the history
of quantum theory), then went on to the quantum theory of atoms
and particles, not mentioning light again. There is a good reason
for this. Quantum theory, as I have described it up to now, was
non-relativistic, assuming the existence of absolute time. But the
study of light certainly needs special relativity, which has after all
the speed of light built into it from the beginning.

We can see by a simple argument that the addition of special
relativity to quantum theory is bound to lead to something quite
new. In the quantum theory of the previous chapter, we always began by specifying the particles contained in the system under study
(for example, in a hydrogen atom, one electron and one proton),
and the number and type of these particles did not change. The
formalism just could not handle any such thing. If we write down
Schrödinger's equation for a hydrogen atom, there is no way we
can use it to study a helium atom: for that, we would have to begin
again. But experiments involving collisions between fast particles
commonly reveal changes in the numbers and types of particles.
For example, a collision between two protons can (if the energy is
high enough) produce extra particles called *mesons*. Some of the
kinetic energy of the protons is used up to make the rest energy of

the mesons, according to Einstein's formula

$$\text{energy} = c^2 \times \text{mass}$$

(where $c$ is the speed of light).

One might think that this type of particle production could be prevented at least if we limited ourselves to collisions that do not have too much energy. But even this is not so. Suppose we ask about the state of the system at some very sharply defined time. Then, according to Heisenberg's uncertainty principle, the energy is quite uncertain. So there may be enough energy for extra particles to be produced for a very short time. Such evanescent particles are called *virtual*. They do not live long enough to be detected directly, but they have an indirect effect on other things that can be measured.

We face, then, two questions: how to make a quantum theory of light and how to make a quantum theory that can allow for particle production. It turns out that a single step leads to the solution of both these problems. This is to make a quantum theory of the electromagnetic *field*. (Later on we will encounter other sorts of field.) In the last chapter, we took Newton's equations of particle motion and found a corresponding quantum theory. In this chapter, we must take Maxwell's equations for the electromagnetic field and find a corresponding quantum theory. This sounds a daunting task: in order to define the position of a particle we need just three numbers, but to define the state of, say, an electric field we need to specify it at *every* point of space (in principle). In other words, a field has an unlimited number of *degrees of freedom*. However, it is not difficult to make a start on "quantum field theory", as I will explain in the next section. Historically, quantum field theory was created soon after quantum theory, in the late 1920s, but, as we shall see, there were worrying snags that were not overcome until the late 1940s.

## 9.2 Fields and Oscillators

Take Maxwell's equations of electromagnetism. We know that they predict the existence of electromagnetic waves: light, radio waves and so on. To begin with, let us think about these waves propagating in free space, leaving aside the electric charges that produced them.

A wave can have any wavelength and any direction (and any polarization). According to the principle of superposition, we can add together such waves to get more complicated solutions of Maxwell's equations. Space around us is busy with all sorts of different electromagnetic radiation: radio, radio-telephones, microwaves escaped from ovens, heat radiation, light and so on. They all proceed independently of each other (at least to a very good approximation): otherwise radio communication would be impossible. So, if we can do the quantum theory of any *one* such wave (with a given wavelength and direction), we can do the quantum theory of any electromagnetic field.

For a given wavelength, the field varies in time with a period given by

$$\text{period} = \frac{\text{wavelength}}{c}.$$

The time dependence is just a sine curve, as in Figure 4.1. This is the same time dependence as in the motion of a pendulum (provided it swings through a small angle). Anything that has a sine curve time dependence, with a definite period, is called a *simple harmonic oscillator*. I shall write just *oscillator* for short.

It follows from all this that, if we can do the quantum theory of an oscillator, we have the necessary building block to do the quantum theory of the field. An oscillator has only one degree of freedom, so we can build up all the unlimited degrees of freedom of the field from a lot of independent oscillators with 1 degree of freedom each.

It turns out that the quantum theory of an oscillator is very simple. The allowed energies are *equally spaced*. They are multiples of

$$\hbar \times (\text{frequency}),$$

with the multipliers

$$\frac{1}{2}, \frac{3}{2}, \frac{5}{2}, \ldots$$

This equal spacing will be crucial in the argument to follow. I do not know of any simple argument why this is so, but one can at least understand quantitatively why the lowest energy is not zero. Classically, an oscillator can be in a state of rest, so that it has no energy. But this would be to specify its position exactly, a task that in quantum theory conflicts with the uncertainty principle (Section 8.6). In the quantum state of minimum energy (corresponding to the $\frac{1}{2}$), there is an uncertainty in the position and in the momentum, such that the kinetic and potential energies are each nonzero and each contributes half the total energy.

Electromagnetic waves carry momentum as well as energy (see Section 6.8). The two are related very simply. In any given volume of space,

$$\text{(energy)} = c \times \text{(momentum)}.$$

Consequently, the quantization of energy of the electromagnetic oscillators implies quantization of momentum too, in multiples of

$$\hbar \times \text{(wavelength)}.$$

Thus

$$\text{(energy quantum)} = c \times \text{(momentum quantum)}.$$

At this point, we must recall Einstein's relation connecting the energy and momentum of a *particle*. This is (see Section 5.11)

$$\text{(energy)}^2 = c^2 \text{(momentum)}^2 + c^4 \text{(mass)}^2.$$

Thus the connection between the energy and momentum of the quantum electromagnetic oscillators parallels that between the energy and momentum of a particle with *zero* mass. As explained in Section 5.11, particles with zero mass are allowed in relativity theory, but they must move with exactly the speed of light.

We can now see a deep parallel between quantized electromagnetic oscillators on the one hand and particles on the other: increasing the energy and momentum of an oscillator by one quantum has exactly the same effect as adding one particle of the same energy and momentum (and zero mass). In fact, we can go further:

- In quantum theory, the electromagnetic field behaves exactly as an assembly of arbitrarily many massless particles. The number of particles of a given momentum and energy

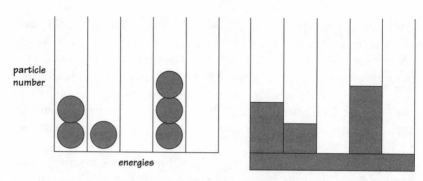

FIGURE 9.1   A schematic indication of the parallel between particles and quantized fields. In each case, five representative energy and momentum states are shown (of the very many that exist). In the particle interpretation, on the left, the numbers of particles "put into" these states are 2, 1, 0, 3, 0, respectively. In the field interpretation, on the right, the corresponding oscillators are in the energy levels labeled by these five numbers. In addition, each of the oscillators has a half-unit of ground state energy.

just corresponds to the energy level of the corresponding electromagnetic oscillator.

The parallel between quantized fields and particles is illustrated in Figure 9.1.

The particles appearing in the quantization of the electromagnetic field are called *photons*. Here are examples to give an idea of orders of magnitude. For a radio wave of frequency 100 MHz (that is, $10^8$ per second), the energy of a single photon is about $10^{-25}$ joule. A transmitter with a power of 1 kilowatt emits $10^{28}$ photons per second. In the laboratory, photons can be produced with energies of up to about $10^{-8}$ joule, approaching the energy of a flying insect. For comparison, the heat energy of a typical air molecule is about $10^{-21}$ joule.

But if a quantized field is nothing but a lot of particles, how can we keep our picture of, say, a magnetic field as a smoothly varying quantity that we can measure at any point of space? The answer is that this classical picture is indeed a very good approximation in some circumstances: when the numbers of particles are very large and indeed uncertain. This is a typical quantum situation: when the number of particles is measured the field is uncertain, and when the field is measured the number of particles is uncertain. This is the "wave-particle duality" that has loomed so large in quantum theory from the first years of the twentieth century. From a modern perspective, it is just one example of the counterintuitive nature of the quantum world.

Photons have spin angular momentum. At the end of Chapter 4, I described a circularly polarized electromagnetic wave: one in which the electric and magnetic fields rotate about the direction of propagation. Not surprisingly, such a wave has angular momentum about that direction. By knowing the spatial distribution of electromagnetic momentum, the angular momentum can be calculated. One finds that (in any given volume of space)

$$\text{angular momentum} = \frac{\text{energy}}{\text{frequency}}.$$

But the energy of a photon is $\hbar$ times the frequency, so the angular momentum of a photon is just $\hbar$. This angular momentum is about the direction of propagation of the wave, that is, about the direction of motion of the photon. The sign comes out positive or negative according to whether the circular polarization is right or left. Thus the spin of a photon about the direction of its motion has value

$$+\hbar \text{ or } -\hbar.$$

We have already, in Section 8.7, discussed the possible spins that particles may have and seen that one possibility was (in any given direction)

$$+\hbar, 0, -\hbar.$$

How is it that for the photon the value 0 is missing? The answer is that in Section 8.7 we implicitly assumed that the particle has nonzero mass. Then we could study it from the point of view of an observer moving with it, so it appears to be at rest. The three possible values of spin then follow. A photon, however, moves with the speed $c$, and there is no possible observer for whom it appears to be at rest, so the argument given does not apply. A right circularly polarized wave appears to have the same angular momentum for *any* observer.

We learn from this that there is a qualitative difference between the spins of particles with zero mass and those with nonzero mass. In this example, there are two and three possible spin states in the two cases. We will see that this simple fact has important consequences for the unification of electromagnetism with other forces of nature.

## 9.3   Lasers and the Indistinguishability of Particles

We saw, in Section 8.13, that particles of a given kind are indistinguishable in quantum theory and may be either fermions or bosons according to whether the wave function does or does not change sign under a swap of two of them. If photons are particles, are they bosons or fermions? The answer is apparent from their interpretation as quanta of oscillations in electromagnetism. Look at Figure 9.1. On the left side, it *might* have been that we could distinguish the six photons – perhaps by writing numbers on them. Then the situation in the figure could be achieved in 60 different ways. (The reader may recognize this as $\frac{6!}{(3!2!)}$.) On the right of the figure, it is clear that this makes no sense. All we can say is that the five oscillators are in their second, first, zeroth, third, zeroth energy states. There is just one such state. The "interchange of two photons" does nothing at all; it does not alter the excitations of the oscillators. Photons are clearly bosons.

We began the previous chapter with the thermal properties of radiation, especially the frequency spectrum in such radiation (Figure 8.1). It should be possible to rederive the results, thinking of the radiation as a collection of photons. Just this was done in 1924 by the Indian physicist Bose. He realized that he needed to assume that photons are indistinguishable. Bose's discovery was taken up and generalized by Einstein. The way of counting distinct states, illustrated here, is now called *Bose-Einstein statistics* (statistics because of its importance in statistical physics, in which the correct enumeration of distinct states is essential), and particles that obey Bose-Einstein statistics are called *bosons*. Thus photons are bosons. But Einstein realized that other particles may be bosons. For example, helium atoms are bosons.

The *laser* is a device that exploits the bosonic nature of photons. Suppose that, in a ruby crystal, say, some atoms in an excited state can fall to the ground state with the emission of a photon with a certain definite wavelength (colour). Suppose most of the photons of this wavelength are kept in the ruby by reflection from mirrors at the ends. Suppose at a given time there is a number $n$ of such photons. A photon may be absorbed by an atom in the ruby in its lowest state, thus sending that atom into its excited state. The probability of this event is, as one would expect, proportional to $n$. When this happens, $n$ changes to $n - 1$. The converse effect is the emission of a photon from an atom passing from the higher state to the ground state. Then $n$ changes to $n + 1$. Einstein realized that the probability that this will happen is, surprisingly, proportional to $n + 1$. The 1 in the $n + 1$ ensures that the process can occur even when no photons are present initially (so $n = 0$). This is reasonable. The $n$ in the $n + 1$ tells us that pre-existing photons make the process more likely, a phenomenon called *stimulated emission*. (*Laser* is an acronym for *light amplification by stimulated emission of radiation*.)

Einstein arrived at this factor $n + 1$ indirectly by considering the equilibrium of the whole system (of photons and atoms). It also follows from the quantization of electromagnetism. There is a sort of symmetry between emission and absorption:

emission: $n \to (n + 1)$, (probability) $\propto (n + 1)$;

absorption: $(n - 1) \leftarrow n$, (probability) $\propto n$.

We can now go back to the ruby laser. Suppose that many atoms are somehow "pumped" into the higher state by some indirect means, which need not concern us here. Photons begin to be emitted. Once this begins to happen, it happens faster and faster, because of the increasing factor $n + 1$. The result is a cascade of photons, all with identical wavelengths, all in fact in a single quantum state. The resulting radiation, as well as being intense, is quite unlike that from, say, a light bulb (or from the Sun), which is a jumble of photons in different states.

The first optical laser was made in 1960, though a microwave version (*maser*), using ammonia molecules and centimetre wavelength radiation, had been devised by Townes earlier. Nowadays, lasers are commonplace, in, for example, compact disc players and check-out scanners.

## 9.4   A Field for Matter

At the beginning of this chapter, we confronted the problem posed by relativity for quantum theory: how to deal with changing numbers of particles. If we accept photons as particles, Section 9.2 has provided the answer for them: quantizing the electromagnetic field automatically leads to a theory that can be interpreted as containing arbitrarily many photons. This naturally raises the question, Can we find a "field" that, when quantized, describes arbitrarily many electrons, or protons or other particles of matter? We cannot find such a thing by looking around us for a classical field that will do the job. The reason for this will become clear shortly.

The required field, together with the equation it obeyed, was discovered in 1928 by the British physicist (he had a Swiss father) Paul Dirac. He had already reconciled the quantum theories of Heisenberg and Schrödinger and formulated quantum theory in its deepest and most general way. He was then, in 1928, trying to reconcile quantum theory with relativity. He did not at first realize that what he had discovered was a field; he thought it was a new sort of Schrödinger wave function. The way I have written this chapter is therefore unhistorical. Dirac did not, in 1928, know about the possibility of the change in the number of electrons. In fact, he predicted it in 1930, and it was confirmed by experiment two years later.

Dirac's field is in some ways like the electromagnetic field, but also very different. I will explain some similarities first, then some differences.

The Dirac field is a field, which means that it is a quantity that varies with position and with time. In fact, it is a set of four quantities (the four *components* of the field), somewhat as the electromagnetic field has six components (three in the electric vector, three in the magnetic vector). Dirac's equations relate the changes with time and position of the components of his field, just as Maxwell's equations relate the changes with time and position of the electric and magnetic fields. (In Section 3.5, I gave a pictorial version of Maxwell's equations. I do not know of a similar version of Dirac's.)

Dirac's equation, like Maxwell's, has wave solutions that are like oscillators, with frequency related to wavelength. When these oscillators are quantized, the theory is seen to describe arbitrarily many particles (e.g., electrons), with their energy and momentum connected according to Einsteins's relation.

Now for some of the differences between Dirac's and Maxwell's fields and their equations.

They are both consistent with the geometry of Minkowski spacetime. The way the electromagnetic fields achieve this was mentioned in Section 5.12. The way the Dirac field does it is more subtle. It exploits a property of space that was mentioned in Section 8.7: the possibility that quantities can appear to change in a non-trivial way when turned through a complete rotation of 360 degrees. In fact, under such a rotation, the components of the Dirac field all change to minus what they were before the rotation. Quantities that behave like this are called *spinors*. The spinor nature of the Dirac field implies that it cannot possibly be a classical quantity that one might measure directly, like a magnetic field, say. Under a rotation through 360 degrees, which gets everything back to where it started, a directly measurable quantity should not be changed. But things like the square of the Dirac field do not change under a rotation through 360 degrees (there are two minus signs), so they may be observable.

Like the electromagnetic field, the Dirac field carries angular momentum. The particles obtained by quantization turn out to have possible values of angular momentum (in any given direction)

$$\frac{\hbar}{2}, \quad -\frac{\hbar}{2}.$$

As explained in Section 8.7, these are just the spin states of an electron.

The energy stored in the electromagnetic field (see Section 3.5) is obviously positive: it is formed by adding the *squares* of the electric and the magnetic fields. The corresponding energy for Dirac's field does not have this property of being positive. It can equally well be negative as positive. At first sight this is a disaster. It looks as if one could put unlimited "negative energy" into the Dirac field, compensated for by unlimited positive energy elsewhere. If a bank allows you an unlimited overdraft, you become infinitely rich. Both these phenomena are contrary to experience. Dirac found the answer to this paradox in 1930. I will defer the explanation until the next section.

The relativistic equation connecting energy and momentum, which comes out of quantizing the Dirac field, is the one with a mass, $m$, in it. Here $m$ (or more accurately the inverse length $mc/\hbar$) appears as an arbitrary parameter in Dirac's equation. It can be put equal to zero, but it need not be. To describe the electron, $m$ would be chosen to match its measured mass. In contrast, Maxwell's equations do not allow such a mass to appear, without doing violence to the structure of the equations.

The electric and magnetic fields are made up of real numbers. Dirac chose the components of his field to be complex quantities, that is, to have a phase angle as well as a length associated with them (or, equivalently, to involve $i$, the square-root of $-1$). It turns out that such a complex field naturally has electric charge associated with it. Out of the components of the field (using products of different components taken two at a time), one can construct a local electric charge density and a local electric current density. Dirac's equations ensure that these are related consistently, so that the rate of change of charge in a region is balanced by the currents flowing in and out.

Thus the particles obtained by quantizing the Dirac field naturally have an electric charge, just as electrons do.

Electrons are spinning and have electric charge. A circulating electric charge produces a magnetism, so it is not too surprising that electrons behave as tiny magnets. One of the triumphs of Dirac's work was to predict (with accuracy of the order of 1 in 100) the strength of this magnetism. A very small magnet has a strength described by a single number, called the *magnetic moment*. Dirac predicted the magnetic moment of the electron to be

$$\frac{\text{charge}}{\text{mass}} \times \left(\frac{\hbar}{2}\right).$$

The measured value is now known to be

$$1.0011596522$$

times Dirac's prediction, and the small correction is now understood.

## 9.5 How Can Electrons Be Fermions?

The quantization of the electromagnetic field leads naturally and apparently inevitably to photons' being bosons. As explained in Section 8.14, we know that electrons certainly are fermions. How can the quantization of the Dirac field possibly give fermions instead of bosons? In answering this question, we shall also solve the puzzle about the lack of positivity of the energy of the Dirac field.

The oscillators of the Dirac field are different from ordinary ones. They do not have an infinite tower of possible energy states. Instead, each oscillator can be in one of just two possible states, with energies

$$-\frac{\hbar}{2} \times (\text{frequency}), \quad +\frac{\hbar}{2} \times (\text{frequency}).$$

The difference between these two energies ($\hbar \times (\text{frequency})$) is the same as for an ordinary oscillator, but the energy cannot be increased above the upper value (nor made more negative than the lower one).

The idea is illustrated in Figure 9.2. This figure is to be contrasted with Figure 9.1. The oscillators in the upper part of the figure each can be in one of only two possible energy levels.

The bottom part of the figure shows how the same situation might be interpreted from a particle point of view. Each box represents a possible particle motion (characterized by its momentum and spin). The boxes have room for one particle only. Each box is either empty or full. This is just what the exclusion principle demands.

What about the negative energies? Negative energies do appear, as expected: in fact there is symmetry between the positive and negative energies. But this is not a disaster. The negative energies are limited below; so each oscillator cannot sink to indefinitely negative energies, thereby releasing unlimited positive energy elsewhere.

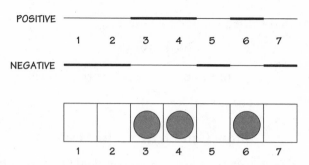

FIGURE 9.2   The top part of the figure represents fermionic oscillators. Seven oscillators (labelled by numbers) are shown. The
bottom line represents their negative energy states, the top line their
positive ones. The solid lines show which state each oscillator is in:
numbers 1, 2, 5, 7 are in their negative energy state; numbers 3,
4, 6 in their positive ones. The bottom part of the figure shows the
same thing from the particle point of view: the seven boxes represent
the seven possible particle energy-momentum assignments, and the
black circles show which of these are occupied by a particle.

This situation is not readily understood according to the particle
picture. One would have to suppose that the unoccupied particle
states all had negative energy, which is not natural.

How is it arranged mathematically that the oscillators of the
Dirac field have the properties described? It is done by a strange
mathematical trick. Mathematicians introduce symbols for other
things beside numbers; for example, symbols to represent operations like rotations. Such symbols must have a consistent set of
rules for manipulating them. There may be things one can do with
the symbols, which are naturally called *addition* and *multiplication*
but may not obey all the same rules as addition and multiplication
of ordinary numbers. We saw in Section 8.4 that quantum theory
needs to use "operators" for which $a \times b$ is not necessarily equal
to $b \times a$.

In the nineteenth century, Hamilton introduced a system of quantities called *quaternions* that he believed would represent naturally
three-dimensional space. Quaternions involve quantities $a$, $b$, $c$,
called *elements*, satisfying

$$a \times b = -b \times a,$$

and so on. Later, Grassmann generalized this scheme to try to represent spaces of any number of dimensions. Grassmann's elements obey the even more strange-looking equations

$$a \times a = 0,$$

and so forth. Neither Hamilton's quaternions nor Grassmann's generalization is now put to quite the direct geometrical uses that their inventors envisaged; but they are very interesting to mathematicians.

Given an ordinary physical quantity, an electric field, say, one can square it, cube it, raise it to the fourth power and so on, thus producing complicated dependence on that quantity. Grassmann variables are much more limiting. Given one Grassmann element, all one can do is multiply it by an ordinary number – squaring it just gives zero. Given two Grassmann elements, $a$ and $b$, there is one more thing that can be done, that is, multiply them together, $a \times b$. Anything more complicated gives zero again.

These are just the mathematical properties needed to express the exclusion principle naturally. Suppose each oscillator of the Dirac field is represented by a Grassmann element. One can make just two different wave functions, for the two states used in Figure 9.2. Any attempt to construct a wave function for a state of higher energy fails because of the property $a \times a = 0$.

The equation $a \times b = -b \times a$ also has a physical interpretation in the properties of fermions. It just corresponds to the change of sign produced by the interchange of the positions of two fermions.

It is therefore assumed that the Dirac field is a Grassmann variable, constructed out of Grassmann elements. This is a radical assumption. It means that the Dirac field is no ordinary physical quantity, such as the magnetic field, that can be measured. But we already knew that this must be true, because of the spinor nature of the Dirac field (changing sign under a 360-degree rotation). All this does not mean that that the Dirac field has no physical significance. Physically measurable things, like energy and electric charge, can be made out of the Dirac field.

In the case of electromagnetism, classical electric and magnetic fields were known long before photons. In contrast, no "classical" Dirac field can possibly exist (for the various reasons explained previously), so the existence of the field had to wait to be inferred indirectly from the existence of electrons.

The use of abstract mathematical quantities, Grassmann variables, in the Dirac field is rather strange. There is, however, an exact parallelism between the mathematics and the physical properties of fermions.

So, if electrons are got by quantizing the Dirac field, they have to be fermions for consistency: otherwise the problem of negative energies would be insoluble. On the other hand, the electromagnetic field was quantized so that photons were automatically bosons. It would also have been inconsistent to make photons into fermions.

These are examples of a general theorem:

- According to relativistic quantum field theory, any particles with integer spin $(0, 1, 2, \ldots) \times \hbar$ must be bosons and any particles with half-integer spin $(\frac{1}{2}, \frac{3}{2}, \ldots) \times \hbar$ must be fermions.

This theorem was first stated and proved in 1940, by Pauli (the formulator of the exclusion principle).

It is remarkable that relativity is needed to prove this theorem. In non-relativistic quantum theory, there was (apparently) *no* reason why electrons (for example) could not have been bosons: the exclusion principle had to be added as another law of nature. The exclusion principle, as we have seen, is crucial to the properties and chemical behaviour of atoms, in which relativity has little direct effect.

## 9.6 Antiparticles

Dirac chose his field to be complex (that is to say, the Grassmann elements were multiplied by complex numbers, not just real ones). A complex number is equivalent to two real numbers, so somehow the Dirac field is twice as complicated as it would have been if it were real. What does this doubling up correspond to physically? From complex fields, one can naturally construct quantities with the properties of electric charge and electric current (automatically satisfying conservation of charge, so that the increase of charge in a region is the total current flowing in). The electric charge does *not* automatically have one sign, negative or positive. So the doubling up means that the particles obtained by quantizing the field can have *both* charges, positive as well as negative.

The charge on the electron is, by convention, called negative. So Dirac's field implied the existence of particles with positive charge, but otherwise just like electrons (same mass, same spin, opposite magnetism). When Dirac realized this in 1930, no such positive particles were known. There was considerable puzzlement. But the positive particles (called *positrons*) were discovered in 1932, accidentally in cosmic rays, by Carl Anderson at the California Institute of Technology.

We may state another theorem:

- In quantum field theory, given any charged particles, there must be *antiparticles* with identical mass and spin but opposite electromagnetic properties.

The antiproton was discovered (intentionally) in 1955 by Chamberlain and Segré at the University of California at Berkeley, using high-energy collisions between protons with enough energy to provide the rest energy $(2(\text{mass}) \times c^2)$ of an extra antiproton and proton.

In spite of the complete symmetry between electrons and positrons, protons and antiprotons, neutrons and antineutrons and so on, ordinary matter contains a vast preponderance of electrons, protons and neutrons. Why this should be so is one of the big questions of astrophysics.

## 9.7  QED

So much for quantized electric and magnetic fields by themselves, or the Dirac field by itself. But these *free* fields are rather trivial. Nothing really changes with time. If certain modes of oscillation are excited at an initial time, they remain like that for all time. Or, from the particle point of view, if there are certain photons (or electrons) moving with certain momenta at one time, they remain like that for all times.

In truth, charged particles and the electromagnetic field interact with one another: charges produce electric and magnetic fields, and these fields in turn exert forces on charges. In the quantum field theory, the (charged) Dirac field must interact with the electromagnetic field. First one must find the form of this interaction. One way to do this is to use the gauge-invariance principle, explained in Section 8.12. This principle referred to the ambiguity in the phase of the electron wave function. In quantum field theory, it is instead the phase of the Dirac field that may be chosen arbitrarily at each point of spacetime. The gauge principle asserts that this arbitrariness must be balanced by a corresponding ambiguity in the electromagnetic FFOCs associated with open curves from one point of spacetime to another. The interaction between the Dirac and electromagnetic fields is then chosen to be the simplest one consistent with gauge-invariance.

Of course, the value of the electron's electric charge comes into this interaction, fixing how strong it is. This electric charge is denoted by the letter $e$ (unfortunately, the same letter used by mathematicians for Euler's constant, the base of natural logarithms, $2.718\ldots$). The charge $e$ has been measured very accurately; it is

$$e = 1.602177 \times 10^{-19} \text{ coulomb.}$$

Of course, this is a small number in ordinary units, because any ordinary macroscopic electric charge is made up of very many electrons. But it does not give one much idea of the strength of the interaction so far as single electrons are concerned.

In relativistic quantum theory, two constants of nature are specially relevant. These are the speed of light $c$ and Planck's constant $\hbar$. The former is woven into the structure of spacetime; the latter is fundamental to any quantum theory. It is natural to use units in which both $c$ and $\hbar$ take the value 1. For example, length can be measured in metres; then the unit of time is $3.335641 \times 10^{-9}$ second (the time it takes light to travel 1 metre), and the unit of mass is $2.21022 \times 10^{-42}$ kilogram. In these units, it turns out that $e$ is dimensionless, that is, a pure number, independently of the system of dimensions used. For historical reasons, the numerical value is usually quoted for

$$\alpha = \frac{e^2}{4\pi} = \frac{1}{137.03599},$$

conventionally denoted by the Greek letter alpha, $\alpha$.

The (rather small) size of this number helps to determine many features of the world we live in: for example, that an atom is a comparatively loosely bound structure, much bigger than its nucleus, with electrons moving slowly compared to the speed of light, and that the typical wavelength of light emitted by atoms is again much larger than an atom. For present purposes, the important thing is that $\frac{1}{137}$ is a *small* number.

The complete theory of quantum electrodynamics (QED), including the interaction between the fields, is very complicated. The states of excitation of all the (many) oscillators of each field can keep changing. In particle terms, electrons can keep emitting or absorbing photons. There is little chance of making exact calculations in such a theory.

But the fortunate smallness of $\alpha$ gives the hope of working things out approximately. Start with the free electromagnetic and Dirac fields. This is a trivial situation. Then take account of corrections of order $\alpha$, that is, corrections of order 1 percent or so. Then include corrections of order $\alpha^2$, one hundredth of a percent, and so on. Some calculations have been done up to order $\alpha^3$.

There is a systematic way of carrying out this type of calculation, which I will describe in the next section.

## 9.8   Feynman's Wonderful Diagrams

Although people had struggled with quantum field theory since the 1930s, systematic ways to develop successive approximations were not found until around the year 1949, by two Americans, Schwinger and Feynman, and an Englishman, Dyson, working in the United States. The styles of these three physicists were strikingly different. Schwinger's work was complicated and difficult for other people to learn how to use. Feynman presented his ideas in an intuitive way, easy to use but harder to place in the context of conventional calculations. Dyson tied everything together in a way that ordinary physicists could assimilate. (I remember beginning as a research physicist by poring over Dyson's beautiful papers.)

Let us start from Feynman's sum-over-histories version of quantum theory. Now we are doing field theory, *history* means, not a motion of a particle, but a succession of values of the fields (at all points of space) – a very complicated thing. For each such history there is a complex number, whose phase is given by the *action* for the fields. The action contains a part for the electromagnetic field by itself, another for the Dirac field by itself, but also the part describing the interaction between the two (proportional to the charge $e$). This complex number is systematically approximated: first a piece independent of $e$, then a piece proportional to $e$, then a piece proportional to $e^2$ and so on.

The first term, independent of $e$, is trivial. The quantum state of the fields does not change. Take the term proportional to $e$. This contains the interaction part of the action, which is given in terms of the fields at some spacetime point $P$. The Feynman sum contains a sum over all such points $P$. It also contains a sum over field histories at times earlier than the point $P$, and another sum for times later than $P$. But in these two sums the interaction between the fields is neglected, and they can be worked out fairly simply.

Going on to the $e^2$ approximation, there are the interaction operating at two points of spacetime, say $P$ and $Q$, and free fields before, between and after the times of these two points. Again the Feynman sums can be done for these free field periods, and then sums over the positions of $P$ and $Q$ remain to be done. Similarly, for the $e^3$ term, three points $P$, $Q$ and $R$ are required, and so on.

Feynman devised a diagrammatic way of representing all this, which has been perhaps the most used tool ever devised in theoretical physics.

Take an example. Suppose a beam of photons is shone onto a substance containing electrons (neglecting the attraction of the atomic nuclei in the substance, which is justified if the photons have high enough energy). What happens? The photon may interact with the electron, be deflected, knock the electron into motion and by so doing itself lose energy. The calculation gives a contribution to this process first at order $\alpha$ in the Feynman sum-over-histories. Probabilities are worked out from a square of this, and so are proportional to $\alpha^2$.

Feynman diagrams corresponding to this are shown in Figure 9.3 Vertices, like $P$ and $Q$, denote the spacetime points where the interactions between the fields take place. The lines denote the effect of the sum-over-histories for free fields, continuous lines for the Dirac

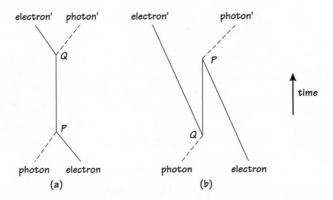

FIGURE 9.3   Feynman diagrams representing a photon in collision with an electron. Continuous lines represent the electron, dashed lines photons. *P* and *Q* are the spacetime points where the interactions take place. Time may be visualized as vertically upwards, as shown. The lines at the bottom of the diagrams represent the photon and electron before the collision; the lines at the top represent them afterwards. The line joining *P* to *Q* represents the sum-over-histories for the free electron field between *P* and *Q*.

field (electrons and positrons) and dashed lines for the electromagnetic field (photons). These are drawn as straight lines, symbolizing the fact that free fields behave simply. In fact, we just have to work out the strength of a wave pulse, emitted from one point, when it arrives at the other. When we have done that, all that remains is to sum over all positions and times of *P* and *Q* in spacetime.

The interpretation of Figure 9.3(a) is clear. The incident photon hits the electron and is absorbed by it. A wave of the Dirac field "propagates" from *P* to *Q*, where it emits a photon again. As a result, the electron recoils and the photon emerging has different direction and wavelength from the original one. The Feynman sum-over-histories yields, when suitably squared, the probability for any such outcome of the collision.

I have shamelessly switched back and forth between particle language and field (wave) language. As explained in Section 9.2, these are just two ways of talking about the same physical reality. However, care must be taken if one thinks about the line *P Q* representing a particle moving from *P* to *Q*. Since there is a limited time

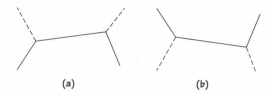

(a)                              (b)

FIGURE 9.4    Examples of Feynman diagrams describing, respectively, (a) the annihilation of a positron by an electron to produce two photons and (b) the production of an electron-positron pair by the collision of two photons.

between $P$ and $Q$, the uncertainty principle forbids one to ascribe a definite energy to that particle.

What about Figure 9.3(b)? At $Q$, the incoming photon produces an electron together with a positron (remember that the quantized Dirac field describes positrons as well as electrons). The Dirac field for the positron propagates to $P$, where it annihilates the original electron, producing a photon again. Electric charge is conserved, because the electron and positron have equal and opposite charges. Energy conservation does not prohibit the creation of the pair at $Q$, because the energy of the particles between the times of $Q$ and $P$ is not sharply defined.

The only difference between the two diagrams in Figure 9.3 lies in the time ordering of $P$ and $Q$. If we agree to sum over *all* positions and times of each of these points (without regard to time ordering), we need only draw one diagram. In fact this is the way Feynman diagrams most conveniently work: a single mathematical expression neatly combines the two contributions. This is just as well. The point $Q$ may be outside the light cone at $P$ (see Figure 5.2). Then it is impossible to say unambiguously which of $P$ and $Q$ is the later. The single Feynman diagram, and the corresponding single mathematical expression, include this possibility. Thus everything is consistent with special relativity theory, with no preferred time direction.

Feynman diagrams are even more flexible than this. Look at the diagrams in Figure 9.4. Here (a) represents a positron colliding with an electron and destroying it to produce a pair of photons and (b) represents the inverse of this. But these diagrams can be got from those in Figure 9.3 by pulling lines from the top to the bottom,

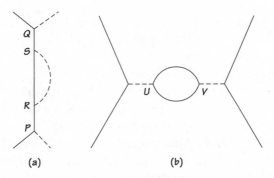

FIGURE 9.5    Two examples of Feynman diagrams with small corrections: (a) is a small correction to Figure 9.3; (b) represents another process: the interaction between two electrons by the mediation of a photon. In this example, an electron-positron pair is created at $U$, lives a short time and annihilates back into a photon at $V$ (or the other way round – as usual, a single Feynman diagram has several possible time orderings).

and vice versa. Thus, if no regard at all is paid to time-ordering, a single diagram (and a single mathematical expression essentially) represents several different physical processes.

All this is summed up in the slogan

A positron is an electron going backwards in time.

The Feynman diagrams in Figures 9.3 and 9.4 are the simplest ones that relate to the collision processes mentioned. But there are other diagrams that give higher powers of $e$, and therefore small corrections. One example (out of many) is shown in Figure 9.5(a). An extra photon is emitted at $R$ and absorbed again at $S$.

By means of this approximation method, calculations have been done in QED with great accuracy. One of the most accurately calculated and measured quantities is the magnetism of the electron. Dirac's equation (see Section 9.4) gives the value $\frac{e\hbar}{2m}$ (where $m$ is the mass of the electron) for the strength of this magnetism, but Feynman diagrams similar to Figure 9.5(b) give small corrections, multiplying Dirac's value by

$$1.0011596521.$$

Experiment agrees to this accuracy, that is, to within 1 part in

$10^{10}$. This is among the best calculated and measured quantities in physics. There can be little doubt that QED is "right", or at least very much on the right lines.

But I must mention a problem. There is often doubt, for any method of successive approximations, whether it does really go on getting closer and closer to the true answer. Given the small value of $e^2$, one naturally expects that each successive correction calculated should be only a small percentage of the one before. But there is no proof that large numbers should not arise, multiplying the small $e^2$: the number of Feynman diagrams rises very rapidly when complicated diagrams are drawn. It is widely believed that eventually, perhaps after about 100 powers of $e^2$, the "corrections" would begin to get bigger again. One must hope that any limitations, which there may in principle be to the method of approximation, do not spoil its accuracy in practical calculations.

## 9.9   The Perils of Point Charges

We have been assuming that the electron is a "point particle" – that it has no size. This certainly appears to be the simplest assumption to make. But is it consistent? This is an old question.

Start with classical electromagnetism. An electric charge carries around an electric field, in which energy is stored (see Section 3.5). How much? To begin with, imagine that the charge is distributed around a small spherical shell, with radius $r$. We can get a point electron by putting $r = 0$ (if possible). The electric field outside the sphere is proportional to the total charge $e$ and has an inverse-square law dependence on distance from the centre. The energy density at any point is proportional to the square of the field. To get the total electric energy, one has to sum over all points of space (outside the sphere). The result, whatever it is, is proportional to $e^2$ and can also depend upon $r$. There is only one way to get something with the dimensions of energy:

$$\text{total energy} = \text{some number} \times \frac{e^2}{r},$$

where the value of the number is unimportant. This energy is called the *self-energy*, because it does not depend on the existence of any other charges.

Now we see the problem. We cannot put $r = 0$. That would produce nonsense. Some people would express this by saying, "The energy of a point particle is infinite".

According to Einstein's special theory of relativity (see Section 5.11), if we divide the preceding energy by $c^2$, we get an equivalent mass. This is the inertial mass stored in the electric field. So the electron seems to acquire a "self-mass", which gets larger and larger as $r$ gets smaller.

What happens in (relativistic) quantum theory? One might guess that the problem should be eased, because the uncertainty principle implies that the electron is not fixed at one point (if its momentum is known). Also, since Planck's constant $\hbar$ may appear, we cannot use the preceding dimensional argument about the self-energy. It might be simply proportional to $e^2 m$ (where $m$ is the mass, and units are used where $\hbar = 1$, $c = 1$).

But unfortunately, calculations show that the self-energy has a piece

$$(\text{some number}) \times e^2 m \times \log\left(\frac{1}{rm}\right).$$

This contains a logarithmic dependence on $\frac{1}{r}$. All one needs to know about this is that it gets bigger and bigger without limit as $r$ gets smaller (though the logarithm grows much more slowly than $\frac{1}{r}$ itself). The point electron again seems to be inconsistent.

The diagram in Figure 9.5(a) illustrates how self-energies arise in Feynman diagrams. The portion between $R$ and $S$ is responsible for the self-energy. The large logarithm just mentioned comes when the spacetime points $R$, $S$ are close together.

This problem was appreciated in the 1930s and drove some physicists to despair. Did it mean that the whole edifice of relativistic quantum field theory was unsound? Should one try to construct a consistent theory of extended particles (a very difficult task)? Some physicists, who had seen the revolutions of relativity and quantum theory, looked forward to another fundamental change (like a Second Coming) that would solve the problem.

History was not so dramatic. By the end of the 1940s, people had learned that, provided they asked the right questions, they could obtain consistent and (as we have seen) spectacularly accurate predictions from the theory. Asking the right questions means, in particular, not worrying about the self-mass of the electron. All we measure is the *total* mass. Perhaps this is made up out of an original "bare" mass (which the electron would have if we switched off electromagnetism) together with the self-mass. But we cannot measure one sort of mass without the other. So let us just accept the total mass as something measured and not inquire further than that. The trick then is to arrange all the calculations so that the answers come out just in terms of this total mass. This procedure is called *renormalization*.

There is a nice property of Feynman diagrams that facilitates the process of renormalization. The self-energy portion, between $R$ and $S$ in Figure 9.5(a), can appear in different places or in diagrams for different physical processes. Wherever it appears, it always gives the same dependence on the interval between $R$ and $S$.

Is renormalization cheating? Whatever the final truth, I think it is certainly the sensible way to proceed until we know more. When we worry about putting the size $r = 0$, we are assuming that quantum electromagnetism is all of the relevant physics at all length scales, however small. Planck's constant connects inverse lengths to momenta, and $c$ connects these to energies, so this is the same as saying all the relevant physics at all energies, however large. We certainly have *not* included all the relevant physics. Apart from anything else, we have omitted the curvature of spacetime, as given by Einstein's theory of gravitation. One expects this to become relevant at the Planck length (see Appendix D). At the present time, people do not know how to work with a quantum version of Einstein's gravity, so nothing much can be done.

I have described the problem of the "infinite" self-mass. There is a second similar problem: the "infinite" self-charge. The diagram in Figure 9.5(b) gives ever-increasing contributions as the spacetime points $U$ and $V$ get closer together. This is connected with the behaviour of the electric charge $e$. To understand why, take an analogy.

A crystal of common salt is made up of regular arrays of positive sodium ions (sodium atoms missing one electron each) and negative chlorine ions (chlorine atoms with an extra electron each), arranged so that each sodium ion is at the centre of a cube of chlorine ions, and vice versa. Left to themselves, the ions maintain these positions as a result of the electric attraction between the two sorts and the exclusion principle that prevents the electron in the ions from over-lapping each other. Now insert an extra small electric charge into the crystal, say, a negative one. The electric field of the extra charge slightly distorts the crystal near it. It pulls the positive ions a little nearer and pushes the negative ones a little farther away. This effect decreases with distance from the extra charge.

Now imagine a sphere drawn around the extra charge, big enough to contain a good number of ions. What is the total electric charge inside this sphere? The numbers of positive and negative ions are not quite equal because the distortion of the crystal has brought in some extra positive ions and pushed out some negative ones. (Actually, averaging over spheres of different sizes, this is only a fractional effect.) So, from a little way away, the extra charge is *shielded* by the distortion of the crystal, that is, apparently reduced. The extra charged particle has a reduced effective charge, if mea-sured from largish distances. Measured very close up (at distances smaller than the spacing between the ions of the lattice), the *bare* charge is revealed. The charge measured from different distances looks something like Figure 9.6(a).

What has this to do with an electron in empty space? The un-certainty principle implies that "empty space" (the "vacuum") is in some ways not so empty. If a measurement were done at a very precise time, the energy would be uncertain. If that uncertainty is about $2mc^2$ (where $m$ is the electron mass), there is nothing to pre-vent an electron-positron pair from being present in the vacuum. We can imagine the vacuum as flickering with such evanescent pairs, each existing for only a very short time (about $10^{-21}$ second). Now put an extra charge, say an electron, into the vacuum. Its electric field distorts the evanescent pairs in the vacuum, a little like the distortion of the crystal in the preceding analogy. The result is sim-ilar: a reduced charge effective at large distances, a larger "bare"

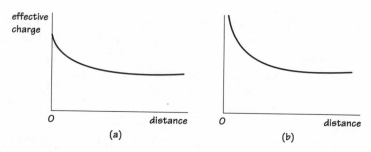

FIGURE 9.6    The effective charge at different distances. (a) What happens in an ionic crystal. The distance from which the charge is measured is plotted horizontally. The *bare* charge is measured very close up. At large distances, the effective charge appears to be smaller. (b) The analogous effect for an electron in free space. In this case the effective charge grows without limit at short distances.

at very small distances ("small" and "large" distances here means compared to $[\hbar/mc] = 3.9 \times 10^{-13}$ metre).

Figure 9.5(b) shows how this happens in Feynman diagrams. The portion of the diagram between $U$ and $V$ represents an electron-positron pair, influenced by the photon (that is, by the electromagnetic field).

In this case, however, unlike in the crystal analogy, the effective charge goes on increasing without limit at smaller and smaller distances; if you like, the "bare charge is infinite". The effect is illustrated in Figure 9.6(b). The problem is like that with the electron mass. The way out is also similar: to express everything in terms of the effective charge at large distances, which is what is most easily measured, and not to worry about the meaning of "bare charge". The value $\frac{1}{137}$, which I have quoted, for $e^2$ refers to the effective charge at large distances.

There is still a potential problem. Suppose one does experiments in which electrons pass very close to one another and are strongly deflected. The relevant effective charge is then bigger than $\frac{1}{137}$, so the method of successive approximations (which depends upon the smallness of $e^2$) is less accurate. In principle, in an extreme enough case, the method might become completely useless. Fortunately, this happens only when the relevant distance is of the order of $10^{-1,000}$ metre, so there is no problem in any practical case. However, the

distance-dependence of effective charges in other contexts turns out
to be a very interesting question, to which we return in Chapter 11.

## 9.10   The Busy Vacuum

There is an important and unanswered question concerning the
"energy of the vacuum". A free field is just a collection of oscillators,
one for each possible mode (that is, each wavelength, direction
and polarization). When quantized, each oscillator has a nonzero
minimum energy of

$$\frac{1}{2}\hbar \times \text{frequency}$$

for bosons and

$$-\frac{1}{2}\hbar \times \text{frequency}$$

for fermions (see Sections 9.2 and 9.5).

A problem arises when one tries to sum these energies over all
the modes. Wavelengths can be arbitrarily small, so frequencies can
be arbitrarily large, and the sum just does not exist (it is "infinite").
To make the situation a little more quantitative, we can do the sum
first for all wavelengths greater than some chosen small length $l$.
The biggest part of the sum then comes out to be some number
times

$$\frac{c\hbar}{l^4}.$$

(This is actually the energy in unit volume of space.) So the energy
gets bigger and bigger the smaller we choose $l$.

This argument applies to free fields only. We know that there are
interactions, for instance, between the electromagnetic and Dirac
fields. Taking these interactions into account, no one knows the
exact answer, but it is unlikely that the preceding large terms should
be cancelled.

But there is one possibility for cancellation. The boson and
fermion contributions have opposite sign. Could they be arranged
so that the minimum energies of the oscillators cancel out or at least
cancel enough so that we can put $l = 0$ and get a finite answer? This

would need rather careful arranging. One must cancel not only the $\frac{1}{l^4}$ terms, but possible $\frac{1}{l^2}$ and $\log l$ terms too.

The minimum requirement for this boson-fermion cancellation to work is that there should be equal numbers of fermions and bosons in nature. This is not the case in QED: there are two sorts of boson (photons with each of two polarizations) but at least four sorts of fermion (electrons and positrons, each with two spin states). However, as we shall see in Section 15.2, there is a rather nice way to impose that the totality of fields in nature should have this property of cancellation.

Let us suppose that the vacuum energy is finite but not zero. The vacuum is busy with a sort of hum of oscillators, a bit like static on a blank television screen. Does this have any physical significance? Or is it an unobservable and meaningless concept? Have we got back to something like the old idea of an aether, filling all space but so subtle as to be almost undetectable?

The vacuum energy is different from most other sources of energy. It does not pick out any direction in spacetime, that is, any standard of rest. If it did, we would hardly call use the word *vacuum*.

Nevertheless, there are two reasons for taking the vacuum energy seriously. The first is a simple experimental matter. Suppose two accurately flat metal plates are arranged parallel and very close to each other. The electric field between the plates has to satisfy special conditions due to the plates' being electrical conductors. So the modes of oscillation of the field between the plates are not the same as the modes in empty space. Therefore the "vacuum" between the plates has a different energy (per unit volume) than it has elsewhere. What is more, this energy depends upon the separation of the plates. If one of the plates is moved a little, the energy changes, so there must be a force between the plates. This force can be calculated. It is attractive and comes out to be some number times

$$\frac{c\hbar}{(\text{separation})^4}$$

per unit area (a finite quantity: the infinite sums cancel on comparing the energy between the plates with that elsewhere). This is very small, about a millionth of atmospheric pressure when the separation is $10^{-6}$ metre, but it has been measured. Note that it is independent of the value of $e$, the charge on the electron. (To prevent any confusion, I should say that it is the electromagnetic field only, and no other field, that contributes to this force between conducting plaes.)

The second reason for worrying about the vacuum energy is more profound. According to Einstein's theory, energy (including of course rest energy, $mc^2$) is the source of gravitation, that is, of spacetime curvature. So the vacuum energy should influence spacetime curvature. Since the vacuum is everywhere, it has no local effect. But it should influence the curvature of the universe as a whole; that is, it should have implications for cosmology, as I explain in Chapter 14.

## 9.11 Conclusion

Quantum theory and special relativity are reconciled in quantum field theory, which predicts that particles should have antiparticles and explains why there are bosons and fermions. QED gives an excellent account of photons, electrons and positrons, allowing remarkably accurate predictions to be made. It is probably not a complete, consistent theory on its own, but its limitations are hidden away at extremely short distances.

But QED is not the whole of physics. What about the *strong* forces between protons and neutrons, which hold them tightly together in atomic nuclei? What about the so-called weak forces, which, for example, allow a neutron to turn (with very low probability) into a proton together with an electron and a neutrino? In Chapters 11 and 12, we come back to these questions and explain how they are related to QED. But first we make an excursion in a very different direction. Its relevance will emerge later.

# ORDER BREAKS SYMMETRY

*How solids and liquids change from disordered to ordered states on being cooled.*

## 10.1 Cooling and Freezing

We are always ignorant about the detailed microscopic state of a macroscopic lump of matter. What we know about it is generally of a statistical nature. Statements about its temperature, pressure, magnetism and so on, are statements about average properties. The entropy is defined as a measure of our ignorance. All this was explained in Chapter 2.

There is one exception. If we could get the lump of matter to the absolute zero of temperature, it would (in principle) be in a single quantum state: the state of minimum energy. The entropy (as well as the temperature) would be zero. This would be very interesting, because we would be studying the quantum theory of macroscopic things, not just of atoms. It is very difficult to get near enough to the absolute zero of temperature to achieve this single quantum state. But fortunately nature does provide us with many examples of interesting large-scale quantum effects that occur when bodies are made cold enough. Some of these have always been familiar; others were total surprises when they were discovered in the twentieth century.

When water is cooled, ice crystals form. This happens at a well-defined temperature (for a given pressure). There is a qualitative difference between water and ice. In a crystal, the molecules are arranged in a regular array, held in place by the forces between them. Knowledge about the positions of molecules at one point tells us something about their positions at macroscopic distances away (say, millimetres). There is *long-range order*. Not so in a liquid.

There is something funny about this. Take a spherical water drop. No special direction is picked out by such a drop. If we rotate it nothing has changed. A spherical ice crystal (assuming it to be a perfect crystal) does define special directions, by the way the molecules are lined up in regular arrays. If we rotate it, it does in general change, because the array is rotated. So freezing has *broken a symmetry*, symmetry under rotation. Something that did not change when rotated has turned into something that does.

What then determines the orientation of a crystal when it is formed by freezing a liquid? Presumably it is chance, influenced by all sorts of accidental and unknowable details about the original liquid drop.

We can sum up what we learn from this example in a slogan:

- Reducing temperature can produce long-range order, thus destroying symmetry.

Here is another example, which I will describe in the opposite direction (raising the temperature instead of lowering it). Take a grain of magnetized nickel. The individual atoms are like tiny magnets (because of the magnetism of some of their electrons). At low temperatures, the atomic magnets are all lined up the same way, so that their magnetic effects add up, making the whole grain into a macroscopic magnet. Warm up this nickel. The thermal fluctuations cause the atomic magnets to wobble about their aligned directions, but still an average (but reduced) overall magnetization remains. The magnetization goes on decreasing, until at 631 K it has dropped to zero. At higher temperatures, it remains zero.

Thus, at any temperature below 631 K, there is some magnetization in some preferred direction. At any temperature above, there is no magnetization and no preferred direction.

This example has several characteristics in common with the melting of an ice crystal. There is long-range order in the magnetized state: knowing about the orientation of electrons in nickel atoms at one position tells us something about their orientation at macroscopic distances away. If the nickel crystal is cut in half, the magnetic orientations of the two halves are the same. There is no such order above 631 K. Below that temperature, the atom orientations pick out a preferred direction, so a rotation makes a difference. Above it, all directions are equivalent (so far as the magnetic orientations are concerned).

In a magnetized material, it is possible to excite waves ("spin waves") in which the directions of the atomic magnets vary slowly from place to place in a wave-like manner. The wavelength of these waves can be arbitrarily long (provided that it is less than the size of the specimen), and their energy is then arbitrarily low. These waves are excitations in which the ordered quantity varies slowly from place to place.

There is a difference between these two examples. The transition in magnetism occurs in a smoother way than in freezing. At 0 K, ice and water can coexist. There is still a sharp distinction between them, and heat is required to turn the ice into water. At 631 K, on the other hand, there is only one state of magnetization of nickel (with zero average magnetization).

In the next sections, I will describe examples of transitions that occur in a similar way to magnetization. But what has all this to do with the forces between particles like electrons, protons and neutrinos? There are two answers to this question. The first is that physicists learnt, from the study of the transitions, about the possible ways physical systems could behave. They had their imaginations stimulated and applied their experience to the theory of particles. The second answer is that there is compelling evidence that the universe as a whole was very hot at early times and has since cooled. Transitions may have occured as it has cooled. From the examples given, we would expect such transitions to do two things: (i) produce long-range order (at least locally) and (ii) break some symmetry that was present at higher temperatures.

## 10.2  Refrigeration

In the nineteenth century, people began to try to liquefy, and perhaps solidify, as many gases as they could. For every gas, there is a temperature, called the *critical temperature*, below which it can be liquefied by application of sufficient pressure. Faraday produced liquid chlorine. Oxygen was liquefied in Paris in 1877, and shortly afterwards in Cracow. (Mendelssohn has given a fascinating account of all this work in his book.) Dewar, working, like Faraday, at the Royal Institution, liquefied hydrogen in 1898.

This would have been the end of the liquefaction story if helium had not been discovered, first indirectly by its spectral lines in the Sun and then, in 1895, in the mineral pitchblende. This gas was the hardest of all to liquefy. It was liquefied in 1908 by Kamerlingh-Onnes in his great cryogenics laboratory in Leiden. Dewar was beaten in this race and gave up cryogenics. The critical temperature of helium turned out to be about 5 K.

In all this work, the methods used to attain low temperatures were based on much the same principles as used in domestic refrigerators. The idea is to go in steps, first lowering entropy, then temperature. For example, compressing a gas, that is maintained at constant temperature by contact with a reservoir which can absorb heat, lowers the entropy of the gas. This is because the gas is confined in a smaller space, so that fewer positions are available for its molecules, but their speeds have not been increased since the temperature has not risen. In the second step, the gas is thermally isolated (so heat cannot flow in or out) and the pressure reduced, say, by pulling out a piston. Molecules bouncing off the receding piston have their speed reduced, so the temperature goes down.

By methods like this, using compressed and expanded gases and liquids, temperatures down to about 1 K were attained. It seemed difficult to go much further, essentially because there was then so little entropy in the molecular motion left to reduce. But there is a way to get temperatures much less than 1 K: magnetic cooling. Again, this goes in two steps: reduce entropy, then reduce temperature.

Some atomic nuclei are magnetic. Their magnetization is only about a thousandth that of the electron (recall that Dirac predicted a magnetic moment inversely proportional to mass, see Section 9.4). There are very small magnetic forces coupling the nuclear magnetization to that of the electrons, but only at very low temperatures are these forces sufficient to overcome the thermal disorder. Thus, at 1 K say, although the atom positions are well ordered (in a crystal), the nuclear spins remain disordered: there is entropy in the spins. If a very strong magnetic field is applied, the nuclear magnets are aligned along the magnetic field direction: the entropy drops and heat is given out. Then the crystal is thermally isolated (so heat cannot flow in) and the magnetic field turned off. Some of the energy in the small remaining atomic motion transfers itself to partially disorder the nuclear magnets; that is, the temperature is reduced.

After the liquefaction of helium, the quest for lower and lower temperature (and entropy) became an end in itself. In fact, this has been one of the great themes of physics, along with the exploration of shorter and shorter distances in particle physics, and greater and greater distances in astronomy. In England in the 1930s, both Oxford and Cambridge started low-temperature research. In Cambridge, a special laboratory was built (with Rutherford's support) for the mercurial Russian Kapitza. In Oxford, Lindemann (later Churchill's influential science adviser and Lord Cherwell) attracted low-temperature physicists from Berlin and Breslau, and liquid helium was produced there in 1932, before it was at Cambridge. Incidentally, when Kapitza returned to the USSR in 1934, he was not allowed to leave again. His equipment was sent out to him from Cambridge.

The first great surprise occurred in 1911, when Kamerlingh-Onnes discovered *superconductivity*: the phenomenon whereby the electrical resistance of many metals and alloys totally vanishes below a certain *critical temperature*, usually a few degrees Kelvin. Electric currents, once started, persist indefinitely. I return to this subject in Section 10.5. First, let us move forward to the 1930s.

## 10.3   Flow without Friction

People have noticed odd things about liquid helium ever since it was first produced.

Below a certain *critical temperature*, about 2.2 K, helium suddenly became a very good conductor of heat, better than metals like copper. What is more, the conduction of heat was associated with flow of the helium, producing, for example, a little fountain. Also, helium remains a liquid down to 0 K, solidifying only if it is subjected to 25 times atmospheric pressure.

In 1938, the science journal *Nature* published several articles about helium. In particular, Kapitza in Moscow and Allen and

Misener in Cambridge announced that helium below its critical temperature flowed through small channels with no detectable resistance. Helium above the critical temperature behaved as other liquids did and came to be called *helium I*. The helium below that temperature, displaying *superfluidity*, was called *helium II*.

Why is flow without friction so surprising? Fluid flow normally experiences friction because the energy of the original ordered flow gets transferred, perhaps first into confused turbulent flow, but eventually into heat energy. The process is one of increasing disorder. It seems an inevitable (and one-way) process.

Fritz London studied classics and philosophy but turned to theoretical physics and worked with Schrödinger in Zurich. In 1933 (with his brother Heinz) he went to Oxford. In 1938 (by then in Paris), he made a daring suggestion. He was familiar with Einstein's work on indistinguishable particles (see Section 9.3). Helium atoms are bosons, because they each contain two electrons, two protons and two neutrons (each of which is a fermion), and an interchange of the positions of two helium atoms produces in the wave function six minus signs; that is, it makes no change.

Consider, as a mathematical idealization, an *ideal gas* of helium atoms. It is easy to work out how, for different temperatures, the atoms are distributed over the possible quantum states. I will describe qualitatively what happens.

To be definite, take helium (at, say, atmospheric pressure) in a one-centimetre cubical box. The wave function of each atom has to fit into the box, and, like a violin string, it can oscillate up and down any number of times between opposite sides of the box. The states are very numerous. How many atoms are there in each state? At "high" temperature (say, above 100 K), the atoms are on average quite energetic and are thinly distributed over many states, with a small chance of there being more than one in any given state. The bosonic property of the atoms is unimportant, and the gas behaves as "ordinary" gases, such as air at room temperature, do.

Now consider the ideal gas at the absolute zero of temperature. There is no thermal disorder (no entropy), and the atoms must certainly distribute themselves so as to get the least energy for the sample of gas. For bosons, this distribution is trivial: each one goes into the same state of least energy. All $10^{22}$ or so atoms are in one and the same state. (Of course, for fermions this would have been forbidden by the exclusion principle.)

What happens at temperatures between these two extremes? Calculations show that, as the temperature is raised from 0 K, the number of atoms in the lowest state goes down, but it remains "macroscopic" (that is, of order $10^{22}$), until it drops to a few (not macroscopic anymore) at 3.1 K and remains that way for higher temperatures. (Remember that the critical temperature for *real* helium liquid is 2.2 K.)

A macroscopic number of particles (bosons) in a single quantum state is called a (Bose-Einstein) *condensate*. Such a condensate is certainly a very ordered state of the fluid. But can we define the order in a more quantitative way? In the case of magnetization, the direction of the atomic magnets was the relevant *order parameter*. The order parameter for a Bose-Einstein condensate in an ideal gas is just that single wave function, which is the same for each of the atoms. It is a sort of "macroscopic wave function" of the condensate.

This definition of an order parameter does not quite work for real liquid helium, in which forces between the atoms are important (otherwise it would not be a liquid). But an order parameter may nevertheless be defined as follows. Take a sample of helium containing a large (macroscopic) number $N$ of atoms (for example, $N$ might be $10^{22}$). Call the wave function of the whole sample, in the state of minimum energy, $\psi_N$. This is a very complicated thing, depending on the positions of the $N$ atoms. Now take a sample containing $N + 1$ atoms, with wave function $\psi_{N+1}$, depending upon $N + 1$ positions. By comparing $\psi_{N+1}$ with $\psi_N$, one can define a quantity $f(x)$, which depends upon just one position $x$ (the position of the "extra" atom). This $f(x)$ is the order parameter. It is the quantum amplitude for adding one extra atom.

In the absence of condensation, $f$ is small, like $1/\sqrt{N}$. If there is a condensate, $f$ is of order 1. The extra $\sqrt{N}$ is similar to a factor that appeared in the theory of lasers (Section 9.3). $f$ is a complex number. Its length is fixed by the degree of condensation. But its *phase angle* is arbitrary, like the direction of magnetization in a magnet. Helium II breaks a symmetry, in that any specimen arbitrarily selects a phase angle.

A state of superfluid flow comes about if the phase of $f$ varies slowly from position to position. (Compare the wave function illustrated in Figure 8.2.) The flow is in the direction of increasing phase.

There is a special case that can be studied qualitatively fairly simply. That is the case of a gas near absolute zero with small forces between the atoms. How does this differ from the ideal gas? The force between atoms causes a little mixing between the lowest atomic state and those states with a small, but nonzero momentum, so the lowest state of the gas is a little more complicated. But also important are the states just a little above the lowest in energy (the "excitations"). For the ideal gas, these contain slowly moving free atoms. For a real gas, the forces between the atoms cause the moving atoms to interact with all the atoms in the condensate. The result is that there is a *minimum* possible speed for the excitations.

The minimum speed is crucial for the existence of frictionless flow. The question is whether the energy of flow could be dissipated by producing low-energy excitations. This is impossible provided the speed of flow is less than the minimum speed of an excitation, just as an aeroplane cannot emit a sonic boom if it is travelling below the speed of sound.

For real liquid helium II, there is a minimum excitation speed, which has been measured to be about 300 metres per second. A qualitative account of the excitations of helium II was given by Feynman in 1954.

Let me summarize the essential features of two examples of transitions to ordered states: magnetism and superfluid helium II.

- A symmetry is broken: by arbitrarily selecting a direction of magnetism and by arbitrarily selecting a quantum phase angle (the phase of the order parameter $f$) in helium.
- There is long-range order; the ordered quantities are the magnetic direction and the quantum phase angle, respectively.
- There are low-energy excitations in which the ordered quantity varies slowly with position: spin waves and superfluid flow, respectively.

## 10.4 Superfluid Vortices

When the order parameter $f$ varies slowly from place to place, there is superfluid flow. Take a closed loop in the helium II. Suppose that the quantum phase angle varies on going round this loop. This is consistent provided that, on going all the way round, the phase has changed by a multiple of 360 degrees (for the phase is an angle, and 360 degrees makes no difference to an angle). Then there is flow round this loop.

Now slightly distort the loop. The amount that the phase changes round the new loop cannot suddenly jump from one multiple of 360 degrees to another: it must stay the same. So there is a related flow round the new loop. But suppose we slowly shrink the size of the loop, by successive small distortions, until it shrinks to a point. In what direction is the flow at this point? And what is the phase angle at this point? We seem to have reached a contradiction. We are saved, however, if the order parameter $f$ goes to zero near this point: where $f = 0$ no phase angle is defined. Thus, if there is a thin core of helium I (where $f = 0$) embedded in the helium II, it is consistent for the phase to change by a certain multiple of 360 degrees, on traversing *any* loop that encircles the thin core, as illustrated in Figure 10.1

Such a flow configuration is called a *superfluid vortex filament*. As the name suggests, the pattern of flow is similar to that of a vortex in an ordinary fluid, as in a bath plug vortex or a tornado. There

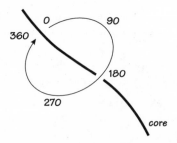

FIGURE 10.1   A vortex filament. The thick line shows the core. The thin line is an example of a closed loop encircling the core of the filament. Possible behaviour of the phase of the order parameter is illustrated.

is a crucial difference, however. The "strength" of a superfluid vortex must be dictated by the whole number that is the multiple of 360 degrees in the phase angle change. One may say that the vortex strength is "quantized". Left to itself (with no outside driving forces), an ordinary vortex slowly decays as a result of friction. A superfluid vortex cannot do this, because its strength (being quantized) cannot change continuously.

In a superfluid vortex, the angular momentum of each helium atom moving around the core is a multiple of $\hbar$ (the same multiple as the multiple of 360 degrees in the phase change). The quantization of angular momentum in multiples of $\hbar$ is no surprise: it is a general property of quantum theory (see Section 8.7). But what is special about a superfluid vortex is that *all* the atoms have the *same* angular momentum.

The actual speeds of flow in a superfluid vortex are quite slow. At a distance of one millimetre from the core, an atom takes about one minute to complete a circle (for the weakest type of vortex). But the speed decreases inversely as the distance.

A vortex filament cannot come to a sudden end within the helium II. It can end on a wall of the vessel containing the helium, or it can curl up into a closed loop (like a smoke ring).

Although vortices are macroscopic objects, they are at least as important as atomic excitations in the damping of superfluid flows. If a cylindrical vessel containing helium II is set rotating, the helium is at first unaffected because of the absence of friction. But if the rate of rotation of the vessel is big enough, vortices form near the edge of the vessel and drift into the helium. In this way, angular momentum is transferred to the fluid (as it would be by friction in an ordinary fluid). In superfluid flow through a small tube, closed vortex loops can be formed, thus slowing down the fluid near the loop.

Although the helium I core of a superfluid vortex is very small, the flow extends to macroscopic distances. Yet its strength is "quantized" (in the sense just explained). The quantization stems from geometric (more accurately, topological) reasoning: from the fact that angles are defined only up to 360-degree changes. These vortices are the simplest examples of a class of structures called *topological solitons*. The word *soliton* somehow comes from the idea that these things can exist on their own, solitarily.

## 10.5 Metals

Before describing superconductivity, in which electric currents flow without resistance, I must say something about ordinary metals, in which of course there is electrical resistance.

Take a simple metal like sodium. Each atom has 11 electrons, but 1 of these is farthest from the nucleus. These outermost electrons are strongly affected by the presence of the other atoms and in fact lose their attachment to any particular atom to wander through the metal. These are the *conduction electrons*. Each sodium atom nucleus, together with its remaining 10 electrons, constitutes a positively charged *ion*, and the ions arrange themselves as a regular *lattice*. The wave functions of the conduction electrons have a symmetry imposed upon them by the regularity of the lattice.

The conduction electrons interact with each other, by their electric repulsion, and interact with the positively charged ions. These two sorts of interaction tend to cancel out, especially at large distances, so the conduction electrons are more like free particles than one might expect.

The most important thing about the conduction electrons is that they are fermions and obey Pauli's exclusion principle. There can be no more than two electrons (two because of two possible spin directions) in any state of motion; this means that the electrons must span a wide range of energies. They fill states with all energies up to a certain maximum, called the *Fermi energy*. This energy is simple to estimate. The volume available to each electron is about the cube of $10^{-10}$ metre (that is, the spacing between the ions). By the uncertainty principle, there is an associated momentum of order $10^{10}\hbar$. This corresponds to a kinetic energy of order $10^{-18}$ joule. It is more helpful to quote the corresponding temperature, which is about $10^5$ K, large compared to room temperature.

At absolute zero, electrons occupy all the states up to the Fermi energy, and none above it. At ordinary temperatures, of a few hundred Kelvin, thermal fluctuations drive some electrons from a little below the Fermi energy to a little above, but the distribution is not very much altered.

Electrons with energy much below the Fermi energy are locked in: they cannot easily change to another state because the states of comparable energy are all full. Electrons with energy close to the Fermi energy are more easily excited, and we can think of them rather like classical free electrons. These electrons, with energy near to the Fermi energy, are fast moving: some hundred times as fast as the speed of sound in the ionic lattice.

Now suppose an electric field is applied to the metal. What happens is most easily pictured from a classical point of view, so long as we restrict ourselves to electrons with energy near to the Fermi energy. Such an electron begins to accelerate under the influence of the electric field. But it does not do so for long. It may hit another electron or an ion in the lattice. In such collisions, energy is transferred from the electron, and this energy ends up as heat. Soon, a balance is attained between the tendency of the field to make the electrons flow and the dissipation of energy as heat. At this point, there is a constant flow of electric current determined by the field (and, according to Ohm's law, proportional to it to a good approximation). If the electric field is turned off, the current quickly fades away.

## 10.6   Conduction without Resistance

We have seen how, in 1908 in Leiden, Kamerlingh-Onnes won the race to liquefy helium, thus making possible temperatures as low as a few degrees Kelvin. In 1911, Onnes found something totally unexpected. When mercury was cooled below 4.1 K, an electric current, once excited, persisted forever (certainly for years) without the necessity of an applied electric field. The majority of metals, and some metal alloys, exhibit this property of *superconductivity*, below a certain critical temperature. The highest critical temperature for metals (or metal alloys) is about 20 K.

In 1933, a new discovery about superconductors was made in Berlin. This, called the *Meissner effect* after one of its discoverers, was that magnetic fields are incapable of penetrating into a superconductor (so long as it remains in its superconducting state). The magnetic lines of force are deflected so as to lie outside. It was known that some materials tend to push magnetic lines of force out to some extent, an effect that can be explained by electrons flowing and producing a magnetic field opposite to the applied one. What was unexpected about superconductors was that a magnetic field was *totally* expelled.

If the applied magnetic field is made too strong, superconductivity ceases; then of course the field can penetrate into the metal. The field necessary to destroy superconductivity goes down as the temperature is raised and drops to zero at the critical temperature itself.

In spite of many attempts, superconductivity was not explained until nearly half a century later than its original discovery. After the discovery of superfluidity in helium in 1938, it was natural to think there might be some sort of similarity between the two sorts of frictionless flow at low temperatures. Superfluidity in helium is connected with the helium atoms' being bosons, and so able to "condense" into a single state. But conduction electrons are fermions, so how can a similar type of explanation possibly work for superconductivity?

Even helium atoms are bosons only in a certain sense. They are made out of fermions – electrons, protons and neutrons. But they behave as simple bosons as long as they are not forced to overlap with each other (when the exclusion principle between the electrons would come into play). Might conduction electrons similarly bind together to form structures that act like bosons? In 1956, the American physicist Leon Cooper argued that pairs of conduction electrons, with energy near to the Fermi energy, and with equal and opposite spins and velocities can loosely bind together to form structures, *Cooper pairs*, with a bosonic character.

It is surprising that electrons should bind together, since they repel one another electrically. In a metal, this repulsion is "shielded" by distortion of the lattice of positive ions. But still an attractive force is required to bind Cooper pairs. Already in 1950, Fröhlich in Liverpool and Bardeen in Illinois had shown how such an attraction could come about, mediated by distortions in the ionic lattice. The electron momentarily distorts the lattice towards itself. Because the electron is fast moving, the shape of the distortion is a sort of long, narrow "wake". This "wake" has an excess of positive charge, which is capable of attracting another electron moving near it in the opposite direction.

Any attractive force, however weak, is sufficient to bind Cooper pairs of oppositely moving electrons. This is surprising. Usually, in quantum theory, an attractive force (acting over a given range) has to have a certain minimum strength in order to bind two particles moving in three-dimensional space. With Cooper pairs, it is different. The constituent electrons are both moving with nearly the Fermi energy, and the reduction in total energy due to the attractive force is much smaller than this (it is in fact about equal to the critical temperature, a few degrees Kelvin). A Cooper pair is big, about $10^{-6}$ metre, that is, thousands of times the distance between the ions in the lattice.

Soon after it was made, the prediction that this attractive force had something to do with superconductivity was confirmed in a dramatic way. Thanks to progress in nuclear technology, different isotopes of metals could be separated. These isotopes are identical with each other, except for having different numbers of neutrons in their atomic nuclei, so all their electrical properties are identical. Yet they were found to have slightly different superconducting transition temperatures. Lattices of ions with slightly different masses have slightly different frequencies of vibrations, and this difference should indeed affect the attractive force and the strength of binding of the Cooper pairs.

A complete theory of superconductivity was developed in 1957 by Bardeen, Schrieffer and Cooper – known universally as the *BCS theory*. The crux of the BCS theory is that Cooper pairs undergo a Bose-Einstein condensation: there are a macroscopic number of such pairs in a single quantum state. So assume now that superconductors have a "condensate" of Cooper pairs. (This is perhaps harder to visualize than a condensate of helium atoms, because the Cooper pairs do overlap each other a lot. Nevertheless, the mathematical forms of the condensate wave functions are similar.) How does this explain the existence of persistent electric currents, and the Meissner expulsion of magnetic fields?

In the case of superfluidity, the Bose-Einstein condensation resulted in a macroscopic "order parameter" for the condensate, called $f$ in Section 10.3. A sample of superfluid helium "chooses" a phase angle for $f$ in an arbitrary way and so has a "macroscopic phase" associated with it. Does the same happen with the Cooper pair condensate in a superconductor? There is a crucial difference. Cooper pairs carry electric charge (actually of amount $2e$, where $e$ is the charge of an electron). The phase of a wave function for a charged particle is subject to the gauge-invariance principle (see Section 8.12): it is ambiguous, and there is a compensating ambiguity in the FFOC (Section 6.5). So it would be meaningless to say that a superconductor chose a phase for the order parameter $f$ of the Cooper pairs.

What can one say instead? Provided that the magnitude of $f$ is not varying with position, there is a simple rule about the phase of $f$ that is consistent with gauge-invariance. The rule states how the phase angle varies from place to place, if there are magnetic fields present:

- In moving along a curve, the phase angle of $f$ changes by an amount determined by the FFOC associated with that curve.

This rule is sufficient to explain the properties of superconductors qualitatively.

First, the Meissner effect – the expulsion of a magnetic field from a superconductor. Consider the change in the phase of $f$ on going round a *closed* curve, lying entirely within the superconductor. The FFOC for a closed curve is just the flux through it, so, by the previous rule, the phase angle change determines this flux. But the phase angle must have got back to where it started, so the magnetic flux must be zero. This is true of any closed curve within the superconductor, so there can be no magnetic field anywhere within the superconductor – the Meissner effect.

Second, persistent currents: for this purpose, it is simplest to consider a specimen of superconductor in the shape of a wedding ring. Apply the rule to a closed curve within the superconductor and going one around the ring (that is, once round the finger wearing the wedding ring). Such a curve is different from one drawn in, say, a ball of superconductor: you cannot, by gradually shrinking it (within the superconductor), make it indefinitely smaller and smaller – it must always go round the ring. In consequence, the phase of $f$ *can* change on going round the curve, but it must change by a multiple of 360 degrees. This means that the magnetic flux through the ring (in the direction of the finger, say) need not be zero: it can be any multiple of

$$\frac{h}{2e}$$

(where as usual $e$ is the electron's charge, and the 2 represents the two electrons in a Cooper pair). This is a very small quantity, about $10^{-15}$ weber (the name of the SI unit of magnetic flux, which is the same as a volt second). Nevertheless, flux changes in these tiny quantized steps have been measured.

If there is such a nonzero magnetic flux, there must be a current flowing round the ring to cause it. It is impossible for resistance to make this current fade away, because if it did so, the magnetic flux would have to die away too, passing through values other than multiples of $\frac{h}{2e}$. Thus superconductivity is inevitable.

Several questions come to mind. What prevents the current from losing energy by exciting electrons to states with energies just above the Fermi energy? The answer is that in a superconductor, there is a small *gap* between the Fermi energy and the least energy of an excited state. This is (roughly) the energy required to break up a Cooper pair. The size of this gap, expressed in terms of an equivalent temperature, is in the region of 1 K. So long as this amount of energy is not available, there is no way the current can transfer its energy into heat. The energies of excitations are illustrated in Figure 10.2.

If a supercurrent flows, why does it not produce magnetic fields within the superconductor, contrary to the Meissner effect (the expulsion of magnetic fields)? The reason is that the currents in superconductors flow in a thin layer near the surface, of thickness about $10^{-7}$ metre (that is, about 1,000 atoms thick). In this layer, the metal is changing from its normal to its superconducting state.

Weak magnetic fields are expelled from superconductors, but if a field is applied of more than a certain critical value (which depends upon how far the temperature is below the critical temperature), superconductivity breaks down and the metal reverts to its

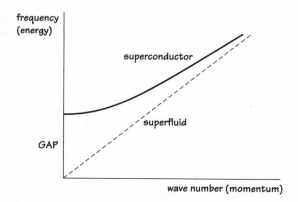

**FIGURE 10.2** The frequencies of excitations in a superconductor plotted against their wave numbers. In quantum theory, this is the same shape as the plot of energies against momenta. The dashed line shows the excitations in a superfluid, where there is no *gap*.

normal state. But there are some materials, mostly alloys, called *type II superconductors*, in which the effect of a magnetic field is more complicated. Relatively low fields can penetrate partially into the material. This happens by the formation of thin *flux tubes* within which there is a transition from superconducting to normal state. Outside the tubes, the magnetic field remains zero. Current flows round the edge of the tube, to confine the magnetic field inside. The phase of the order parameter, f, increases by (a multiple of) 360 degrees round any closed curve encircling the flux tube. Not surprisingly, the flux in such a tube is quantized in units of $\frac{h}{2e}$.

Flux tubes in superconductors are the analogues of vortex filaments in superfluids (see Section 10.4).

There exists a very beautiful way to see the reality of the phase of the order parameter $f$ in a superconductor. This is called the *Josephson effect*, after Brian Josephson, who predicted it in Cambridge at the age of 22. Take two superconductors separated by a very thin layer (in the region of $10^{-9}$ metre) of insulator. The wave function of a Cooper pair in either superconductor does not drop to zero abruptly at the edge of the insulator, but has a "tail" extending across the insulator. This means that Cooper pairs may "tunnel" through the insulator: that is, a current can flow from one superconductor to another. If the phases of the two order parameters are different, such a current does indeed flow. This arrangement is called a *Josephson junction*.

Now suppose that a fixed voltage difference is established between the two superconductors, so that there is an electric field in the insulating layer between them. We know that a magnetic field makes the phase vary with position. In a similar way, an electric field makes the phase vary in time. So the phase difference between the two superconductors varies proportionately with time. Thus there is a periodically varying, *alternating* current caused by a *steady* voltage. The frequency of the alternations is related to the voltage difference by

$$\text{frequency} = \frac{2e}{h} \times \text{voltage},$$

the same factor $\frac{2e}{h}$ that we have encountered before.

Josephson junctions can be used to do very accurate measurements. For example, they yield very accurate values of the ratio $\frac{e}{h}$. If a ring of superconductor is interrupted by two Josephson junctions, the magnetic flux through the ring can be measured with great accuracy, by deducing it from the change in the phase of the order parameter on going once around the ring.

Since 1986, excitement has been caused by the discovery of a new class of superconducting materials with high transition temperatures, approaching 100 K. These are not metals, but complicated compounds. For example, some are composed of the elements yttrium, barium, copper (all metallic) and oxygen. Their crystal structure is a complicated, layered one, with copper and oxygen atoms concentrated in two-dimensional planes. Above their transition temperatures (where they are not superconducting) they behave differently than metals. There is as yet no well-established theory, analogous to the BCS theory, of the superconductivity of these materials.

## 10.7   Conclusion

Materials sometimes exist in states with *long-range order*, such as the alignment of atomic magnets in a magnetized substance. A subtle form of long-range order exists in superfluids and superconductors, in which the order parameters are quantum phase angles. The long-range order breaks a symmetry. For example, a piece of magnetized material "chooses" a direction of magnetization, breaking the symmetry between all directions in space.

In the next two chapters, we see these ideas inspired new theories about elementary particles.

# 11

# QUARKS AND WHAT HOLDS THEM TOGETHER

*How protons and neutrons and other baryons are made of quarks,*
*bound together by a subtle generalization of electromagnetism.*

## 11.1 Seeing the Very Small

Lenses were used for microscopic purposes in the second half of the seventeenth century. The Dutch microscopist, van Leeuwenhoek identified blood capillaries, red blood cells, spermatazoa and bacteria. In the nineteenth century, the cell structure of plants and animals was established.

But optical microscopes, using light of wavelength about four to eight times $10^{-7}$ metre, cannot resolve smaller distances than these and certainly cannot be used to study atoms or even large molecules. To do better, one must use electromagnetic radiation of shorter wavelength, like X-rays, or beams of particles that have, according to quantum theory, wave functions with wavelengths determined by the momenta of the particles. X-rays were discovered by Röntgen in Würzburg in 1895. Their ability to penetrate opaque bodies, like hands, was of course the initial cause of excitement, but it is not our concern here. In 1912, von Laue in Berlin showed that X-rays incident on a crystal produce (on a photographic plate) a pattern of spots. The spots (that is, positions of maximum intensity) appear where the waves scattered by individual atoms add up constructively (where there phases are equal). This is the phenomenon of interference described in Section 4.6. It works because the wavelength of X-rays is comparable to the atomic separation in crystals.

From the X-ray "diffraction patterns" (of spots), it is possible to deduce the arrangement of atomic positions in the crystal. The Braggs, father and son (William and Lawrence), developed these methods. In the 1950s in Cambridge and London, X-rays were used to find the atomic structure of large organic molecules like haemoglobin (done by Perutz) and DNA (Franklin, Wilkins, Crick and Watson).

If an electron falls in an electric field through a potential difference of 10,000 volts, the wavelength of its wave function is about $10^{-11}$ metre, less than $10^{-4}$ of the wavelength of visible light and less than the size of an atom. Electron beams can be guided and "focused" by magnetic fields and thus produce an "electron microscope" (invented in 1932 by Ernst Ruska of Berlin). Electron microscopes can produce detailed pictures of, for example, viruses.

An even more remarkable instrument was invented in 1982 by Gerd Binnig and Heinrich Rohrer, the scanning tunneling microscope. A tiny metal tip is placed very near (within a few times $10^{-10}$ metre of) a conducting surface, and its position delicately controlled. At this closeness, the wave functions of electrons in the tip and in the surface overlap a little. If a voltage difference is established between the tip and the surface, an electric current can flow (the electrons "tunnel" across the gap). The strength of the current depends very sensitively on the size of the gap. By scanning the tip over the surface and monitoring the current, a relief map of the surface is constructed, detailed enough to show individual atoms. Striking computer-generated "pictures" are produced, showing the arrangement of atoms on the surface.

## 11.2   Inside the Atomic Nucleus

In Section 8.2, I described how Rutherford and his colleagues discovered the atomic nucleus in about 1911. Rather than light or X-rays or electrons, they used alpha particles as their tool. These have an associated wavelength less than $10^{-14}$ metre, small enough to respond to the atomic nucleus.

By 1932, the compositions of atomic nuclei were understood. They were made of *protons* and *neutrons*. The proton is the (positively charged) nucleus of hydrogen, the lightest atom. The neutron is similar but is electrically neutral. The proton's mass is about 1,840 times that of the electron, and the neutron's about 2.3 electron masses more. The lighter atoms usually have an equal number of neutrons and protons, but heavier ones have more neutrons.

There must be some force holding the protons and neutrons together in the nucleus. This force is certainly *strong* compared to the electric force binding electrons in an atom: the energy needed to disrupt a nucleus is about a million times more than that to ionize (remove an electron from) an atom. The strong force also acts over a short range, about $10^{-15}$ metre: much outside that distance only the electric, inverse-square law, force remains.

In 1934, Yukawa in Japan suggested that the quantum theory of the strong nuclear force should require a particle, now called the *pion* (from the Greek letter $\pi$, pi), just as the electromagnetic forces are connected with the photon. Because of the short range of the nuclear force, Yukawa predicted that the pion, unlike the photon, should have a mass given by a simple uncertainty principle estimate

$$\text{mass} \approx \frac{\hbar}{c \times \text{range}},$$

that is, a few hundred electron masses.

Particles with all the expected properties were discovered in Bristol in 1947. The masses were about 270 electron masses. The pions were of three kinds, positively charged, negatively charged and neutral. (The neutral one is about 9 electron masses lighter than the charged ones.) Yukawa's prediction seemed to be brilliantly fulfilled.

In retrospect, Yukawa's arguments seem less than decisive. In the same year, 1947, new particles a bit like pions but heavier, and new particles a bit like protons and neutrons began to be found. The first group of particles, bosons with spin a whole number times $\hbar$, are called *mesons* (from the Greek for "middle"). The second group, fermions with spin $\frac{1}{2}\hbar$, $\frac{3}{2}\hbar$, ... and so forth, are called *baryons* (from the Greek for "heavy"). From the 1950s onwards more and more such particles have been discovered. Now the American Institute of Physics issues a booklet with some 100 pages of data about subnuclear particles. We are dealing with something at least as complex as the periodic table of elements. Do these "elementary particles" have an internal structure just as atoms do?

The answer to this question began to emerge in the late 1960s from experiments done at Stanford in California with a machine called a linear accelerator. This used electric fields to accelerate electrons (in a straight line, hence "linear") until they had energies up to 40,000 times their rest energy ($mc^2$), that is, speeds below the speed of light by about 1 part in $10^9$. The electron beam was directed at a target of liquid hydrogen, in order to probe the structure of the protons in the hydrogen. The wave function of an electron moving with that energy has a wavelength of $10^{-17}$ metre. This is about a hundredth of the expected size of a proton, so the electrons should be able to resolve the internal structure (if any) of the proton.

When a proton is struck by a high-energy electron, the proton may just recoil as a whole or other particles like pions may emerge (shaken out of the proton, as it were). In the first class of events, it was found that the electrons were deflected through only small angles. This is just what one expects if the proton is a large (compared to the electron wavelength), diffuse object: the electron will plough through, only slightly affected. In the second class of events (where extra particles came out), the electrons were deflected much more. The experimenters were surprised and made careful checks to be sure they had made no mistake. In retrospect, we see that history was repeating itself. More than half a century before, Rutherford had probed the atom and found the nucleus within it: now the proton itself was being probed and something small being found within it. The scale of the experiment had been reduced some thousand times: that is, the energy increased some thousand times. If an electron happens to hit a small constituent (with a size comparable to the electron's wavelength) in the proton, it may be strongly deflected.

Feynman, with characteristic informality, named these constituents *partons*. What were these partons? What were their electric charges, masses, spins and so on? By the mid-1970s, these questions had been partially answered. The most important partons were *quarks*. The name *quark*, as recondite as *parton* was simple, was taken by Murray Gell-Mann from a passage in *Finnegan's*

*Wake.* (The fact that quark is a type of German cheese seems to be irrelevant.)

The quark model of baryon and meson structure was originated by Gell-Mann and Zweig in 1964, but at first the claim was only that hadrons appeared *as if* they had a quark substructure: the actual existence of quarks was left in some doubt. The debate over the literal existence of quarks in some ways echoed the controversy, some hundred years before, over the existence of molecules.

Like, for instance, electrons, quarks are fermions and have spin one-half $\hbar$. They have electric charges, but these have fractional values in terms of the charges on electrons, protons, mesons and so on. Quarks belong to pairs, with charges

$$+\frac{2}{3} \text{ and } -\frac{1}{3}$$

times the charge on the positron (or on the proton).

Moreover, three of these pairs of quarks are known, differing in their masses. The names given to the six quarks are *up, down; charm, strange; top, bottom.* These names are just about as arbitrary as the names of pet cats. Abbreviating the names by the first letters, the quarks may be arranged like this:

$$\begin{pmatrix} u \\ d \end{pmatrix} \begin{pmatrix} c \\ s \end{pmatrix} \begin{pmatrix} t \\ b \end{pmatrix}.$$

In each of these doublets, the upper members ($u$, $c$, $t$) have charges $+\frac{2}{3}$ and the lower members ($d$, $s$, $b$) have charges $-\frac{1}{3}$ (in units of the charge on the positron). The lightest quarks are $u$ and $d$, and protons, neutrons and pions are mainly composed of just these.

Electrons and all baryons each have antiparticles. Particles and antiparticles have most properties in common, but they have opposite electrical charges. The antiparticle of the electron is the positron (discovered in 1932). The antiproton was discovered in 1955. Similarly, each quark has an antiquark of opposite charge.

Baryons are made of three quarks, and mesons are made of a quark and an antiquark. In this way, the charges of baryons and mesons come out to be whole numbers. For example, the proton consists of *uud*, the neutron *udd*, the positive pion of *ud̄* (the bar on top of a letter for a particle denotes its antiparticle). So, for example, the charge on the proton comes out as

$$+\frac{2}{3} + \frac{2}{3} - \frac{1}{3} = +1,$$

and the charge on the positive pion as

$$+\frac{2}{3} - \left(-\frac{1}{3}\right) = +1.$$

Particles containing $s$ or $\bar{s}$ quarks began to be discovered in the late 1940s. They came to be called "strange" because they had properties that were not at first understood. The $s$ quark is heavier than $u, d$ and hence "strange" particles are a little heavier than ordinary protons, neutrons and mesons. "Strange" particles may be produced in pairs in high-energy collisions, when there is enough energy to provide the rest masses of a pair of quarks $(s/\bar{s})$.

The other three quarks are even heavier. The $c$ quark has about 1.5 proton masses, the $b$ about 4.5 and the $t$ nearly 200. One consequence of this is that these heavy quarks, inside mesons and baryons, move fairly slowly compared to the speed of light, so composites of heavy quarks are in some ways easier to understand than those of light ones. For example, the bound states of $c$ together with $\bar{c}$, called *charmonium*, have a pattern not totally unlike the states of the hydrogen atom (bound states of an electron with a proton) – though the forces are different in the two cases.

In the proton and neutron, the three quarks do not all have their spins aligned in the same way: that is why the total spin is $\frac{1}{2}$ not $\frac{3}{2}$. There are baryons with spin $\frac{3}{2}$. For example, there is a *uuu* system, with charge twice that of the proton. But its mass is a little more than the combined mass of a proton and a pion, so in fact it decays very quickly (in a few times $10^{-20}$ second). It is not really a particle, but its existence affects the distribution in energy of protons and pions emerging from collisions: the distribution may be peaked at the energy corresponding to the *uuu* system.

Apart from their masses, all the quarks behave similarly. If there is a *uuu* system, one may expect the existence of others in which some of the $u$ quarks are replaced by $s$ or heavier quarks, for example, a *sss* system, with spin $\frac{3}{2}$ and charge $-1$. If a decay of this particle were possible, it would have to be into a baryon and a meson with a total of 3 $s$ quarks between them. It turns out that there are no such baryon and meson light enough, so the $sss$ system should be a genuine particle. (In fact it does decay via weak interactions, but only after about $10^{-10}$ second, in which time it can move more than 20 millimetres.) This particle is called $\Omega^-$ (the capital Greek letter omega, with a minus charge). Its existence was spectacularly predicted by Gell-Mann in 1962 (by a more quantitative version of the preceding argument), and it was discovered within the next two years.

Figure 11.1 shows the pattern of certain 3-quark compounds: those made of $u$, $d$ and $s$ quarks, and with all three spins aligned

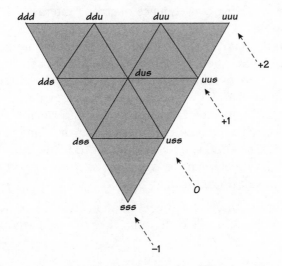

FIGURE 11.1   Some baryon compounds of three quarks, in which the three spins are all aligned in the same direction. Electric charges are indicated by the dashed arrows. The number of $s$ quarks increases down the diagram, and consequently the compounds get heavier towards the bottom of the diagram. All the compounds except the one at the bottom are not true particles, but show up indirectly in energy distributions of hadrons emerging in collisions. The $sss$ compound at the bottom was predicted as a true particle, by Gell-Mann in 1962, to complete this diagram.

in the same directions. The bottom entry is the $\Omega^-$. The 3-quark compounds with spins not all aligned the same way can be arranged in somewhat similar patterns made of equilateral triangles.

In spite of all this, there were good reasons why people were cautious about the existence of quarks. Here are some of these problems.

(i) Free quarks have never been found. A free quark would be recognizable by its fractional charge.

(ii) The parton model works *too* well. The quarks seem to behave as if they were almost *free* particles, during the short intervals involved when they are struck by high-energy electrons. How is this fact to be reconciled with the strong forces between baryons and mesons, presumably a consequence of strong interactions between quarks?

(iii) Quarks have spin $\frac{1}{2}$ so they must be fermions, subject to Pauli's exclusion principle. Yet there are particles, such as the $\Omega^-$, consisting of three identical quarks (three $s$ quarks in this example) with their spins all aligned in the same direction. In such cases, the exclusion principle prevents any two of the quarks from being at the same place. This makes a bound state very difficult to form. In fact the quark model seems to work just *as if* the quarks were bosons not fermions.

Unexpectedly, there is a theory that neatly overcomes each of these difficulties, and this is the subject of the next section.

## 11.3   Quantum Chromodynamics

Let us begin with the third of the three problems listed. There is a brute force way to get round this, though it seems to fly in the face of Occam's razor. Suppose that there are not 6 different quarks but 18, with each of the 6 types ($u$, $d$, $c$; $s$, $t$, $b$) in 3 different varieties. These 3 varieties are called *colours* – an arbitrarily chosen word. Let us call the 3 colours red ($r$), blue ($b$) and yellow ($y$). Then the up quarks, for example, have 3 colours, say $u_r, u_b, u_y$, and similarly for the other 15. The 3 colours of up quark have identical masses, charges and spins, and similarly for the 3 colours of down quark, and so forth.

The fermion problem is now solved if we assume that the three $s$ quarks in, for example, the $\Omega^-$ each have different colours, so they are $s_r$, $s_b$ and $s_y$. Then there is no conflict with the exclusion principle. The general rule is that every baryon is made of three quarks, one of each colour, and the wave function changes sign when any two colours are interchanged. In this way, the force of the fermionic property (the change of sign when any two quarks are interchanged) is entirely taken up by the different colours.

As an explanation, colour would not rate very highly if it were not linked to other properties of quarks. It turns out that colour is an ingredient of the theory of the forces between quarks. How this works is a longish story, which I will now begin.

Sections 8.12 and 9.7 have emphasized the importance of *local gauge-invariance* in quantum electrodynamics (QED). This principle may be stated as follows. The action for a Dirac field (of, say, an electron) is unchanged when the phase angle (in the sense of the phase of a complex number) of that field is changed by any amount that is the same at all points of space and time. If we demand instead invariance by a change of phase that may *vary* in space and time, this requires the Dirac field to be coupled to electromagnetism. The coupling is via a FFOC (see Section 6.5), and this changes to compensate for the changes of phase.

In 1964, Yang and Mills in the United States invented a generalized form of gauge-invariance. At the time, this seemed to most people to be a pretty piece of mathematics with no discernible application to physics. By a remarkable coincidence, a graduate student in Cambridge, Ron Shaw, invented exactly the same thing independently but (sadly for him) a few months later. Often discoveries in science are made independently by different people because they are "in the air". I do not think there was much sign, prior to 1964, of generalized gauge theories' being "in the air".

The generalization begins by generalizing the gauge-invariance of a Dirac field. Take as an example the three colours of any one quark, say the $u$ quark for definiteness. (Yang, Mills and Shaw did not have this particular physical application in mind.) Since the three colours have identical properties, the action is invariant, not only under changes of phase of each colour separately, but also under mixings of the different colours. This full set of transformations, called a *group* of transformations by mathematicians, is quite complicated. We shall need to know two things about it.

Firstly, the order in which transformations are performed matters. For example, first changing the phase of $b$, then mixing a little of (the new) $r$ into the new $b$ is not the same as first mixing a little $r$ into $b$, then changing the phase of the new $b$. A group in which order matters like this is called *non-Abelian* (after the Norwegian mathematician Niels Abel, 1802–1829). As another example, the group of rotations of a solid object (in three dimensions) is non-Abelian.

Secondly, it is interesting to know how big (in some sense) is our group acting on the 3 colours of quark field. One way to answer this is to look at very small transformations and count how many independent kinds there are. First, there are small changes of the phase of each colour separately. There are 3 like this. Second, there are transformations that mix a small amount of any one colour into any other. There are 6 like this, making a total of 9. However, a change of phase of all 3 colours by the same amount is a special, and more trivial, case, and it is convenient to exclude this kind of transformation. The number arrived at is then 8, and mathematicians say that 8 is the "dimension" of the group. (This group was introduced to physicists, in a quite different physical context, by Gell-Mann and by the Israeli physicist Ne'eman, and they called this theory the "eightfold way", referring to Buddhist teaching. Colour was irrelevant in this earlier application of the mathematics.)

Now we try to extend this invariance to make it local, with different transformations at each point of space and time. We need to couple the coloured quarks to some generalization of the electromagnetic field. Not surprisingly, 8 separate fields are needed. These are called *gluon* fields (because they provide the "glue" between quarks). We may say that there are 8 different "colour charges", and there is one of the 8 gluon fields associated with each charge. Of course, in electromagnetism there are just one electromagnetic field and just one charge (the electric charge).

When the electromagnetic field is quantized, the particles that result are photons. They move, of course, with the speed of light; or equivalently they have no mass. They have spin $+\hbar$ or $-\hbar$ in their direction of motion. Similarly, we expect the 8 gluon fields to be associated with 8 massless particles, with the same spin states.

Now comes the important point. Because the group is non-Abelian, the transformations of the group act on the gluon fields themselves and mix them up. We may say that the *gluons have colour charge*, similarly to the way in which the quarks themselves have colour charge. Thus each gluon *both* has colour charge *and* produces colour fields. This is a much more complicated situation than in electromagnetism. Electrons have electric charge and produce electromagnetic field, but the photon itself is electrically neutral and does not produce electromagnetic field.

So, *local, non-Abelian gauge-invariance*, in this particular example related to quark colour, gives rise to a complicated theory with 8 gluons and interactions between the quarks and the gluons and between the gluons and themselves. This is called *quantum chromodynamics* (reverting to words of Greek origin), or QCD, by analogy with QED.

But now we see why Yang, Mills and Shaw could see no application for their mathematics. Gluons are forced to be massless, just as the photon is, but no one has ever directly seen any massless particles except photons (and perhaps neutrinos, and, maybe one day, gravitons). We seem only to have added to our difficulties, by piling unobserved gluons on top of unobserved quarks.

Before coming to the (possible) resolution of this problem, let us look at the second dilemma in the list at the end of the last section: why do quarks behave as if they were almost free for short intervals? The strength of the forces in QED is characterized by the dimensionless number

$$\alpha = \frac{e^2}{4\pi\hbar c} \approx \frac{1}{137} = 0.0073,$$

where $e$ is the electric charge of an electron. The strength of the QCD forces is determined by a similar dimensionless number, called $\alpha_S$ ($S$ for "strong"). As one might expect, this is bigger:

$$\alpha_S \approx 0.12.$$

We saw in Section 9.9 that the strength of QED really depends upon the distances being probed. The value quoted, $\frac{1}{137}$, applies for large distances: at shorter distances the number gets larger. This is because the electron-positron pairs momentarily present in the vacuum *shield* the charge seen at large distances, that is, make it smaller than at shorter distances. Around the year 1973, it was realized that the *opposite* is the case in QCD: the strength decreases at short distances. This solves the problem (ii) (at the end of the last section), because, in the high-energy experiments, the electrons pass very close to the quarks, and the quarks then behave briefly as if the forces between them were weak.

But what was wrong with the intuitive argument (in Section 9.9) about shielding, suggesting that the force strength should always decrease with distance? Calculation shows that quark-antiquark pairs in the vacuum do indeed shield the colour charges as expected, but gluons in the vacuum have the opposite effect, which predominates (as long as there are not too many quarks). This behaviour of gluons depends upon their colour charge's not being distributed uniformly along their spin directions. In the colour field of a quark, the gluon spins are aligned, and this causes a net movement of colour charge opposite to that which would produce screening.

This "antishielding" effect is unique to non-Abelian gauge theories (at least among reasonable field theories). Its discovery made physicists believe in QCD. It meant that they could calculate the predictions of QCD for high-energy scattering and make quantitative tests with experiment, in much the same way (though with less accuracy) as had been done in QED.

The corollary of the weakening of QCD at short distances is its strengthening at longer distances. When forces are strong, calculation is difficult: one cannot proceed by working out small corrections to free fields. So at last we are up against the third problem: to find out how QCD behaves when quarks get farther apart, and in so doing explain why isolated quarks and gluons are not found.

There is no single, decisive solution to this problem. But, from various clues, a fairly convincing quantitive picture has emerged, which I will now try to explain.

Take as an example a meson made of a charm quark $c$ with a charm antiquark $\bar{c}$. As remarked in Section 11.2, these heavy quarks move fairly slowly, and the system is not much different from a hydrogen atom. There is an attractive colour Coulomb field

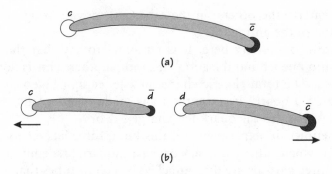

FIGURE 11.2    The disruption of a quark-antiquark system. Quarks are denoted by open circles, antiquarks by black blobs. In (a) the $c$ and $\bar{c}$ are moving apart, and the QCD fields are concentrated in a tube joining them. In (b) a $d$, $\bar{d}$ pair has been created, cutting the tube and producing two mesons that are moving apart.

between the quark and antiquark. Now suppose we try to pull the $c$ and $\bar{c}$ apart. Suppose, for some reason yet to be explained, the colour-electric field between the $c$ and $\bar{c}$ becomes concentrated in a thinnish tube between them (unlike what would happen to the electric field between a proton and an electron). This state of affairs is indicated in Figure 11.2(a).

The energy stored in the colour field is proportional to the length of the tube. If the $c$ and $\bar{c}$ get far enough apart, the stored energy becomes more than the rest energy of another quark-antiquark pair (say, $d$, $\bar{d}$). Then this pair can "pop out of the vacuum", producing the arrangement in (b) of the figure. Now the two mesons, $c\bar{d}$ and $\bar{c}d$, are free to separate. Our attempt to tear the original $c$ and $\bar{c}$ apart has been frustrated: we have made two mesons instead.

The belief is that this sort of behaviour is inevitable. If any meson or baryon is subject to disruption that might knock quarks out, what happens instead is that tubes of colour field form and then break up into other mesons or baryons. Something like this is observed. High-energy collisions between electrons and positrons may momentarily form a photon that can sometimes produce a quark-antiquark pair. These are not seen: instead two "jets" of mesons and baryons may be emitted in roughly opposite directions. These jets are believed to be the remnants of the colour tube, breaking into fragments, some (in one jet) moving roughly in the direction of the quark, the others (in the other jet) moving roughly in the direction of the antiquark.

There is a subtlety here. It is tempting to say that the quark turns into one jet and the antiquark into another. This is not quite accurate. The total electric charge in a jet cannot be equal to the charge on a quark, because the jet is made of integrally charged particles whereas the quark has charge $-\frac{1}{3}$ or $\frac{2}{3}$.

Sometimes, in experiments of this kind, three jets emerge. This happens when either the quark or the antiquark first emits a gluon. These three partons are then attached by colour tubes that sever to produce the three jets. The distribution of the directions of the jets affords indirect evidence for the presence of a gluon at intermediate times.

All this assumes that colour fields, when "stretched", shrink into relatively narrow tubes. Why should this happen? The theory behind this is inspired by analogy with superconductors, in particular with the formation of filaments of magnetic flux in some superconductors (see Section 10.6). It is believed that the QCD vacuum is analogous to a superconductor, but with the role of electricity replaced by colour magnetism, and of magnetism by colour electricity. So colour *electric* lines of force are generally expelled from the QCD vacuum and concentrated into thin tubes. The best evidence for the truth of this picture comes from large computer calculations, approximating the solution of the equations of QCD.

The analogy goes further. Superconductors have a critical temperature, above which there is a transition to conduction as in a normal metal. The QCD vacuum likewise probably has a critical temperature, above which QCD behaves more like QED. The transition temperature is probably of the order of $10^{12}$ K (in energy terms, this is about the rest energy of a pion). Above this critical temperature, quarks and gluons should behave more as ordinary free particles. Experiments are under way to produce (for a short time) a hot blob of quark-gluon matter by colliding atomic nuclei at high energies.

Note the radical step that has been taken. An analogy has been drawn between the *vacuum* of particle physics and a superconducting *material*. The vacuum means just the absence of any particles or radiation, that is, in quantum field theory, all the field oscillators

(Section 9.2) in their ground states. However, this way of talking about independent oscillators presupposes that the interactions between them are small. In QCD (but not in QED), this approximation seems to be qualitatively misleading. Hence the vacuum may have more complicated properties, as suggested by the superconduction analogy.

## 11.4   Conclusion

QCD explains the forces between quarks. Three ideas have gone into the understanding of QCD: generalization of local gauge-invariance, antishielding, and the non-trivial nature of the vacuum suggested by analogy with superconductors. In the next chapter, all three ideas will be used again, but with important variations.

# UNIFYING WEAK FORCES WITH QED

*How weak forces, responsible for beta decay, and so on, and electromagnetism have a heavily disguised unity.*

## 12.1 What Are Weak Forces?

Newton, Faraday and Einstein each believed that there must be some sort of unity in the forces of nature. Faraday, having unified electricity and magnetism, conducted experiments to try to find a connection between these forces and gravity. From 1922 until his death in 1955, Einstein was trying mathematically to unify gravity with electromagnetism. In the event, the force that was eventually unified with electromagnetism is one that was unknown to Newton and Faraday, and perhaps not considered very important by Einstein. This is the so-called weak force. In previous chapters I have ignored the weak force, but now I must explain what it is.

In 1896 (a few months after the discovery of X-rays) Henri Becquerel in Paris discovered the radioactivity of uranium. Studied further by Marie and Pierre Curie, the nature of radioactivity was finally elucidated by Rutherford and Soddy in Montreal during the early 1900s.

In radioactivity, an atomic nucleus emits "rays", unpredictably but with a definite probability per unit time, often changing into a different nucleus in the process. By a historical accident, three totally different processes are lumped under the heading "radioactivity". One is the emission of alpha-particles, the nuclei of helium atoms, each made up of two protons and two neutrons. This happens as a result of the strong, nuclear forces. The only reason that it happens slowly (slowly compared to the time for an alpha particle to traverse the nucleus) is that the alpha particle's wave function has only a very small "tail" outside the nucleus. The second radioactive process is the emission of a photon by the nucleus. This is no different in principle from the emission of photons by the electrons in atoms, but the photons (called *gamma rays*) have more energy (comparable to the rest energy of an electron).

For the purposes of this chapter, the interesting sort of radioactivity is the emission of electrons (or sometimes positrons) from the nucleus. These are not "in" the nucleus before they are emitted. They are a consequence of a new type of transformation of particles called *beta decay*. (The three types of radioactivity were termed *alpha*, *beta* and *gamma* by Rutherford, before their nature was known.)

The basic process is the decay of a neutron, inside the nucleus:

$$\text{neutron} \rightarrow \text{proton} + \text{electron} + \text{antineutrino}.$$

Here the antineutrino is the antiparticle of the neutrino; both have spin $\frac{1}{2}\hbar$, zero electric charge and very little or no mass. Note that the total electric charge remains fixed (zero) during this decay.

The decay of the neutron can be observed for a free neutron (emerging from a uranium reactor, for example). The mean lifetime of a neutron is 887 seconds, an extraordinarily long time by the standards of nuclear physics: for example, $\hbar/(c^2 \text{ neutron mass})$ is about $10^{-24}$ second. This long mean life is partly due to the fact that the neutron is only just heavy enough for the decay to be possible: the kinetic energy released in the decay is less than 0.1 percent of the rest energy of the neutron. But the main reason is the intrinsic smallness of the "force" involved. In the account of beta decay given by Fermi in 1933, the quantity characterizing the force was given as about

$$10^{-5} \times (\text{neutron mass})^{-2}.$$

This is a small number, and hence the force is called *weak*. But notice that we have a small number because we have expressed it in terms of the neutron mass. This mass is the largest that appears to be relevant: if we used a smaller mass, like that of the electron, the coefficient would be even smaller (more like $10^{-11}$). But Fermi's beta decay constant is not like the number $\alpha = \frac{1}{137}$ characterizing QED. The latter is dimensionless and so is unambiguously a small number. On the other hand, the weak force is truly weak only so long as very large energies are not involved.

Why should one use the word *force* in speaking about something like neutron decay? According to quantum field theory (Chapter 9), any field that creates a particle can equally destroy its antiparticle. This means that, in the decay process, we can switch the antineutrino from the final state and replace it by a neutrino in the initial state. In this way, we get from the decay process the reaction

$$\text{proton} + \text{neutrino} \leftrightarrow \text{neutron} + \text{positron},$$

which looks more like a collision leading to scattering, for instance, the scattering of an electron on a proton. The difference here is that two of the particles have changed their nature (in particular, changed their charges): a proton into a neutron and a neutrino into an electron. If we are prepared to envisage a force that, as well as deflecting particles, changes their nature, we will be happy to use the word *force* (or *interaction*) here.

That a neutrino, as well as an electron, is emitted in beta decay was postulated in 1930 by Pauli. (The Italian diminutive ending points to Fermi's involvement.) If a neutron turned into just a proton and electron, angular momentum could not be conserved since there are just three $\frac{1}{2}\hbar$ spins involved. Further, the electron should have a unique energy, whereas in fact the energy varies from decay to decay.

Neutrinos have no (direct) electromagnetic interactions or strong ones. They interact via the weak force only. This means that (except at very high energies) they are extremely difficult to detect. An (anti)neutrino was not detected directly until 1956. Antineutrinos from a nuclear reactor (the experimenters, who had worked at Los Alamos, had thought of using a nuclear bomb) were incident on large tanks of water, containing some cadmium chloride. Occasionally an antineutrino hit a proton, producing a positron and a neutron. The neutron was absorbed by a cadmium nucleus, which then emitted photons that could be detected. The positron annihilated with an electron, making another detectable photon. The experiment was done some 12 metres underground to reduce false signals due to cosmic rays. About $10^{17}$ antineutrinos passed through the tanks per second. The expected signal (that is, sequence of photons) was seen about three times per hour.

The experimenters, Reines and Cowan, sent a telegram to Pauli, who replied, "Everything comes to him who knows how to wait".

We now know that the epithet *weak* is misleading. The strength of the "weak" forces turns out to increase with the energy of the participating particles. At energies of about a hundred times the rest energy of a proton, the "weak" forces (for example, the preceding reaction of a neutrino with a proton) become comparable with electromagnetic ones. This energy is some 10,000 times the energy emitted in nuclear beta decay.

Thus the "weak" forces are energy-dependent ones. The reason for this is that they are *short-range* forces. I will explain this. Electrostatic forces and gravity have no intrinsic range. These forces decrease with distance, but only by the natural, geometric inverse-square law (because the lines of force spread out). Short-range forces have an additional sharp (exponential) decrease with distance outside some characteristic range, of order $10^{-18}$ metre in the "weak" case. Why do short-range interactions appear to be weak at low energies? If the wavelength of a particle's wave function is much larger than the range, the force has little effect on it (a bucket does not scatter radio waves much). But when the energy is high, the wavelength can become comparable with the range, then the force shows its true strength. Energies high enough for this were not achieved on Earth until the 1980s.

I will continue to write *weak force*, but *short-range* would be a better description.

I have only so far mentioned one example of a weak interaction. In fact there are hundreds of processes ascribable to the weak force. First, there are those associated with *leptons*. This word comes from the Greek for "light" (it is also a Greek coin), but nowadays it means a fermion that does not participate in strong interactions. At present, six leptons (together with their antiparticles) are known. Three of them are negatively charged: the electron; the muon (from the Greek letter mu, $\mu$), whose mass is 207 times as great; and the tau (another Greek letter, $\tau$), whose mass is 3,484 times the electron's (that is, getting on for twice the mass of a proton). The other three leptons have no electric charge and are in fact three different neutrinos.

Each neutrino is associated with one of the charged leptons, and they are denoted by the Greek letter $\nu$ (nu) with a suffix denoting which lepton. So we may arrange the leptons in a pattern

$$\begin{pmatrix} e \\ \nu_e \end{pmatrix} \begin{pmatrix} \mu \\ \nu_\mu \end{pmatrix} \begin{pmatrix} \tau \\ \nu_\tau \end{pmatrix}$$

(here $e$ stands for electron). By a weak process, an electron can change into its neutrino, $\nu_e$, but not into either of the others. Similarly, a muon can change into $\nu_\mu$ and a tau into $\nu_\tau$. Each neutrino remembers its origin. But the only way to tell one neutrino from another is by seeing which lepton it can produce (or is produced from).

Each neutrino has, at most, a very small mass. The masses may be exactly zero, in which case they must move with the speed of light (like the photon). But this is not known. All we have are upper limits on the masses. The electron's neutrino has mass less than $10^{-5}$ of the electron mass. The limits on the $\nu_\mu$ and $\nu_\tau$ are less stringent: 0.2 and 34 electron masses, respectively.

An example of a weak process involving two of the lepton doublets is

$$\mu \rightarrow \nu_\mu + e + \bar{\nu}_e.$$

(A bar denotes an antiparticle.) Note that the last two particles are together in one lepton doublet, and the other two particles are from another doublet. This process is the decay of the muon, which gives the muon a mean lifetime of about $2.2 \times 10^{-6}$ second. The muon decays faster than the neutron because more energy is released in the decay, but the muon decay rate is still very small by nuclear standards.

My statements in the last but one paragraph, strictly speaking, presupposed all the neutrino masses to be zero. If they are not, there are two ways to characterize a neutrino state: by the lepton it accompanies *and* by its mass. There may be a mismatch between these two characterizations. Then, a neutrino born from an electron may be a superposition of states with different masses. Since these propagate with different speeds, after some distance the superposition is different, and the amount of electron-producing component in it may be less than 1.

Now about the weak interactions of hadrons (strongly interacting particles), like the beta decay of the neutron with which we began this section. The weak interactions of hadrons result from the weak interactions of their constituent quarks, such as

$$d \to u + e + \bar{v}_e,$$

or, equivalently,

$$d + v_e \leftrightarrow u + e.$$

(Here $d$, $u$ stand for the down and up quarks with charges $-\frac{1}{3}$ and $\frac{2}{3}$, $e$ stands for electron and $\bar{v}_e$ for an antineutrino.) Note that electric charge balances on either side of each of these two transitions.

There are similar weak transitions between the $s$ and $c$ quarks, and between the the $b$ and $t$ quarks. This fact reminds us that, in Section 11.2, the quarks were arranged in three doublets, reminiscent of the three doublets of lepton displayed here. We recall that the quark doublets are

$$\begin{pmatrix} u \\ d' \end{pmatrix} \begin{pmatrix} c \\ s' \end{pmatrix} \begin{pmatrix} t \\ b' \end{pmatrix}.$$

As with the lepton doublets, the charge difference between the two members in any doublet is 1. As with the lepton doublets, weak interactions involve transitions between members of the same doublet.

I have made a small but important refinement in writing out these quark doublets, calling the lower members $d'$, $s'$, $b'$. The reason is the same as something mentioned earlier about neutrinos. The $d$, $s$, $b$ quarks are defined by the (different) masses that they have. The $d'$, $s'$, $b'$ states are defined to be the weak interaction partners of, respectively, $u$, $c$, $t$. There is a mismatch between these two definitions. Consequently, for example, $s$ is not pure $s'$ but contains an admixture of about 5 percent (in terms of probabilities) of $d'$ (and a smaller admixture of $b'$). Consequently, transitions like

$$s(d') \to u + d' + \bar{u}$$

are possible (where $s[d']$ means the $d'$ admixture in the $s$). In this way, particles containing $s$ quarks, called *strange particles*, can decay weakly into hadrons without strange quarks. As an example, a neutral particle called $\Lambda$ (made of $sdu$) decays into a proton together with a negative pion or a neutron together with a neutral pion, with a consequent mean lifetime of about $2.6 \times 10^{-10}$ second.

## 12.2   The Looking-Glass World

One of the most important principles in physics is that the laws of nature do not pick out any special positions or directions in space, nor any special times. In classical physics, this principle led to the conservation of momentum, angular momentum and energy (Sections 6.8, 6.9). The same is true in quantum physics.

There is another possible principle: that the laws of physics should be the same after reflection in a mirror. Some things look the same in a mirror; others do not. A plain teacup does; a corkscrew does not. Things that do not have *handedness*, left or right (the word *chirality*, from the Greek for "hand" as in *chiropractor*, is also used).

Here is an instance in which the laws of nature are the same under reflection. When an electric current flows in a straight wire, there is a magnetic field and the lines of magnetic force are circles round the wire. The direction of the field is such that it appears *clockwise* to someone looking along the wire in the direction of the current (see Figure 3.8). In a mirror, the magnetic field will appear to go *anticlockwise*. This fact in itself does not mean that the laws of nature change on reflection; it just means that we define magnetic field in a way that involves a standard of handedness (a clock).

Now suppose there is a second current-carrying wire parallel to the first. There is a force on this wire, whose direction is given in terms of the field by another rule (see Figure 3.7), which again involves a standard of handedness. The final result is that the direction of the force (attractive if the currents are in the same direction) does look the same in a mirror. The physical conclusion is unchanged on reflection.

Newton's and Einstein's laws of motion and of gravity, Maxwell's laws of electromagnetism and the equations of QCD are all unchanged on reflection. It is natural to believe that *all* the laws of nature should have this invariance, and most physicists did so believe until about 1956.

In classical physics, reflection invariance has an important difference from translation invariance and rotation invariance: there is no conserved quantity, analogous to momentum and angular momentum, connected with reflections. The reason for this difference is connected with the following observation.

If you want to compare a teacup and a rotated teacup, you can use two identical cups. Alternatively you can take one cup, start with it in one position and rotate it into a new position. But, given a corkscrew, there is nothing you can do to it (short of bending it) to turn it into its mirror image. You can compare a corkscrew with its mirror image, but you cannot change it into its mirror image.

In quantum theory, it is different. Discontinuous changes are not forbidden. The possible states of an atom form a discrete set, yet transitions (jumps) can occur from one such state to another. Likewise, quantum theory allows, in principle, a right-handed corkscrew to flip over into a left-handed one. For a macroscopic thing like a corkscrew, the probability of such a flip is to all intents and purposes zero. But in the microscopic domain, handedness flips can readily occur.

Therefore, in quantum theory, if there is mirror symmetry, then there *is* a conserved quantity. This quantity is called *parity*. But it is a very different thing from the other conserved quantities, momentum, angular momentum and so forth. If the operation of reflection is done twice, one gets back to where one started. When the hairdresser holds a mirror behind your head, and you see your image reflected twice, your parting does appear on the correct side. Consequently, parity has the property

$$(\text{parity})^2 = 1,$$

that is,

$$\text{parity} = +1 \text{ or } -1.$$

No other values are possible.

Figure 12.1 illustrates the working of reflections and parity. On the left is illustrated a photon with its spin pointing in its direction of

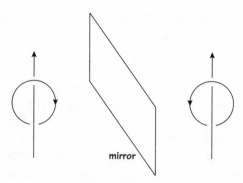

FIGURE 12.1    The reflection of a circularly polarized photon (on the left of the diagram). The reflection (on the right) is a photon circularly polarized in the opposite sense.

motion: that is (from the standpoint of classical electrodynamics), it is a photon of circularly polarized light, with the electric field rotating clockwise about the direction of motion. The reflection in a mirror (placed as shown in the diagram) is a photon with its spin in the opposite direction, that is, a left circularly polarized photon.

In quantum theory, there exist states got by superposing the two oppositely circularly polarized states. If the wave functions are simply added, the resulting state has parity $+1$: it is unchanged on reflection. (This state represents a plane polarized photon, with the electric field always parallel to the mirror.) If the two wave functions are subtracted, the resulting state has parity $-1$.

Atomic physics uses quantum theory applied (mainly) to electromagnetic forces. So atomic physics naturally has mirror symmetry, and parity is a conserved quantity. The same seemed to be true of nuclear physics. For example, the particles called pions – spinless particles with masses about 270 times the electron mass – have parity equal to $-1$. The neutral pion decays into two photons, and the polarizations of these two photons are correlated so that their combined parity comes out to be $-1$.

When weak interactions (beta decay, and so forth) were studied, from the work of Fermi onwards, it was taken for granted, almost universally, that mirror symmetry would apply in this domain too, being in fact a universal law of nature. Then, in the mid-1950s, nature sprang one of its surprises. Certain particles called *kaons* were found to decay, by weak interactions, sometimes into two pions and sometimes into three. The decay into two pions showed the parity of the kaon to be $+1$ (since $(-1)^2 = +1$), but the three-pion decay suggested a parity of $-1$. (The argument is actually a little complicated, since the parity of a three-pion state depends upon their mode of motion, as well as upon the intrinsic parity of the individual pions. But the dominant mode of motion is very simple and does have mirror symmetry.)

Two Chinese-U.S. physicists, Tsung Dao Lee and Chen Ning Yang, analyzed all that was known about beta decay, and so on, and found that, up to then, no other experiments were sensitive to whether mirror symmetry held good or not. Perhaps the weak interactions were not reflection symmetric. Yet a third Chinese-U.S. physicist (this time a woman), Chien-Shiung Wu, an expert on beta decay, set out to do a test. One way to find something that changes under reflections is to involve spin directions as well as directions of motion (as Figure 12.1 indicates). Wu did this by studying the beta decay of atomic nuclei whose spins had been aligned by the application of a high magnetic field at low temperature (otherwise thermal agitation would randomize the spin directions). Call "up" the direction relative to which the spins appeared to be clockwise. Then, she found that the electrons emitted in the decay more often went near the "down" direction than the "up" one. In a mirror, this spin-direction correlation would appear oppositely (as in Figure 12.1), so its existence proved that weak interactions did not have mirror symmetry.

This experiment is surely one of the great crucial experiments in the history of physics. The result astonished most physicists. Two as eminent as Pauli and Feynman are said to have bet that Wu would not find mirror symmetry to be broken.

It soon became clear that mirror symmetry is often, in a well-defined sense, "maximally" broken in weak interactions. An example of what this means comes from neutrinos, which have no other interactions except weak ones (and gravitation). Neutrinos were found to have their spins *always* oriented so as to appear to be anticlockwise relative to their direction of motion. (This spin-direction correlation is called *left-handed helicity*.) Thus neutrinos are quite unlike photons, in which the spin can be clockwise or anticlockwise (as on the two sides of Figure 12.1). The mirror image of a neutrino is not a possible state of a neutrino.

(These statements presuppose that neutrinos move exactly with the speed of light. If not, one could consider a neutrino at rest, and then there would be no direction of motion with which to correlate its spin. There is now some evidence that neutrinos have very small, but not exactly zero rest masses; so they cannot move with exactly the speed of light. If this is so, the last paragraph requires slight qualification, but it remains a good approximation to the truth.)

At this point, another broken symmetry of nature enters the story. Many particles have antiparticles with almost identical properties except that their charges are opposite. The antiparticle of the electron is the positron. These two have identical mass and spin, but opposite electric charge. The change of a particle into its antiparticle is called *charge conjugation*. If charge conjugation is done twice, the particle gets back to the way it started. In this respect, charge-conjugation is like reflection. It is natural to speculate that the laws of nature should be invariant under charge conjugation. Indeed, so far as strong and electromagnetic interactions are concerned this seems to be the case. Soon after the discovery of mirror symmetry breaking, it was realized that weak interactions break charge conjugation invariance also. A convenient way to illustrate this is again from neutrinos.

Neutrinos have distinct antiparticles, called just *antineutrinos*. Because neutrinos have no electric charge, the difference between them and antineutrinos is not obvious. But they are different. In nuclear beta decay, neutrinos appear with positrons but antineutrinos come with electrons. The neutrinos and antineutrinos react differently with matter. An antineutrino can react with a proton to produce a positron and a neutron. A neutrino can react with a neutron to produce an electron and a proton. But an antineutrino cannot do the former, nor a neutrino the latter.

It is found that, whereas neutrinos have left helicity, antineutrinos have right helicity. Thus particle and antiparticle have different properties, and charge conjugation invariance does not hold. This is just one example of charge conjugation breaking. It occurs in all weak interactions, whether neutrinos are involved or not.

These properties of neutrinos and antineutrinos suggest a very nice hypothesis: that although mirror and charge conjugation invariance are each separately broken in weak interactions, there should be invariance under the combined transformation

(reflection) × (charge conjugation).

This is called *CP* for short (*C* for charge conjugation, *P* for parity). Thus *CP* invariance, if true, means that if you look at a neutrino in a mirror, you see something with the properties of an antineutrino. More generally, the mirror image of any process in our world looks like the corresponding process in an "antiworld" (where matter is predominantly made of antiprotons and antineutrons). If Wu had done her experiment in an "antiworld", she would have seen positrons moving "up" (relative to the antinucleus spin) rather than electrons moving "down".

The hypothesis of *CP* invariance seems to provide a beautiful substitute for simple reflection invariance. If true, the laws of nature would not provide a standard of "handedness". If a being from another galaxy told us that the neutrino defines "left-handedness", we would have to ask it how we knew what it called neutrino and what antineutrino (or what electron and what positron). It could not tell us, short of sending a sample.

Great physicists, Einstein and Dirac, for example, have assured us that beauty is a good indication of the likely truth of a scientific theory. By this token, *CP* invariance should have been true. The looking-glass world would have been a world of antiparticles. But brute nature was recalcitrant. In 1963, at the Brookhaven National Laboratory in the United States, an experiment was performed that showed that *CP* invariance was *not* exact. The particles used were again kaons, but this time the neutral ones were important. There are two of these with very nearly the same mass, but one has a mean lifetime of about $10^{-10}$ second and the other $5 \times 10^{-8}$. It is the latter, longer-lived, one that is relevant. If *CP* were a good symmetry, this should be forbidden from decaying into two pions. The experiment showed that it did rarely decay into two pions, with a probability of about 0.3 percent (the dominant decay modes were into three particles).

There is another fact about the decays of this long-lived kaon, which shows more directly the violation of *CP*. Two of the decays are into

positive pion + electron + antineutrino

and into

negative pion + positron + neutrino.

These two are related by charge conjugation. If *CP* were an exact symmetry, these two decays should occur with equal frequency. (Mirror symmetry – the *P* part of *CP* – is irrelevant here, because any configuration and its mirror image are included to get the total probability of decay.) In fact, the probabilities of the two decays differ by some 0.6 percent.

Now the extraterrestrial being *can* communicate to us a standard of handedness, embedded in the laws of nature. It says, "Measure the two decay rates (just mentioned) in the more frequent mode, one of the particles is a positron (not an electron), and the accompanying neutrino is left handed".

Within weak interactions, the violations of mirror symmetry and charge conjugation invariance are large (often maximal). But the violation of *CP* is very small: so small that it only shows up in a few specially favourable cases.

There is a third discrete transformation of importance in science: the change in the direction of time, *time reversal*. Newton's laws of motion are invariant under time reversal. This is because they refer to acceleration, which is rate of change of rate of change of position. Each "rate of change" gets a minus sign under time-reversal, but as there are two of them, there is no overall change. Given any motion possible under Newton's laws, there is another one in which everything runs backwards. If we were shown a video of the motions of the planets and their satellites, we would be hard put to know whether it were being run backwards or forwards. Maxwell's laws of electromagnetism have the same property of time reversal invariance (magnetic fields change sign under time reversal).

Many phenomena in nature emphatically do not appear to be time reversal invariant. A cup of hot tea in air at room temperature cools down. The converse process (heat flowing into the hot tea) is not observed. In general, entropy increases. (See Section 2.7.) Yet if we could follow the motion of a few molecules of air and water for a very short time, they could all be consistently reversed. I will return to this famous paradox in Chapter 14. In the meantime, let us confine ourselves to the time reversal invariance, or otherwise, of the microscopic laws of physics.

In quantum field theory, there is a mathematical theorem (based on very plausible assumptions) called the *CPT* theorem. This says that there must be invariance under the triply combined transformations

$$\text{charge conjugation} \times \text{reflection} \times \text{time reversal.}$$

According to this theorem, if *CP* invariance fails, then so must time reversal invariance.

I will give a very incomplete indication of the reason for the *CPT* theorem. If we stand in front of a mirror, back and front are interchanged in our image. This is reflection. If we "about turn" (that is, rotate through 180 degrees about a vertical axis), back and front are interchanged, but *also* left and right. This is a rotation *not* a reflection. In a reflection, one direction (or more generally, an odd number of directions) is reversed. But reversal of two directions can be accomplished just by a rotation, and rotational invariance is not in doubt.

Now think of spacetime as some sort of four-dimensional space. The combination of space reflection with time reversal (*PT*) does reverse *two* directions in spacetime. So we might expect that for this double reflection (which is a sort of "rotation") invariance would be guaranteed. This argument, as it stands, is wrong, because time is essentially different from space in Minkowski spacetime (Section 5.5). There is not really any continuous series of small transformations (rotations or Lorentz transformations) that reverses the direction of time. However, the argument can be patched up, and in the process charge conjugation has to be introduced as well as space reflection and time reversal.

Accepting the truth of the *CPT* theorem, and given the experimental fact of *CP* non-invariance, it follows that weak interactions cannot be exactly invariant under time reversal either. There is indeed indirect evidence for this, but, because of the smallness of the effect, nothing that stands out as obviously being not time reversal invariant. Much of the experimental data in particle physics is about the results of collisions. For example, two particles $A$ and $B$, with given velocities and spin directions, collide and turn into two others, $C$ and $D$, with measured velocities and spins:

$$A + B \rightarrow C + D.$$

If time-reversal invariance were exact, the probability for this process would be equal to that for

$$C' + D' \rightarrow A' + B',$$

where $A'$ is the same as $A$ except that the velocity and spin direction are reversed, and similarly for $B'$, $C'$, $D'$. Because of the lack of time-reversal invariance, these two probabilities should in general be slightly different, but direct evidence of this kind does not yet exist.

What is the reason for this untidy, small breaking of $CP$ and time reversal invariance? No one knows for certain. I will return, in Section 12.5, to current thinking on the subject.

Before ending this section, I mention three questions:

(i) There are many *molecules* in nature that do not possess mirror symmetry, that is, have handedness. Molecules are held together almost entirely by electromagnetic forces, which do have mirror symmetry. So why the handedness?

(ii) The known matter of the universe is largely composed of protons, neutrons and electrons, and not their respective antiparticles, so it is not $CP$ symmetric. Is this anything to do with the lack of $CP$ symmetry of the laws of nature?

(iii) Does the lack of time reversal invariance shown in the large by increase of entropy have any connection with the lack of that invariance in the laws of microscopic physics?

It is thought that the answer to (ii) is yes. I return to this in Section 14.5. As to question (iii), to my knowledge, no one has been able establish such a connection. Here, I will discuss (i) a little more.

The nineteenth-century scientist Pasteur's first great discovery was that solutions of tartaric acid (a by-product of wine making) rotate the plane of plane-polarized light. This is a handed effect, so the molecules of tartaric acid must exist in two forms with opposite handedness. The acid produced by a living organism always has one handedness, but synthesized in the laboratory it is a mixture of both. Many organic molecules have these properties. The two forms may taste and smell different.

The handedness is easy to understand, given that organic molecules contain carbon atoms, which often combine with four other atomic groups placed roughly at the corners of a tetrahedron. The corners of a tetrahedron can be labelled in two distinct ways, which are mirror images of each other.

Pasteur was excited by his discovery. He wrote: "Life is dominated by asymmetrical actions. I can imagine that all living species are primordially ... functions of cosmic asymmetry".

If a handed molecule exists, the mirror symmetry of electromagnetism implies that its mirror image must also exist. There is no problem with this. But quantum theory goes further and in principle allows a handed molecule to flip from one form to another, so that a handed molecule should not be a truly stable form. This appears to be a contradiction, until it is realized that the *rate* of this transition in some molecules is so slow (compared to the age of the universe, say) as to make the probability of flip utterly negligible.

There remains a question still. All life on Earth seems to use molecules of the same handedness. Is this an accident of evolution, resulting from handedness of the molecules that the first life happened to use? Could it possibly have anything to do with the lack of mirror symmetry of the weak interactions? Is it (somewhat as Pasteur envisaged) due to polarized light (polarized by the indirect action of magnetic fields) bathing the clouds from which stars (and their planets) form and in which some carbon-containing molecules are synthesized?

## 12.3   The Hidden Unity of Weak and Electromagnetic Forces

In the mid-1960s, weak and electromagnetic forces were shown to be two sectors of a unified whole, now called *electroweak* interactions. This was an insight comparable to the unified understanding of electricity and magnetism in the nineteenth century (Chapter 3). On the theoretical side, the two main architects of the theory (working independently) were the American Steven Weinberg and the Pakistani Abdus Salam, working mainly in London. As we shall see, they tapped various currents of ideas that were about at the time.

Very soon after this theoretical work, plans were afoot in the European laboratory CERN, in Geneva, to test the predictions of the new theory. The bold idea was to store antiprotons by making them circulate in a magnetic field, in a ring-shaped tube at a very high vacuum. Antiprotons, of course, do not exist naturally on Earth and have to be produced in high-energy collisions between, say, protons. Antiprotons annihilate rapidly in contact with ordinary matter. To store sufficiently many required great technical ingenuity. Protons could be made to circulate in the opposite direction in the same tube. Where the two beams intersected, proton and antiproton would annihilate with sufficient energy to produce, in 1981 for the first time, the two new particles, called $W$ and $Z$, predicted by the electroweak theory.

At first sight, weak and electromagnetic forces seem as different as chalk and cheese. Here are some of the apparent obstacles to unification:

(i) Weak forces, unlike electromagnetic ones, can change one particle into another, for instance, a neutrino into an electron.

(ii) Weak forces involve handed states (of neutrinos, for instance): electromagnetism has mirror symmetry.

(iii) Weak forces, unlike electromagnetic ones, have a short range.

All these obstacles can be overcome. I will divide the presentation into parts. First I will describe a halfway house in which (i) and (ii) are overcome, but (iii) is just ignored. That is, I will pretend temporarily that weak forces do *not* have short range and describe an imaginary world in which they are no different from electromagnetism in respect to range. Then later I show how a short range is imposed.

## 12.4 An Imaginary, Long-Range Electroweak Unification

We want to enlarge electromagnetism so as to include weak forces. Electromagnetism is a gauge theory (Section 8.12), and we have already seen, in Section 11.3, that a natural generalization is to non-Abelian gauge theories. This means generalizing the changes of phase angle of the fields of charged particles to a bigger set of transformations. In Section 11.3, these generalized transformations mixed up the three colours of each type of quark. For the electroweak theory, we have to specify what fields are mixed up by the generalized gauge transformations.

In order to do this, start with the doublet

$$\begin{pmatrix} \nu_L \\ e_L \end{pmatrix},$$

where $\nu$ and $e$ are the neutrino and electron fields, and the suffix $L$ indicates that the left-handed parts only are to be used. ("Left-handed" here demands some explanation, since the helicity of a particle with mass depends upon the velocity of the observer who measures it. I will just say that the left-handed part of a field is such as would produce a left-handed particle *if* its mass were zero.) Non-Abelian gauge transformations based upon this doublet include phase changes of $\nu_L$ and $e_L$ and also the mixings of a little $e_L$ into $\nu_L$, and vice versa, four kinds of transformation in all. Of these, three interlock with one another. The fourth, the transformation of $\nu_L$ and $e_L$ by an equal phase, is independent.

Take the first three for the moment. In the non-Abelian gauge theory, they have associated with them three "photons". Call them $W_+$, $W_-$, $W_0$, where the suffices denote their electric charges. These are the analogues of the eight gluons that cause the strong forces (Section 11.3).

Now let us bring in the $u$ and $d$ quarks. Arrange them (along with the electron and its neutrino) in doublets:

$$\begin{pmatrix} \nu_e \\ e \end{pmatrix}_L \begin{pmatrix} u_r \\ d'_r \end{pmatrix}_L \begin{pmatrix} u_b \\ d'_b \end{pmatrix}_L \begin{pmatrix} u_y \\ d'_y \end{pmatrix}_L.$$

The suffices $r$, $b$, $y$ distinguish the three colours of quarks. The suffix

$L$ indicates the left-handed states of the particles. (The prime on $d'$ was explained at the end of Section 12.1.) The $W$ particles interact with each of these four doublets in identical ways. They are connected with gauge transformations that transform between upper and lower members of each doublet. (The gluons of the strong forces, in contrast, are connected with gauge transformations of the three quark colours of each quark doublet.)

In addition, there are the muon and tau lepton and their neutrinos and the $c, s$ and $t, b$ quarks. These just repeat the previous pattern twice more, making three *generations*, as they are called. The generations are identical except for the masses.

There are altogether 12 doublets, 4 in each generation. Each of the 12 doublets is assumed to transform in the same way under the non-Abelian gauge transformation and therefore to interact in the same way with the three "photons", $W_+$, $W_-$, $W_0$.

We can now see that the actions of the charged $W$ particles are just right to induce weak forces (except for the range!). Figure 12.2 shows the Feynman diagrams for three examples of weak processes. It would be nice if $W_0$ could be identified with the true physical photon. Unfortunately it cannot. It interacts with the neutrinos, which have no electric charge, and its interactions with charged particles (via their left-handed parts) do not have mirror symmetry. These two faults can be put right, if not in a very elegant way. Introduce another gauge symmetry, changing the phases of all the left-handed parts and also the right-handed parts of the charged particles. There is an associated, electrically neutral "photon", called $Y_0$. The three $W$s and the $Y_0$ are independent of each other. They have their own independent interaction strengths.

The $W_0$ and the $Y_0$ are both neutral "photons". It turns out that neither of them is identified directly with a physical particle: rather two different superpositions of them are so identified. One superposition gives the true photon. The other is a neutral particle called $Z_0$. The interactions of $Y_0$ have been so arranged that the photon does not interact with neutrinos (the $Y_0$ contribution cancels the $W_0$ one) and interacts with charged particles in the appropriate way, with mirror symmetry (equally to right- and left-handed parts).

The $Z_0$ generates new types of weak interactions, not having mirror symmetry but nevertheless in general involving right- as well as

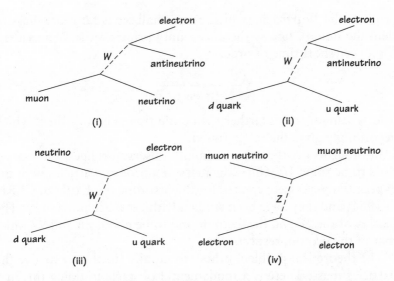

FIGURE 12.2   Feynman diagrams for four examples of weak processes. Time runs from left to right. Continuous lines represent leptons or quarks. Dashed lines denote $W_+$, $W_-$ or $Z_0$ particles. Diagram (i) gives the decay of the muon, (ii) is responsible for neutron decay and (iii) for a possible result of a neutrino collision. (iv) is the scattering of the muon neutrino $\nu_\mu$ on an electron.

left-handed states. These weak processes are not so readily observed as those due to the charged $W$ particles, but one example, which has been observed, is shown in Figure 12.2(iv).

## 12.5   The Origin of Mass

The model just described is totally fictitious. There is nothing in it to make the weak interactions short-ranged. The "$W$ and $Z$ particles", like the photon, have no mass. Somehow, this range has got to be imposed on the theory. This is the same as giving the $W$ and $Z$ particles masses, for the range is connected with the mass by

$$\text{range} = \frac{\hbar}{c} \frac{1}{\text{mass}}.$$

One can see this by an argument using the uncertainty principle. Take Figure 12.2(i) as an example. There is no energy available to produce the $W$ particle (which as we shall see is far more massive than the muon), but by the uncertainty principle the $W$ may exist for a very short time of order

$$\frac{\hbar}{c^2 \times (\text{mass of W})}.$$

The $W$ cannot travel farther than $c$ multiplied by this time, which accordingly gives the range stated.

In order to get the experimentally observed ranges, the masses need to be some 100 times the mass of a proton. Particles with the expected masses were created for the first time (on Earth) at CERN in 1981, and they have been studied with great accuracy since. The masses of the $W$ and the $Z$ are found to be about 86 and 97 times that of the proton, respectively.

The theoretical problem is how to modify the theory to give the particles mass. There is a fundamental obstacle to doing this in a naive way. This obstacle lies in a *qualitative* difference between spin 1 particles without mass and with mass. The photon, the known example of a massless spin 1 particle, has *two* possible spin states: aligned with the motion or opposite to the motion (corresponding to right- and left-circularly polarized light). Because the photon always moves with the speed of light and cannot be overtaken, each of these spin states looks the the same to any observer.

A spin 1 particle with mass is quite different. One may study it from the point of view of an observer with respect to whom it is at rest. Then it follows from basic principles of quantum theory that the allowed values of the spin in any given direction are (in units of $\hbar$)

$$-1, 0, +1,$$

that is, *three* values. To go from a massless to a massive spin 1 particle, one must somehow provide one extra spin state. How can this be done?

How this question was answered has a complicated history, which has not, to my knowledge, been written in detail. The crucial clue came from the theory of superconductivity (see Section 10.6). I will mention just two people who came into the story. The first is the imaginative Japanese-U.S. physicist Yoichiro Nambu, who first (in around 1960) imported into particle physics ideas from superconductivity theory. The second is the British theoretician Peter Higgs, who, working in Edinburgh, put the idea into its definitive form in the mid-1960s. Higgs's name, as we shall see, is attached to a predicted but yet undiscovered particle.

A magnetic field, as we saw in Section 10.6, penetrates only a small distance into a superconductor. In other words, magnetism acquires a short range inside the superconductor. If a short range is linked to mass, and a massive photon must have a third spin state, how does superconductivity provide that third spin state? The answer lies in the *order parameter f* that characterizes the superconducting state, the macroscopic "wave function" of the Cooper pairs. This $f$ is a complex number, whose phase is chosen arbitrarily by the superconductor. When a magnetic field is applied to a superconductor, the phase of $f$ varies with position and time, and this degree of freedom provides the third spin state. The mass of the "photon" inside the superconductor depends upon the magnitude of $f$.

To mimic this mechanism in particle physics, we must assume that the vacuum (that is, "empty" space) has a sort of "order parameter". Higgs showed how to do this by postulating the existence of four spinless particles, which interact with the W and Z particles somewhat as the leptons and quarks do. One of these spinless particles has charge +1, one −1, and two have zero charge. In the imaginary, massless world (described in the previous section) these would be four massless and spinless particles, just tacked on to the other particles.

Now suppose that one of the neutral spinless particles undergoes Bose condensation, just as the Cooper pairs in a superconductor do. The choice of which particle does this is arbitrary, just like the choice of phase in a superconductor. The making of this choice is usually called *spontaneous* symmetry breaking, but *vacuum symmetry breaking* would be a better term. The laws of the electroweak theory do not distinguish among the four spinless particles, but nature has chosen one, in an arbitrary way, to undergo Bose condensation. The vacuum is "full" of those particular particles. This process does not do any obvious violence to our idea of the vacuum: it is the same everywhere and at all times (or, at least, over some large regions of space and long periods of time).

The fate of the other three spinless particles is to provide the third spin states for the $W_+$, $W_-$ and $Z_0$ particles when they acquire mass. The magnitudes of these masses are determined by the magnitude of the order parameter. The first spinless particle, the one undergoing Bose condensation, also gets a mass and is a real physical spinless, neutral particle, called a *Higgs particle*. It has not been discovered yet; that means that it must have at least about 100 times the mass of a proton. Unfortunately, the electroweak theory does not make a definite prediction what the mass should be. Nevertheless, the theory demands that Higgs particles should exist and should not be too heavy, probably not more than about 200 proton masses. The search for these particles is one of the great experimental challenges in particle physics at the present time. A machine is under construction at CERN that will cause oppositely moving proton beams to collide against each other at energies equivalent to several thousand proton masses. This collider, to begin producing results in 2005, has as one of its main aims the discovery of Higgs particles (if they have not been seen elsewhere before).

(We have now used analogies from superconductivity theory twice. In this chapter, the analogy is quite close: the Higgs particles correspond to the Cooper pairs. In the previous chapter, on QCD, the analogy was more indirect: it was an analogy with an imaginary "magnetic superconductor", from which electric lines of force rather than magnetic ones are expelled. In both cases, the role of the analogy has been to inspire guesses about how the non-Abelian gauge theories might behave. These guesses were then to be confirmed by more or less complete calculations.)

A criterion for a good scientific theory is that it should make many predictions with few adjustable parameters (that is, numbers like the charge and mass of the electron, and so on). The electroweak theory contains three basic parameters, which one may choose to be: the unit $e$ of electric charge, a similar number that fixes the strength of the $W$ interactions and the mass of the $W_+$, $W_-$ particles. Given these, the $Z$ mass and interaction strengths are determined, and so are an enormous variety of weak processes.

So far, so good, but sadly the masses of all the leptons and quarks (and the Higgs) are also adjustable parameters. Theory at present does very little to explain why these are what they are. All the masses are thought to be produced by the interaction of the spinless particles concerned with the Bose condensate of Higgs particles, but this just means that all the interaction strengths to the Higgs are adjustable parameters.

The electroweak theory has a clean part and a messy part. The clean part consists of the spin 1 particles (photon, Ws and Z) and their interactions, which are all powerfully constrained by the principle of gauge-invariance. The messy part consists of the interactions of the spin 0 particles (the Higgs and the three others), which obey no such guiding principle and can be different for each lepton and quark.

These Higgs interactions do more than provide masses for the leptons and quarks; they may also cause transitions among leptons or among quarks. This explains the distinction, made at the end of Section 12.1, between $d, s, b$, which have definite masses, and $d', s', b'$, which occur in the electroweak doublets. Experiment shows these mixing effects to be fairly small. Why this should be is not well understood.

We can now return to the mystery of $CP$ (reflection times charge conjugation) and time reversal non-invariance, explained at the end of Section 12.2. A possible explanation of this is the following. Neither of these invariances is built in from the start, as a law of nature. But gauge-invariance forces the "clean" part of electroweak interactions to have exact $CP$ and time reversal invariance. The "messy" Higgs interactions have no such constraint. Nevertheless, it turns out that they have to be quite complicated before the violations can show up. Mixing between quarks has to occur, and there need to be three (or more) quark doublets (as there are). From this point of view, the mystery is not that $CP$ violation occurs, but that it is such a small effect. This smallness is explained in part, but only in part, by the smallness of the quark mixings.

Do similar mixing effects occur in the leptons? The answer to this depends upon whether the three neutrinos all have exactly zero masses or not. If they do, there can be no interaction between neutrinos and Higgs particles, and so no mixing. At the time of writing, there is evidence for large mixing between the neutrinos. There are two separate pieces of evidence, one concerning electron neutrinos and the other muon neutrinos.

The Sun emits electron neutrinos. They are inevitably produced in the nuclear fusion that gives the Sun's energy. In fusion, hydrogen (that is, protons and electrons) converts to heavier elements, which contain neutrons. To do this, some protons must combine with electrons, producing neutrons and neutrinos. Solar neutrinos can, with great effort, be detected on Earth. Only about half the expected number are seen. One possible explanation of this is that mixing, during the journey from Sun to Earth, has converted half of the electron neutrinos into muon neutrinos (which would not register in the experiments).

The Earth is bombarded with cosmic rays: high-energy protons and other nuclei. When these particles strike atoms in the upper atmosphere, they can produce pions, which decay (before reaching the ground) into muons and neutrinos. The muons likewise decay into electrons and neutrinos. The result is a mix of energetic muon and electron neutrinos, in the ratio of 2 to 1. Both of these types can be detected. The ratio of muon to electron neutrinos found is not 2, but only about 1.3. This deficiency too can be explained, invoking mixing with tau neutrinos.

If these conclusions are correct, some neutrinos must have some masses, but they need be only very small, say, about $10^{-7}$ of an electron mass. This would mean that the mixing takes place very slowly and so requires the neutrinos to travel large distances before becoming effective. That is why the mixing may be difficult to measure in the laboratory.

If these experiments, and their interpretation, prove to be right, the mixing of neutrinos is dissimilar to the mixing of quarks. For the quarks, there are large mass differences, but rather small mixing; for neutrinos, very small mass differences but large mixing.

## 12.6 GUTs

The heading GUTs stands for *Grand Unified Theories*. Their aim is to unify the electroweak theory with the gluon theory of strong interactions, subsuming them into a bigger gauge theory. In the electroweak gauge theory, particles within doublets (of quarks or of leptons) transform into one another. In the gluon theory, quarks of different colours (in colour triplets) transform into one another. In a GUT, particles within larger multiplets are assumed to transform into one another. For example, there might be a five-fold multiplet, consisting of a colour triplet together with a weak doublet. In this model, there would be 24 "photons", including the 8 gluons, the photon and the $W$ and $Z$ particles, but with 12 extra spin 1 particles, called collectively $X$ *particles*.

In such a GUT, there is a single, common interaction strength. The strong interaction strength, and the two weak ones, must all come from this single number, and so must be related. The relations come out to be

$$\text{gluon strength} = W \text{ strength} = \frac{8}{3} \times (\text{electromagnetic strength}).$$

(The fraction $\frac{8}{3}$ originates from the sum of the squares of the electric charges of all the particles in one generation:

$$1 + 3[(2/3)^2 + (1/3)^2],$$

where the 3 refers to the three different colour quarks.)

The measured values of the three strengths are

$$0.12, \quad 0.03, \quad \frac{8}{3} \times 0.008 \approx 0.02,$$

which are clearly not equal to each other. However, as explained in 11.3, interaction strengths are not really constants: they vary slowly and predictably with the energy of the process under study. (The numbers just quoted are appropriate at the rest energy of the $Z$ particle.) So, at what energy should we test the GUT relation? There must be Higgs particles giving a mass to the new $X$ particles, as well as to the $W$ and $Z$. The mass of the $X$ must be very high, or else its effects should have already been observed. Only at energies above the rest energy of the $X$ is the full symmetry of the GUT revealed. It is found that the three interaction strengths come much closer to satisfying the preceding two GUT relations, if they are worked out at energies of some $10^{15}$ or $10^{16}$ times the rest energy of the proton. So the $X$ must have such an enormous mass. (But there is still a significant discrepancy: see Section 15.2.)

It is an astonishing claim that we can say something about physics at these energies, when experiments (terrestrial ones, at any rate) have not been done above about 1,000 times the rest energy of the proton. How can we make a leap of 12 or so powers of 10? One part of the answer to this question is that the interaction strengths depend on energy only very weakly, logarithmically to be precise. This means that the variation is not proportional to the energy itself, but to the number of powers of 10 in the energy. So we should think not of $10^{16}$ but of 16. From this point of view the leap is not so great. But still, it needs some courage to be confident that no essentially new and unexpected physics remains to be discovered between about $10^3$ and $10^{16}$ proton rest energies.

The Planck energy, the energy at which the (unknown) effects of quantum gravity are expected to become important, is $10^{19}$ proton rest energies. So the GUT energy is not so far from the Planck energy, especially on a logarithmic scale. It should be possible to calculate with GUTs without worrying about quantum gravity, but there is not much to spare.

An important virtue of GUTs is that they explain the *quantization of charge*. So far as is known, the electric charge of the electron is *exactly* equal to three times that of the down quark, and in fact *all* charges are simple whole number multiples of the latter. Within electromagnetism, or even within electroweak theory, this fact receives no explanation. There would be no contradiction if a particle were to exist with charge 1.27 (or any other random decimal) times the charge of the electron. In a GUT, however, all the spin 1 particles (photon, W, Z, X) belong together in a set that mixes within itself under the symmetry of the GUT. This fact ensures that their electric charges are simply related. Then the interactions of the spin 1 particles with lepton and quarks (and Higgs particles) must, to be consistent, also be simply related.

It is not clear how much weight to give to this explanation of the quantization of charge; other explanations may exist.

Probably no experiment done on Earth will ever achieve these GUT energies. But there may be indirect ways in which the GUT idea can be tested. The most striking is the prediction of proton decay. As stated, GUT multiplets contain leptons and quarks together. A quark can change into a lepton if an X particle is emitted. It follows that, with the aid of X particles, quarks can decay into just leptons, and so the proton can likewise decay: protons, in ordinary matter, should not last forever. Because of the huge mass of the $X$, the predicted rate of decay is very slow: protons should live, on average, for something like $10^{32}$ to $10^{34}$ years. Surprisingly, these very low probabilities of decay are on the edge of being measurable: 1,000 tonnes of matter contains about $10^{33}$ protons, and so 1 should decay every month or so. The rate of proton decay to, for example, positron plus pion is known, from past experiments, to be less than 1 every $10^{32}$ years.

Unfortunately, the exact proton decay rate depends upon the details of the GUT model, so the experimental bound on the decay rate does nothing more than rule out certain models.

According to present theories of cosmology, the universe was very hot very early in its history. When its temperature was high enough, some of the effects of GUTs may have been relatively important. It may be, therefore, that it is to astrophysics and cosmology that we must look for the answers to some of the questions (about neutrino masses, proton decay and so on) raised in this chapter.

## 12.7 Conclusion

Electromagnetism is successfully unified with the diverse set of processes called *weak*. A simple principle controls very many interactions. The method of mass generation is understood in principle, but the details remain unexplained. The discovery of a Higgs particle, when and if it is made, will confirm our understanding.

The ambitious, but theoretically appealing, GUT unification awaits experimental tests.

# GRAVITATION PLUS QUANTUM THEORY – STARS AND BLACK HOLES

*Gravity is very weak compared to the other forces of nature, but when enough particles are assembled together, gravity dominates.*

## 13.1 Black Holes

Let us first pick up some ideas from where we left them in Chapter 7.

When, in 1915, Einstein worked out the consequences for the Solar System of his theory of gravity, he used an approximation. He started with the Newtonian predictions and found corrections of relative size about $10^{-8}$, neglecting further corrections of about $10^{-16}$. This approximation is certainly good enough for the Solar System: the neglected terms are much too tiny to be measured.

The German astronomer and physicist Karl Schwarzschild had been serving in the German army on the eastern front during 1915, and there he contracted a fatal skin disease. He died in May 1916. While he was ill, he found the *exact* metric due to the Sun (that is, outside the Sun and assuming the Sun to be exactly spherical). Evidently there was at least one person who quickly understood Einstein's theory.

What is Schwarzschild's metric like? First, we must decide what sort of map of spacetime to use. I will describe the map originally used by Schwarzschild himself, though like any map it has some drawbacks. On the page, we plot time (as defined by a distant

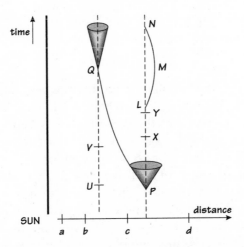

**FIGURE 13.1** A Schwarzschild map of spacetime near the Sun. Time is plotted vertically, and distance from the Sun in one particular direction horizontally (other directions are obtained by imagining the diagram rotated around the world-line of the Sun). The diagram indicates some features of the metric: $UV$ and $XY$ represent equal proper-times, and $ab$, $bc$ and $cd$ represent equal proper-distances. Light cones are shown at $P$ and $Q$. $PQ$ is part of the world-line of a light flash moving directly towards the Sun. $LMN$ is the world-line of a particle projected at $L$ directly away from the Sun and returning to the same spatial position at $N$. This diagram is schematic only and does not have realistic scales.

observer) vertically and "distance" from the Sun horizontally, as in Figure 13.1. The world-line of the centre of the Sun is a vertical line. Because everything is spherically symmetric (that is, the same in all directions), the metric depends only upon time and "distance", and all the information is contained in this two-dimensional diagram. But we can imagine rotating the diagram about the Sun's world-line, thus getting three out of the four dimensions of spacetime. We can think of the transverse directions (directions round the Sun, as opposed to away from it) as being at right angles to the page.

I have put "distance" in quotation marks, because the "distance" used by Schwarzschild is not actually the proper-distance from the Sun. Rather it is defined so that the circumference of a circle round the Sun (say, the length of a circular planetary orbit) is given exactly by

$$2\pi \times (\text{"distance"}).$$

In curved space, this would not be the true circumference if "distance" were the true distance (see Figure 7.1). This way of plotting "distance" in the map has the advantage that transverse distances (at right angles to the diagram) are just what one expects from the diagram, with no distortion to be allowed for.

Now about the metric in the Schwarzschild map. I will be content to point to two features of it. First, take a straight vertical line in the map, like $UV$ or $XY$ (representing bodies somehow held at rest relative to the Sun). How fast does proper-time elapse along such a line? (That is, how would a clock on such a world-line graduate it?) The answer is that proper-time runs slower (compared to the vertical time scale) nearer the Sun. So, for example, $UV$ and $XY$ represent equal increments of proper-time. (This fact in itself is sufficient to give the Newtonian approximation to planetary motion.) Second, how does proper-distance relate to the horizontal "distance" scale on the map? The answer is that it gets less farther from the Sun. For example, $ab$, $bc$ and $cd$ represent equal proper-distances. (We are assuming that the Sun is not changing with time, so the metric is time-independent too.)

These two facts about the metric tell us that the light cone is distorted in the map. Getting nearer the Sun, the light cone stretches in the time direction and squashes in the "distance" direction, as indicated by the two examples at $P$ and $Q$ in the figure. (Nothing happens in the transverse direction.) The behaviour of the light cones will turn out to be the most crucial feature of the Schwarzschild metric (in this map).

In Figure 13.1, the curve $LMN$ shows the world-line geodesic of a particle projected directly outwards at the event $L$ and returning to the same spatial position (at a later time) at the event $N$. According to the action principle (Section 7.5), the geodesic should maximize the proper-time from $L$ to $N$. One can see very roughly how it does this, by steering (in the region of $M$) out farther away from the Sun where proper-time progresses faster (compared to the time plotted vertically in the diagram). Also shown (as the curve

*PQ*) is part of the world-line of a light flash directed towards the Sun. As it is a light flash, the curve must touch the light cone at every event on it.

Figure 13.1 is only schematic. It is not drawn accurately. It wildly exaggerates the distortions in the actual Solar System. Also, it only represents the situation *outside* the surface of the Sun. Inside, the metric is more complicated, depending on the structure of the Sun.

Let us imagine, hypothetically for the moment, that a star similar to the Sun could be compressed to a very small size (less than about 3 kilometres in radius). Then the Schwarzschild metric is applicable down to this very small "distance", and spacetime is strongly distorted. This is illustrated in Figure 13.2. It turns out that there is a critical "distance", called the *Schwarzschild radius*, given by (using units in which the speed of light is 1)

Schwarzschild radius = 2 × (gravitational constant) × (mass)

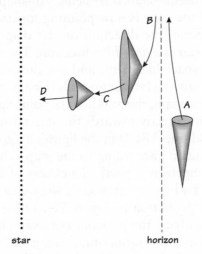

FIGURE 13.2   Schwarzschild map of spacetime near a very small star. Crossing the horizon, the light cone "flips", so that time-like and space-like directions are interchanged. *ABCD* is part of the world-line (a geodesic) of a body falling directly in through the horizon towards the star. Although this world-line "goes off to infinity" near *B* on this map, the proper-time from *A* to *D* is a sensible finite quantity. Once the body is through the horizon, its *future* is *inwards* towards the star.

at which extreme things happen. Here *mass* means the mass of the star. For a mass equal to that of the Sun (about $2 \times 10^{30}$ kilograms), the Schwarzschild radius is about three kilometres.

Approaching the Schwarzschild radius from outside, proper-time slows without limit compared to the vertical time scale on the map, and proper-distance shrinks without limit compared to the horizontal "distance" scale. The spherical surface at the Schwarzschild radius from the Sun has the expected area

$$4\pi \times (\text{Schwarzschild radius})^2.$$

This surface is called the *horizon* (for a reason that will become clear).

Approaching the horizon, the light cone stretches and shrinks without limit, as indicated in Figure 13.2. Passing from just outside the horizon to just inside, the light cone "flips" in direction and goes from being needle-shaped to being fan-shaped, as in Figure 13.2. The axis of the cone is now pointing *towards* the Sun. That is, inside the horizon, the direction on the diagram towards the world-line of the star is a *time-like* direction! The vertical direction in the diagram is space-like. Time and one direction in space have exchanged roles. Moving farther inwards (from $C$ to $D$), the flipped light cone narrows again, the axis always pointing inwards.

If a particle falls directly towards the star through the horizon, as along the world-line $ABCD$ in the figure, it "goes off to infinity and comes back again" according to the map. This extraordinary behaviour at the horizon is partly a sickness of the map used. It is a little like the bad representation of the north and south poles in the cylindrical projection in Figure 7.4. Other maps are more appropriate for studying the horizon. For example, in spite of the look of the diagram, the proper-time along the world-line $ABC$ is finite. That is, a clock carried by the in-falling particle behaves sensibly.

However, the Schwarzschild map does correctly point to one remarkable physical property: once anything has got inside the horizon it cannot get out. Inside the horizon, the inevitable progression of time towards the future means that the particle must continue "inwards" towards the star. The authors (Misner, Thorne and Wheeler) of a well-known textbook are eloquent (insertions in square brackets are my paraphrases):

> The explorer in his jet-powered spaceship prior to arrival at [the horizon] always has the option to turn on his jets and change his motion from [in-fall] to [escape]. Quite the contrary is the situation when he has once allowed himself to fall inside the [horizon]. Then the further motion ["inwards"] represents the passage of time. No command that the traveler can give his jet engine will turn back time. The unseen power of the world which drags everyone forward willy-nilly from age twenty to forty and from forty to eighty also drags the rocket [inwards from the horizon to later points inside]. No human act of will, no engine, no rocket, no force... can make time stand still. As surely as cells die, as surely as the traveler's watch ticks away "the unforgiving minutes," with equal certainty, with never one halt along the way, [the traveler moves inwards inside the horizon].

The body that has fallen through the horizon cannot reverse its path. Neither can it communicate with the world outside. If it sends out a light or radio signal, that must move along the light cones, and so (as Figure 13.2 shows) move inwards too. Signals emitted outside the horizon can move outwards, but those emitted inside can only move farther inwards. Light emitted at the moment of crossing the horizon remains on the horizon, moving neither in nor out.

The horizon and the spacetime inside it are called a *black hole* (the term was coined in 1968 by John Wheeler and has passed into the language, used metaphorically as well as in its technical sense). The idea of a black hole was even advanced two centuries ago for example by the British astronomer John Michell (briefly professor of geology at Cambridge) and by the great French mathematician Laplace: suppose an object could be dense enough so that the escape velocity from it were equal to the velocity of light; then no light could leave it. Of course, a consistent theory of light *and* gravity was required to make the idea definite. Einstein himself tried to argue that black holes could not form.

The Schwarzschild map in Figure 13.2 gives a misleading impression from the point of view of the space traveller falling through the horizon. But it gives a better impression of what an observer from the outside world sees. Such an observer *never* sees the traveller quite cross the horizon: rather, the traveller seems to get ever closer to the horizon, as indicated by the portion $AB$ of the world-line in Figure 13.2. The portion $BCD$ is not observed from the outside world. (Also, less and less light from the spacecraft reaches the outside world, so it appears to dim very fast as it approaches the horizon.)

We are talking about a hypothetical "star" with about the mass of the Sun, but with a radius less than three kilometres. Suppose it was once bigger than three kilometres but contracted to this size. Then the matter of the star itself is in the same situation as the spacecraft that fell through the horizon: it too is moving inescapably inwards. This collapse towards the centre happens very fast – in a few microseconds of proper-time (the time it takes light to go a few kilometres). What happens to the matter of the star when it is crunched into the centre? What happens to the space traveller falling in afterwards? The traveller is certainly soon torn apart. Gravitational tidal forces increase without limit. Mathematicians talk about a *singularity* at the centre. But this is just a word to describe our ignorance. The effects of quantum theory on gravity are usually very tiny, but near the "singularity" these effects certainly become important. Quantum gravity is a very difficult subject, about which little is known. So no one is sure what finally happens to the in-falling matter.

The Schwarzschild metric, on which all this discussion has been based, assumed exact spherical symmetry, that is, that all directions from the centre were exactly equivalent. Could it be that the preceding conclusions depend crucially on this very special assumption? In 1963, a more general solution of Einstein's equations was found by Kerr, corresponding to a rotating black hole. This is more complicated than Schwarzschild's metric, not being rotationally symmetric. Light cones not only get distorted and tipped as in Figure 13.2; they also tip in the direction of rotation. There are a horizon, from which *no* light can escape, but also, outside that, another surface inside which anything (including light) must orbit in the same direction as the black hole is rotating. If something is sent in orbiting the other way, it gets turned round. It is as if space itself were rotating, sweeping everything round with it. There is a maximum angular momentum that a black hole of a given mass can have.

An important characteristic of a black hole is a lower limit to the size of a stable orbit for an object (a "planet") moving in its gravitational field. For a non-rotating black hole, this limit is twice the radius of the event horizon. For a maximally rotating black hole, it is smaller: equal to the radius of the horizon itself. The total energy, including the negative gravitational binding energy, of a body in such an orbit can be as little as $\frac{1}{\sqrt{3}}$ of the its rest energy ($mc^2$). In consequence, rotating black holes can be very efficient machines for converting mass into energy (much more efficient than nuclear fusion, in stars or bombs). Suppose a stream of matter swirls into a black hole. It loses energy by radiation (because the matter will certainly become ionized, that is, consist of electrically charged particles) until it gets to the orbit of minimum size, whereupon it will plunge down through the horizon and be "lost". Thus up to

$$1 - \frac{1}{\sqrt{3}} = 0.42$$

of the rest mass energy may be radiated before that mass itself disappears through the horizon. This is one of the ways in which black holes may be showing up in the universe.

It is also possible, in theory, for a black hole to carry an electric charge, in which case the metric is again different.

It is believed, remarkably, that the spacetime *outside* any collapsed object is *uniquely* determined by just three quantities: the total mass, the total angular momentum (or spin) and the total electric charge (if any). Apart from these three numbers, the spacetime outside contains no traces of the detailed history of the prior collapse. Let me quote an expert (Chandrasekhar in *Truth and Beauty*) on black holes:

> This is the only instance which we have of an exact description of a macroscopic object. Macroscopic objects, as we see them all around us, are governed by a variety of forces, derived from a variety of approximations to a variety of physical theories. In contrast, the only elements in the construction of black holes are our basic concepts of space and time. They are thus, almost by definition, the most perfect macroscopic objects there are in the

universe. And since the general theory of relativity provides a single unique two-parameter family of solutions for their description, they are the simplest objects as well.

But do black holes really form? Can matter collapse sufficiently for this to happen? These are the questions to which we now turn.

## 13.2  Stars, Dwarves and Pulsars

In 1930, a 20-year-old Indian theoretical physicist, Chandrasekhar, returning on a ship to England, realized that any star above a certain mass (about 1.4 times the mass of the Sun) must eventually collapse and form a black hole. This conclusion was hard for people to accept. The prestigious Cambridge astrophysicist Eddington forcefully attacked Chandrasekhar's reasoning, bringing to bear really rather spurious arguments against it – a sad episode in the lives of two great men. Shortly afterwards, Landau in Russia came to the same conclusion as Chandrasekhar, but his work was not known in the West until later. In 1939, Oppenheimer and Snyder in Berkeley, California, studied "gravitational collapse" in more detail.

In order to understand Chandrasekhar's argument, let us briefly describe how stars "work".

Stars are objects for which gravitation is as important as the electromagnetic and other forces that control the structure of atoms. The ratio of the gravitational to the electric force between two protons is very small, about

$$10^{-36}.$$

Nevertheless, unlike electromagnetism, gravity is always attractive, so the smallness of this ratio can be overcome if enough particles are assembled together.

Here is a very rough argument. Take an object containing a large number, $N$, of protons and electrons. Each proton will have electric potential energy due to one or two neighbouring particles (all the others tend to cancel out), but gravitational potential energy due to *all* the other protons. The potential energies go down as the typical distances between particles. In the electric case, this distance is between neighbouring particles. In the gravitational case it is of the order of the total size of the object, that is, $\sqrt[3]{N}$ as great (since $N$ goes as the cube of the size). Thus the ratio of the total gravitational to the total electric potential energy (for all $N$ particles) is, roughly,

$$N \times \frac{1}{\sqrt[3]{N}} \times 10^{-36} = N^{\frac{2}{3}} \times 10^{-36}.$$

This becomes of order 1 for

$$N = 10^{54}.$$

If there are more particles than this, gravity begins to overcome electromagnetism.

The largest planet in the Solar System, Jupiter, does indeed contain about $10^{54}$ protons. It does consist of more or less ordinary atoms. Much bigger, and gravity crunches the atoms up. Something else is needed to support the star. A star similar to the Sun has about $10^{57}$ atoms and needs something else.

That something else is nuclear fusion. The temperature near the centre of the star rises enough so that nuclei can approach near enough (overcoming the electrical repulsion) to fuse, thus releasing nuclear energy, which, escaping through the star, holds it up against further gravitational collapse. About $\frac{2}{3}$ of 1 percent of the rest energy is released in this way. First hydrogen converts into helium, then helium into carbon and oxygen, and then, if the star is big enough, into other heavier nuclei. The fusion stops when iron is reached, since this is the most tightly bound nucleus.

This process has a limited lifetime, about $10^{10}$ years, before the supply of nuclear fuel is exhausted. What happens then depends upon the mass of the star. The star begins to collapse, and so heats up. (It may seem strange that, when nuclear fusion stops, the star at first begins to heat up. This effect is due to the release of gravitational potential energy that the star had when it was larger.) This heat energy is dissipated in radiation and in blowing off outer layers. There are two states in which the remaining core of the star may be left. They each turn out to be "cold", in the sense that their temperatures are unimportant compared to the other energies involved.

The first type of "cold" star is a *white dwarf*, in which the atoms have been crushed too close together to retain their identity. What remains is a fluid of atomic nuclei and electrons. This is prevented from collapsing by, simply, the exclusion principle (Section 8.13) operating between the electrons (the exclusion principle for the nuclei is not important in this case). The electrons are rather like the conduction electron in a metal (see Section 10.5). They fill all available quantum states up to a certain energy, the Fermi energy. Expressed in terms of an equivalent temperature, this is in the region of $10^9$ K. (The star is "cold" because its temperature is much less than this.)

Is the Fermi motion sufficient to support the star against gravity? To answer this, we work out how the Fermi energy depends upon the radius, $R$, of the star. The uncertainty principle relates the momentum of an electron to the dimensions of the space available to it. So the momentum goes as $R^{-1}$. If the electron is moving slowly compared to the speed of light, its energy is proportional to the square of the momentum, and so the Fermi energy depends upon the radius as $R^{-2}$. Gravitational potential energy, on the other hand, is negative and goes as $-R^{-1}$. Gravitation tends to cause the star to collapse. The Fermi motion gives rise to pressure, tending to resist the collapse. For small enough $R$, the Fermi pressure must dominate and prevent collapse.

Stars with masses similar to the Sun's can settle down as stable white dwarves in this way, with a radius some hundredth of the Sun's (that is, similar to Earth's). The first such object was observed in 1914 by W. S. Adams, as one component of the double star Sirius. The explanation was given by the Cambridge theoretical physicist Fowler in 1926, in the same year that Fermi and Dirac worked out the proper formulation of the exclusion principle for electrons (see Section 9.5).

But the qualification I have slipped in, "for electrons moving slowly compared to light", was the point seized upon by Chandrasekhar in 1930. For a particle moving with nearly the speed of light, the energy is approximately $c$ times the momentum, rather than being proportional to the square of the momentum. The preceding argument shows that the Fermi energy then depends upon the radius as $R^{-1}$ (instead of $R^{-2}$), that is, the same dependence as the gravitational potential energy. In this case, there is no simple argument to fix the value of the radius $R$. But the maximum possible number of particles $N$ is now determined by making the total energy, Fermi plus gravitational (the latter being negative) zero, giving

$$\hbar c N^{-\frac{1}{3}} \approx GN(\text{proton mass})^2$$

(where $G$ is the constant of gravitation), that is, a maximum value for $N$ of, roughly,

$$\left(\frac{\hbar c}{G(\text{proton mass})^2}\right)^{\frac{3}{4}} \approx 10^{57},$$

a little more than the number of protons and neutrons in the Sun. This is the *Chandrasekhar limit* on the number of electrons in a white dwarf. Beyond this number, gravity overwhelms the exclusion principle.

The quantity

$$\sqrt{\frac{\hbar c}{G}} \approx 2 \times 10^{-8} \text{ kilogram}$$

is the Planck mass. It is the quantity with the dimensions of mass that one can make out of the three fundamental constants of nature $\hbar$, $c$, $G$. The Chandrasekhar limit is given by the three-half power of the ratio of the Planck mass to mass of the proton.

The essential ingredients of Chandrasekhar's argument are just four: Newton's inverse-square law of gravity (Einstein's theory of gravity makes little difference in this case), Einstein's connection between energy and momentum, the uncertainty relation between position and momentum and the exclusion principle for electrons as stated by Pauli, Fermi and Dirac. The 20-year-old Chandrasekhar certainly understood all these things.

It is ironic that Eddington was fascinated by the large numbers appearing in physics but failed to see the most important implication of the numbers just mentioned.

I have mentioned two large numbers: the upper limits to the number of particles in a planet and in a white dwarf star. The ratio of these two is determined by yet another dimensionless number

$$\left(\frac{e^2}{\hbar c}\right)^{\frac{3}{2}} \approx \left(\frac{1}{137}\right)^{\frac{3}{2}},$$

where $e$ is the charge on the electron (see Section 9.7).

But nature can make more compact objects than white dwarves. These are neutron stars, made predominantly of neutrons, a bit like enormous and very neutron-rich atomic nuclei. Neutrons are fermions, so the exclusion principle tends to prevent collapse, just as it does for the electrons in white dwarves.

Whereas white dwarves have sizes of a few thousand kilometres, neutron stars have sizes in the region of 10 kilometres. The reason for this is that, for a given momentum, the energy of a particle is inversely proportional to its mass (so long as it is moving slowly compared to the speed of light). Therefore, for a given size star, the Fermi energy of a neutron is some 2,000 times less than an electron. So, for the gravitational energy to reach the Fermi energy, a neutron star must be smaller.

But Chandrasekhar's limit applies to neutron stars just as much as to white dwarves, once the neutrons' speed approaches that of light. The maximum number of neutrons is about the same. In fact, the strong nuclear forces between neutrons ought to be taken into account as well as the exclusion principle, so the actual upper limit on the mass of a neutron star is not known exactly; there is good reason to believe this limit to be less than 6 solar masses.

The gravitation near a neutron star is enormous. The gravitational potential energy of a particle on the surface can be 10 percent or more of its rest energy. To escape, it would have to be accelerated to about half the speed of light.

Neutron stars were predicted in 1939 by Oppenheimer and Volkov. What was not predicted was the dramatic manner in which they would be found. In 1968, a group using a radio telescope in Cambridge discovered the first "pulsar". Pulsars are sources of radiation, flashing very regularly, and fast, from less than 1 to 1,000 times per second. The rate can be so regular that it can sometimes be measured with an accuracy of 1 part in $10^{12}$. (But the rates do slow down very gradually.)

In spite of the name *pulsar*, the flashing is not due to pulsation but to rotation. A neutron star can rotate very fast without breaking up, and thus account for these short periods. The star has an intense magnetic field, with the magnetic axis not coinciding with the axis of rotation. Electrically charged matter falling into the neutron star is deflected towards the magnetic poles, and it produces intense radiation concentrated near the direction of the magnetic axis. As the star spins, the beam of radiation is swept round, like a lighthouse beam, so that an observer on Earth sees the intensity vary periodically.

How are neutron stars born? The story is believed to be as follows. A large star undergoes nuclear fusion for some 10 million years, until it has developed a core of iron (and nickel) that can undergo no more fusion. The core collapses in less than a second. Electrons and protons in the core are removed by the process

$$\text{electron} + \text{proton} \rightarrow \text{neutrino} + \text{neutron}.$$

This process needs energy to be available (since the neutron is heavier than the combined masses of proton and electron), but there is an enormous amount of gravitational energy available from the collapse. The neutrinos from this and other reactions stream out through the core and the outer layers of the star. Because neutrinos interact so weakly with matter, they remove energy very efficiently. Nevertheless, they undergo some reactions in the star and help to blow off the outer layers, at speeds of tens of thousands of kilometres per second, in a great explosion called a *supernova*, very bright at first but quickly fading. The nuclei of atoms such as carbon and oxygen (needed for life) are scattered into space. The tiny, spinning neutron star, perhaps a tenth of the mass of the original, remains behind.

The sudden appearance of supernovas, very bright new objects in the sky that flare up suddenly and begin to fade in a few weeks, has been observed for hundreds of years. Chinese astronomers saw such a "guest star" in 1054 A.D., and its remnants (a diffuse, glowing gas cloud) are the Crab Nebula. Tycho Brahe observed a supernova in 1572. In 1987 a supernova was seen (170,000 light years away) in a galaxy close to the Milky Way. Of the $10^{58}$ neutrinos released, $10^{15}$ passed through each square metre at the Earth. Neutrino detectors (which by a fortunate chance had recently been constructed) caught 22 of these, so slight is their interaction with matter.

Can any myth have described a birth more wonderful than the true birth of a neutron star?

### 13.3   Unleashing Gravity's Power: Black Holes at Large

So the only stable "stars" are objects like Jupiter, made of ordinary matter, and white dwarves and neutron stars supported against collapse by the exclusion principle. Anything heavier than the Chandrasekhar limit (not much more than the Sun's mass) must collapse to form a black hole. Is there evidence for any such thing? If there is matter available to be sucked into a black hole, we expect the production of enormous amounts of energy, as mentioned, up to 40 percent of the rest mass of the in-falling matter. The matter will attain very high temperatures and emit a wide range of electromagnetic radiation. The signatures of black hole activity are expected to be high intensity, high temperatures, fast variations and smallness of source.

Many astronomers now believe that there is good evidence for black holes. Some of it has come from the new satellite-born X-ray and gamma ray telescopes. There are, in our galaxy, binary stars emitting X-rays, in which the heavier component (the black hole?) has several solar masses and is thought to be accreting matter from its smaller companion.

Quasar is an acronym for "quasi-stellar radio source"; however, quasars are now known to emit electromagnetic radiation of all kinds. The first quasar was discovered in 1962. They are now known to be outside our galaxy, billions of light years away. They emit a prodigious amount of energy, hundreds of times as much as the whole of our own galaxy. Their brightness can vary in days. This implies that the size of the source cannot be more than a few light days, say, about the size of the Solar System. These may be black holes with masses of millions or even billions of suns. They may be accreting matter from a galaxy in which they sit or from another colliding galaxy.

Even more startling than quasars are "gamma-ray bursts", first discovered in the 1960s by an American satellite intended to monitor nuclear weapons tests in space. The bursts may last for only a few seconds. Since their distribution in the sky is uniform, they almost certainly originate outside our galaxy. They may be caused by the coalescence of two neutron stars to form a black hole, or by accretion of matter into a black hole from an orbiting star. In the latter case, the period of the orbit can show that the central object is too heavy to be anything else but a black hole.

In each these examples, the black hole is far from black. It is pouring out radiation. The reason is that the system in question is not a black hole on its own: it is accompanied by a supply of matter for it to feed on. The radiation is from the gravitational energy released as this matter falls towards the black hole.

## 13.4　The Crack in Gravity's Armour

Several times in the history of physics progress has been made by probing the interface between two subjects that sat uneasily together. Planck initiated quantum theory by combining statistical physics with electromagnetism (Section 8.1). Einstein discovered his special theory of relativity (in part at least) in order to reconcile Newtonian dynamics with electromagnetism.

In 1974, Stephen Hawking put his finger on a crucial juncture between quantum theory and Einstein's theory of gravity, two theories that had hitherto seemed disjoint. The implications of Hawking's insight are by no means worked out, but surely it will prove to be one of the keys to a future theory in which gravity is reconciled with quantum theory and with the other forces of nature.

Hawking's discovery was that a black hole is not black: it glows. What is more, it glows in a very definite and simple way. Each black hole has a definite temperature associated with it, and the radiation emerging is just the same as the radiation leaving an oven at that temperature (see Section 8.1). It is distributed over frequencies according to the Planck distribution, as in Figure 8.1. It is amusing that the thermal radiation from an oven is sometimes called "blackbody" radiation, because it is assumed that, if the oven were cold, it would absorb *all* radiation incident on it, not reflecting any. So black holes, appropriately, emit black-body radiation.

There are several different ways by which people have argued for the existence of Hawking radiation. The overall picture is so nice, and consistent, that experts are convinced. I will give a simple, but by no means rigorous argument that suggests why Hawking radiation should exist. The argument brings together two elements: from quantum theory, the uncertainty principle (Section 8.4), and, from Einstein's theory of gravity, the way in which particles (or light) move in the curved spacetime of a black hole.

We have seen that a particle that falls through the horizon of a black hole continues on an orbit that takes it inexorably towards the centre. There are also possible orbits that lie entirely *within* the horizon, which also move in to the centre. A particle in one of these orbits has a total energy (as measured by an observer nearby) that is *negative*. This just means that the gravitational potential energy, which is as usual negative, out-balances the (positive) kinetic energy and mass-energy ($mc^2$). We, from outside the horizon, could never see a particle in one of these negative energy orbits.

Heisenberg's uncertainty principle implies that, if we specify the time rather precisely, then the energy must be uncertain. This means that, for short time intervals, the conservation of energy does not forbid the appearance of extra particles, so the vacuum flickers with evanescent particles (Section 9.10). Suppose a pair of such particles – photons, let us say – appears near a black hole, one just outside the horizon on an escaping orbit (with positive energy) and one just inside the horizon in one of the negative energy orbits. The positive and negative energies may very nearly cancel each other out so that the total energy is as near to zero as we please. In this case, the uncertainty principle allows this pair of photons to persist as long as we wish. The eventual effect is that the outside photon is far away from the black hole, whereas the inside photon, by virtue of its negative energy, reduces the energy (and so the mass) of the black hole.

This argument as it stands gives us no idea of the typical energy of the photons radiated. But there is another argument that suggests what this should be. For this, we must remember that photons can be interpreted as quantized waves in the electromagnetic field

(Section 9.2). If the preceding argument were translated into the wave picture, what sort of wavelength would we guess to be important? The simplest answer to this question is something of the same sort of size as the black hole horizon itself, that is,

$$\text{wavelength} \approx \frac{G \times (\text{mass})}{c^2}.$$

This guess turns out to be right.

Unfortunately, these simple arguments do not tell in detail the energy distribution of the Hawking radiation. But more complete mathematical arguments show that the radiation should have all the properties of thermal black-body radiation, with a distribution given precisely by the Planck curve in Figure 8.1. The "temperature" defining the Hawking radiation is

$$\text{temperature} = \frac{8\pi\hbar c^3}{G \times (\text{mass})}.$$

Since the radiation is an effect of quantum theory, we expect the temperature to be proportional to $\hbar$. Given that it depends also only upon $c$, $G$ and the mass, the preceding formula is forced, except the number $8\pi$.

To me, it is amazing how the (quite simple) *geometry* of Schwarzschild spacetime leads to the same Planck distribution previously deduced from the *random* distribution of energy amongst different modes of motion.

When a pair of photons is momentarily formed as a quantum fluctuation, the two particles are in a single quantum state – they are *entangled*, in the language of Section 8.11. But if one of the particles is lost to us behind the horizon of a black hole, we have only partial information about the other. This is presumably why the description of the Hawking radiation is of a probabilistic nature, couched in the language of heat.

Because the wavelength of Hawking radiation is comparable to the size of the black hole, it has no meaning to say exactly where the radiation comes from, any more than it has to say whereabouts on a bell the sound comes from. The black-body radiation is what should be detected by an observer far from the black hole and at rest relative to it. What happens nearer the black hole is a subtler question.

The energy emitted in the Hawking radiation is compensated for by a decrease in the mass of the black hole. Black holes "evaporate".

Take a black hole of mass similar to that of the Sun's. The wavelength at the peak of the Hawking radiation distribution is a few kilometres. The temperature is tiny, about $10^{-7}$ K, and so is the rate of "evaporation".

Clearly, the Hawking radiation in this case could never be observed. For any possibility of an observable effect, we have to assume that very small black holes exist. These cannot be formed by collapse of stars, but they might possibly be left over from early stages of the universe. No radiation from such hypothetical black holes has been identified, but this is not to say that it does not exist at some level.

As an example, a black hole of $10^{12}$ kilograms ($10^{-18}$ of the Sun) would have a radius of $10^{-15}$ metre and a temperature of $10^{11}$ K. It would totally "evaporate" in about $10^{10}$ years, the age of the universe.

But we must be careful. The argument for Hawking radiation presupposes a *fixed classical* Schwarzschild spacetime, then analyzes the quantization of the electromagnetic field in that background. As the black hole evaporates faster and faster, there comes a time when the mass is changing as fast as the field is oscillating. Then the approximate analysis totally fails. A proper account would require a quantum treatment of the changing spacetime as well as of electromagnetism. No one yet knows how to do this. The final fate of an evaporating black hole is a mystery.

When does a black hole enter this unknown region? There is only one mass we can construct from $\hbar$, $c$ and $G$, that is, the Planck mass

$$\sqrt{\frac{\hbar c}{G}} \approx 2 \times 10^{-8} \text{ kilogram.}$$

A black hole of this mass has a size of order $10^{-35}$ metre. When a black hole gets as light and small as this, we do not know what happens next.

I have been talking as if Hawking radiation consisted just of photons. In fact the argument applies to any quantized field. Neutrinos and gravitons must also be radiated. If the temperature is high enough (that is, the black hole small enough), electrons and positrons and other particles with mass should be emitted too.

Rotating black holes emit Hawking radiation too. For a given mass, the temperature is less, and it is zero when the spin is maximal.

## 13.5  Black Hole Entropy: Gravity and Thermodynamics

So black holes are like hot bodies in having a temperature. Even more remarkably, a complete thermodynamics of black holes exists, just like the thermodynamics of ordinary heat. The key property is *entropy*. It turns out that a simple geometrical property of a black hole can be identified with its entropy. That is the area of the horizon. More precisely,

$$\text{entropy} = \frac{1}{4}\frac{c^3}{\hbar G} \times (\text{area}).$$

For example, the entropy of a solar mass black hole is about $10^{76}$. (Note that in this formula for black hole entropy there is no logarithm, as there is in some of the equations in Section 2.7.)

Entropy should never decrease. Here are two examples. Suppose two black holes merge to form a single one. There is a geometrical argument to show that the area of the final black hole is at least as big as the sum of the areas of the two original ones. (Here the black holes are assumed big enough that Hawking radiation is negligible.) When a black hole does emit Hawking radiation, its area, and so its entropy, decreases. This is compensated for by the entropy of the emitted radiation.

Here again we encounter a wonderful correspondence between purely geometrical things, on the one hand, and concepts concerned with randomness, on the other. But entropy may give the key to understanding this. Entropy is a measure of missing information. If a black hole is formed by the collapse of a star, we do seem to lose a lot of information. All the intricate information about the structure of the star – its composition, distribution of density and so on – has gone, swallowed up beneath the horizon (at least as far as we observers outside the horizon are concerned). The only information left is a mere pair of numbers: the mass and the spin.

Is this information lost forever? Or is it stored inside the horizon somehow? Hawking radiation seems not to bring any information out again. The radiation has a random thermal distribution, characterized by just one quantity, the temperature. Two black holes formed from quite different collapsing stars would (if their masses and spins were the same) give the same Hawking radiation. But no one knows what happens in the last stages of evaporation. Is the information finally released?

## 13.6 Quantum Gravity: The Big Challenge

Hawking's inspiration about black-hole evaporation told us a little about what is to be expected of a union of quantum theory and gravitation, but a complete theory of quantum gravity is still missing. The situation is a little like that in the early 1900s, when Planck had combined statistical physics with radiation theory, but a complete quantum theory of radiation (or even particles) was not yet developed.

Given any classical dynamical system, there exist methods for trying to make it into a quantum theory, methods for "quantizing" it. One such method, used in this book, is Feynman's sum over histories (see Section 8.8). What happens if we try to apply this to Einstein's theory of gravity?

First, recall how the method works in the quantum theory of particles and in quantum field theory. For particles, we first of all associate a phase factor (calculated from the classical action) for each history of the particles. Then we sum this phase factor over all histories that start from some given position and end at another. This gives the quantum amplitude for going from the first position to the second, from which the rate of change of a wave function can be found.

In quantum field theory, we do something similar, except that particle position is replaced by a state of the field.

To apply this method to gravity, we first have to decide what is the analogue of a particle position or of a state of the field. The answer is, a slice through spacetime (at some given time, if you like). This slice is a three-dimensional space, curved in general. Call this briefly a *3-space*. So we will be concerned with the history of a

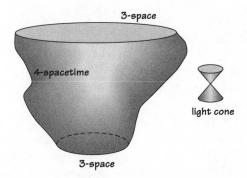

FIGURE 13.3  Feynman's sum over histories applied to quantum gravity. For the purpose of drawing the figure, two space dimensions are omitted, and it is assumed that everything can be embedded in a flat Minkowski spacetime (three-dimensional in the figure). A light cone in this flat spacetime is shown. We seek the amplitude for going from an initial 3-space (represented by the lower curve in the figure) to another (the upper curve). One needs all 4-spacetimes (represented by the surface in the figure) connecting the two 3-spaces.

3-space, deforming from one given 3-space to another. Such a history is nothing but a 4-spacetime, with the first 3-space as one boundary and the second as another. This is illustrated in Figure 13.3

The amplitude for going from one 3-space to the other is given by summing a phase factor (calculated from the action of classical gravity theory) over all 4-spacetimes with the two 3-spaces as boundaries. So far, so good, but what are the snags?

First of all, there is a surprising feature of quantum gravity, which takes a bit of getting used to. This is that the amplitude just defined does *not* depend upon any times. The reason is that, in order to calculate the time between the two 3-spaces, one would need some fixed spacetime to do it in. But we sum over all spacetimes, so we automatically sum over all time differences too. In the same way, a wave function in quantum gravity is something that depends just upon a 3-space, not upon any time variable.

We are used to having a spacetime as an arena in which particles move, fields change and so on. But in quantum gravity, spacetime itself is subject to quantum uncertainties. If one could do quantum gravity successfully, some classical spacetime should emerge as a very good approximation in ordinary situations (but not at scales comparable to the Planck length).

The second problem arises when we try to define precisely the glib phrase "sum over all 4-spacetimes". What geometries are allowed? Should we include spacetimes with holes in? With twists in? Joined up in funny ways?

Suppose we confine ourselves to the most innocent sort of space-times. They can still be varied continuously. How do we enumerate them all? What would one mean by enumerating all possible shapes of a cup? What do we mean by *sum* in such circumstances? In practice, we must approximate the sum by a discrete set of "representative" spacetimes. We can try to make this approximation better and better by choosing more and more of these representative geometries. But how to choose a representative sample? QCD, and other gauge theories, actually have similar problems to these, but they are much less acute.

The problem of carrying out Feynman's sum over histories has a more technical aspect. The histories each have phase factors associated with them, complex numbers each of length 1 but with varying phase angles. What is meant by the sum of an indefinite number of such quantities? A simple model of the sort of thing that might occur is the endless series

$$1 - 1 + 1 - 1 + 1 - 1 + \cdots$$

If you stop this series anywhere, you get either 1 or 0, depending on where you stop. So the endless series appears to be ambiguous. But if you modify it slightly to, for instance,

$$1 - (.99) + (.99)^2 - (.99)^3 + \cdots,$$

then the sum is unambiguously $1/1.99$: by adding enough terms of the series you can get as near to this value as you please (to get to within 1 percent, you need to add about 400 terms). The point about the number 0.99 is just that it is less than 1.

One can try to play a similar mathematical trick to do the Feynman sum over histories. In ordinary quantum theory, and in quantum field theory, there is a clear way to do this. In quantum gravity, on the other hand, no one (to my knowledge) understands how to perform the trick in a reliable way.

Even if we were confident about the answers to all these questions, it would still be very difficult to calculate with quantum gravity in practice. In quantum field theory (electroweak and QCD, for example), a reasonably good approximation method is available (see Section 9.7), depending upon the smallness of dimensionless constants like $\frac{1}{137}$ in QED. Quantum gravity has no such dimensionless number. The strength of the interactions is determined by the gravitational constant

$$G = 6.67 \times 10^{-11} \text{ (metre)}^3 \text{ (kilogram)}^{-1} \text{ (second)}^{-2}.$$

It is impossible to construct a dimensionless constant by combining $G$ with the two fundamental constants $\hbar$ and $c$. One can construct, for example, the Planck length

$$\sqrt{\frac{\hbar G}{c^3}} = 1.62 \times 10^{-35} \text{ metre}.$$

Because this is so small, gravity is normally a very weak force between individual particles, but at very high energies (or very short distances) it inevitably becomes strong. In fact, in just those circumstances when quantum gravity is interesting, gravity is strong and calculations are very difficult. These circumstances include the "singularity" at the centre of a black hole and the very early universe.

Ordinary field theories are almost entirely insensitive to what goes on at very high energies or very small distances. In fact, the coupling strengths in gauge theories actually decrease slowly as energy increases. Quantum gravity behaves oppositely: the strength increases with energy, and one can no longer evade the problems of high energies and short distances.

There are two possible conclusions that might be drawn from the difficulties of quantum gravity. One is that our physics is correct (and complete), but that we are just not clever enough at calculating with it. The other is that some new physical theory is needed – a theory that of course must agree with Einstein's in situations in which quantum effects are negligible. I think that the majority of theoretical physicists at the present time favour the latter conclusion (see Section 15.3).

There is just one thing about quantum gravity which can be said with confidence: there should be *gravitons*, that is, massless particles that are the quanta of (weak) gravitational waves. Classically, a gravitational wave is a small ripple in an otherwise almost flat spacetime. These waves are expected to be excited by motions of (large) masses. In electromagnetism, a single electric charge oscillating back and forth is sufficient to emit waves. There is nothing directly analogous to this for gravity: a single mass cannot move back and forth without another mass to balance it. But two masses moving alternately together and apart, or rotating round their centre of mass, do excite gravitational waves.

There is indirect evidence for the existence of gravitational waves. A binary pulsar (a pulsar closely orbiting a companion star) is known whose orbital period (of about eight hours) has been observed to decrease by 2.7 parts per billion ($10^9$) per year. When orbits speed up, they shrink and their total energy (kinetic plus potential) decreases. The effect, then, is attributed to the emission of energy in gravitational waves.

If a gravitational wave is incident on a ball of matter, it tends to deform into an ellipsoidal shape. The orientation of this ellipsoid depends upon the polarization of the wave. There are two independent polarizations possible. If these are chosen to be circular polarizations, the ellipsoidal shape rotates either clockwise or anticlockwise (about the direction of propagation of the wave).

Devices have been, and are being, built to attempt to detect gravitational waves from extreme astrophysical events, but enormous sensitivity is required, and also large size is needed to detect low-frequency waves.

There is little doubt that quantum theory demands that gravitational waves occur in discrete quanta, gravitons, analogously to the photons that are quanta of electromagnetic radiation. Because of the more complicated nature of the polarization, gravitons have spin $\pm 2\hbar$, as compared to the spin $\pm \hbar$ of photons. Circularly polarized gravitons have spin aligned either in the direction of motion or oppositely to it.

Gravity waves from astronomical sources would have quite low frequencies, so the energy of a single graviton (given by $h$ times frequency) would be minute. There seems no prospect of observing a single graviton.

## 13.7  Something from Nothing

If one had a theory of quantum gravity, the most ambitious thing to try to do would be to find "the wave function of the universe" (this phrase is the title of a well-known paper by Hartle and Hawking). Feynman's principle, as illustrated in Figure 13.3, relates the values of wave function for two different 3-spaces, but it does not determine a wave function uniquely. In ordinary applications of quantum theory, this is to be expected. There is no unique wave function of a hydrogen atom: it can exist in many different states. Given the wave function at some initial time, Feynman's principle determines how it changes with time.

But presumably there is a unique "wave function of the universe". Can Feynman's principle be adapted to determine it? It has been proposed, for instance by Hartle and Hawking, that the principle should be as illustrated in Figure 13.4. Here there is a 4-spacetime with only *one* boundary, a 3-space. For the purpose of drawing

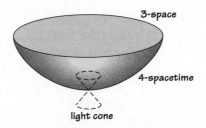

FIGURE 13.4   The Hartle-Hawking proposal to find the "wave function of the universe". The wave function depends upon the variable 3-space. The dependence is obtained by summing a phase factor over all 4-spacetimes each with that 3-space as its single boundary. In this figure, two spatial dimensions are omitted, so that the 3-space becomes a one-dimensional curve and the 4-spacetimes becomes two-dimensional surfaces. It is also assumed (for the purposes of the figure) that everything can be embedded in Minkowski spacetime.

the figure, two spatial dimensions are omitted, and it is also assumed that everything can be embedded in a higher-dimensional, flat Minkowski space.

There is a crucial point about this proposal. To see what it is, I have drawn the light cone at one point in the Minkowski space. The surface representing the 4-spacetime lies entirely outside this light cone; that is, it is a *space-like* surface. Light could not propagate in it. It is not really a 4-*spacetime* at all, but a 4-*space*. Hartle and Hawking argue that it is a correct use of the Feynman principle, in this case, to sum over such 4-spaces, even though they are not directly physical.

The problems involved in this programme are daunting indeed. In addition to the questions of the right quantum gravity and the right use of Feynman's principle, the mystery of the interpretation of quantum theory (see Section 8.11) is most acute when talking about the universe as a whole. What does one do with the "wave function of the universe" if one knows it? Does it tell us the probability that we should find ourselves living in a particular universe with given properties and parameters?

## 13.8 Conclusion

When enough particles are assembled together (as in a star), account must be taken of gravity as well as the other forces of nature. Above the Chandrasekhar limit, gravity is irresistible, and collapse to a black hole takes place. In the presence of a black hole, quantum theory predicts the emission of Hawking radiation, characterized by a definite temperature associated with the black hole. The final stages of gravitational collapse, as well as the very early stages of the universe, require a complete quantum theory of gravity, which is not at present understood with any certainty. Nor is the entropy of a black hole (the information "lost" at its birth) well understood. Many people expect that string theory (Section 15.3) offers the most promising approach to these profound problems.

# 14

# PARTICLES, SYMMETRIES AND THE UNIVERSE

Modern cosmology envisages a universe that, at its early moments, was very hot and very small. In these extreme conditions, experience on Earth gives little guide to the behaviour of matter. We must turn to elementary particle physics in an attempt to divine what matter was like. Conversely, as mentioned at the end of the Chapter 12, particle physics has raised questions that are probably beyond laboratory experiments to answer. Maybe evidence from the universe at large can help.

There are two general features of modern particle physics that make application to the early universe particularly interesting.

- Both the gluon theory of strong interactions (Chapter 11) and the electroweak theory (Chapter 12) employ some analogies with superconductivity (Chapter 10). As the temperature of a superconducting substance is raised, there is a phase transition above which superconductivity ceases. There are good reasons to believe that something similar should happen to the vacuum of particle physics. At the high temperatures of the early universe, the Bose condensation of Higgs particles should go away, revealing the gauge symmetry in an unbroken form (as in Section 12.4). Likewise, confinement of colour should go, and the gluons behave similarly to photons. Thus the physics of the early universe may have been, in some sense, simpler than physics now.

- In non-Abelian gauge theories, coupling strengths decrease as the typical energies explored increase (Section 11.3). In the early universe, the temperature is the relevant energy, so very early, when the temperature is very high, the particles may behave nearly like *free* ones. Then, in spite of the extreme conditions, physics may not be as difficult as might have been supposed.

## 14.1 Cosmology

Up to the time of Galileo and Kepler, most people believed that the universe was limited in size, with the fixed stars on a sphere not much bigger than the Solar System. Descartes, on the other hand, believed in a boundless universe. Newton (by comparing the brightness of stars with that of the Sun) estimated the distance of Sirius to be a million times the size of the Earth's orbit (an overestimate). Ever since the number and distance of stars were appreciated, people seem naturally to have contemplated an unchanging, boundless, uniformly star-filled universe.

But there is a crucial objection to this picture, which was put to Newton and Halley by a physician, William Stukely, in 1721 (when Stukely was 34 and Sir Isaac was 79). This objection is known as *Olber's paradox*, after another physician, a German, who wrote over a hundred years later. The paradox is this: where does all the starlight go? Imagine a big box anywhere in a boundless, uniform, unchanging universe. Through any face of the box, an equal amount of radiation must stream one way as the other. But light is continuously being emitted by the stars inside the box. So the net effect is for the density of radiation inside to build up. Space must eventually be full of radiation at the same temperature as the surfaces of the stars, as in a white-hot furnace (at which stage, stars would absorb as much radiation as they emitted). Fortunately this is not the case; therefore at least one of the assumptions made must be wrong. Modern cosmology escapes from Olber's paradox by denying that the universe is unchanging.

During the 1920s, astronomers realized that many of the nebulae observed are galaxies, assemblies of billions of stars, together with clouds of dust and gas, like our own Milky Way. Observations made especially by Edwin Hubble, working with the great telescopes at Mount Wilson and Palomar Mountain, revealed that the galaxies are receding from us with speeds roughly proportional to their distance. The constant of proportionality is usually called *Hubble's constant* $H_0$, though *constant* is not a good epithet because the expansion rate probably changes with time. Its value is

$$H_0 = 15 \text{ to } 30 \text{ kilometres per second per } 10^6 \text{ light years}$$

(but this number is not very accurately known). Thus the universe is not static, so it evades Olber's paradox: we are indeed bathed in radiation from earlier times, but the expansion of the universe stretches its wavelength and thus cools it down (from thousands of degrees to a few degrees).

At first sight, Hubble's discovery may sound as if it gives a special position to us on Earth: everything is receding from *us*. But any other observer in the universe would see the same thing. All the galaxies are receding from each other, like the currants in a cake that is rising during baking.

There is no reason why the behaviour of such an "expanding universe" should not have been studied by using Newtonian dynamics and gravitation: Newton himself would not have found that hard. But, to my knowledge, no one contemplated anything except static cosmologies until after Einstein had derived his theory of gravitation in 1915.

The recession of the galaxies was discovered by virtue of the *red shift* (that is, increase in wavelength and decrease in frequency) of light from distant galaxies. This may be interpreted (in part, at least) as a Doppler shift, analogous to the fall in pitch of the whistle of a departing train.

From Hubble's constant and the speed of light, we can construct a distance

$$\frac{c}{H_0} \approx 10^{10} \text{ light years} \approx 10^{26} \text{ metres}$$

that we expect to be characteristic of the scale of the universe. Also, there is a time

$$\frac{1}{H_0} \approx 10^{10} \text{ years}$$

that is representative of the time scale of the universe.

In 1917 (a decade before the discovery of the recession of the galaxies), only two years after completing his theory of gravity,

Einstein applied that theory to the universe as a whole. He assumed the universe should be static and showed that this contradicted gravity theory unless (as he proposed at that time) it was modified in some way. Indeed, Newton had realized that a static, uniform distribution of stars was unstable against gravitational collapse.

But Einstein's theory of gravity opens up possibilities for cosmology that Newton could not have dreamed of. Space (or spacetime) as a whole can be curved. Space can be curved so as to close up on itself, like a spherical surface but generalized to be three-dimensional instead of two-dimensional. Of course, when we think of a spherical surface, such as a balloon, we visualize it "embedded" in ordinary three-dimensional space. To generalize this to one higher dimension, we might like to imagine a three-dimensional space embedded in some four-dimensional space. No one can really "see" this: it is enough to realize that it is a mathematical possibility.

A spatially closed universe is an attractive solution to some of the philosophical dilemmas of cosmology. It has no boundary (any more than the Earth's surface has a boundary), yet it is finite. We do not have to worry about some improbable "edge of space", but neither do we have to strain our minds thinking about infinity. However that may be, it is not yet known whether space is closed or not.

In 1922, in Leningrad, Alexander Friedmann pointed out that Einstein's equations had solutions in which the universe was expanding: the size of the closed spherical space increasing with time. In 1927, in Louvain, Abbé Georges Lemaître rediscovered Friedmann's solutions but went further to link the expansion of space to recession of the galaxies and to claim that the universe must initially have been very small. This cosmology was later termed by Fred Hoyle the "big bang". Hoyle meant this disparagingly, but the name has stuck.

In order to do anything so bold as to study the universe as a whole, one must make some simplifying assumptions: usually that the universe is, on a large enough scale, *homogeneous* and *isotropic*: that is, the same everywhere and the same in all directions. These properties can hold in some statistical sense only. Galaxies occur in clusters of sizes of over a million light years, the clusters occur in superclusters of sizes of over 10 million light years, and there may be larger structures too. By comparison, the size of the observable universe is about 10,000 million light years.

On these assumptions, the large-scale geometry of spacetime is specified by just one quantity that varies with time. That quantity is the radius of the universe, which I will denote by $R$, or, because it depends upon time, by $R(t)$. The fractional rate of increase of $R$ with time (that is, the rate of increase divided by $R$) is called $H$. The value of $H$ at the present time is Hubble's $H_0$. The dynamics of the universe is governed by a single equation of Einstein's, involving $R$ and $H$. In Einstein's theory, the source of gravity (that is, curvature) is not in general mass, but rather is energy, so the density of energy, of all kinds, in the universe is relevant. In accordance with the assumption of homogeneity, this density is smoothed out over the universe and is taken to be the same everywhere (though changing with time).

With these preliminaries, Einstein's equation, relating curvature to energy distribution for a closed universe, is remarkably simple, and I will venture to write it out:

$$H^2 + \frac{c^2}{R^2} = \frac{8\pi}{3}\frac{G}{c^2} \times \text{(energy density)}.$$

Here, $G$ is the gravitational constant that determines the strength of gravity, and $c^2$ is the square of the speed of light (converting the energy in the last term into an equivalent mass). The factor $\frac{8\pi}{3}$ need not bother us too much. The left-hand side of the equation is the relevant part of the curvature of spacetime. The first term is that part of the curvature connected with time as well as space. The second term is the curvature of space itself (at any given time).

Einstein's equation as written assumes (as Einstein did) the universe to be closed. There is nothing in the mathematics to insist that this should be so. Space could be infinite. It could have uniform "negative" curvature. It is not difficult to visualize a two-dimensional surface with negative curvature in a small region. At the centre of a saddle there is negative curvature. That means that a "circle" drawn there has a circumference more than $2\pi$ times its radius. But there is no surface embedded in ordinary three-dimensional space that has the *same* negative curvature everywhere.

Nonetheless, a space of constant negative curvature is mathematically consistent. (It can be embedded in four-dimensional Minkowski spacetime, but that is not much help in visualizing it.) Such a space, with constant negative universe, has to go on forever, not close up on itself. Mathematicians call it *open* as opposed to *closed*. In such a space, the spatial curvature term $(c^2/R^2)$ in Einstein's equation just changes sign.

There is a third possibility: that space is flat and open. Then of course there is no spatial curvature term in Einstein's equation.

We can include all these three possibilities in a single form (*Einstein's cosmological equation*):

$$H^2 + k\frac{c^2}{R^2} = \frac{8\pi}{3}\frac{G}{c^2} \times \text{(energy density)},$$

where the number $k$ is just

$$k = 1, \ 0, \ \text{or} \ -1,$$

according to whether the curvature is positive (closed universe), zero (flat open universe) or negative (open universe).

I shall continue to refer to $R$ as the "radius of the universe", although this is strictly a correct phrase only if the universe is closed.

(As it happens, the last equation can be loosely interpreted as energy conservation in Newtonian physics. To make this interpretation, take a typical galaxy with a mass $M$, and rewrite Einstein's equation as

$$\frac{M}{2}(HR)^2 - \frac{GM}{R} \times \frac{4\pi R^3}{3}\frac{\text{(energy density)}}{c^2} = -\frac{kc^2 M}{2}.$$

Then the first term is the kinetic energy of the typical galaxy, the second term is its gravitational potential energy in the universe and the right-hand side is the constant value of the total energy, which may be negative, zero or positive – and can be scaled to one or other of the three values by an appropriate definition of $R$. A closed universe has negative total energy, somewhat analogously to the negative total energy of a closed planetary orbit.)

We can try to estimate the size of the first and last terms in Einstein's cosmological equation, as they are in the present epoch of the universe. The first requires Hubble's constant now ($H_0$).

The last term requires a knowledge of the total energy density from all sources. This total is hard to determine. Let us express the contributions in terms of the *critical density*, defined as the density that would balance the first ($H^2$) term by itself, and so allow the universe to be spatially flat ($k = 0$). The visible matter (in stars, and so on) gives us only a lower bound, because there may be dark matter that does not shine. But there is a way to estimate the number of baryons (protons and neutrons) in the universe, whether they are visible or not. This (as we shall see in the next section) relies on the production of helium by nuclear fusion when the universe was about a minute old. The proportions of these nuclei formed depend upon the baryon density. The conclusion from this argument is that the baryon density is only a small percentage of the critical density.

There is also evidence for more "dark matter" in the outer parts of galaxies, of an unknown, non-baryonic nature, which would bring the energy density term up to about 20 to 30 percent of the "critical density". This matter is required to provide enough gravitational attraction to account for the speed of revolution of stars near the edges of galaxies, and to bind galaxies into clusters.

*If* we have accounted for all the energy in the universe, Einstein's cosmological equation requires that the spatial curvature term is present with a minus sign (the infinite, negative curvature universe) and is comparable in size to the $H_0^2$ term.

It is quite remarkable that the terms in Einstein's equation should be of the same order of magnitude *now*. To understand this we must ask how we expect the energy density to vary with $R$ as the universe develops. This requires some knowledge of what there is in the universe. There are three different types of energy that come most readily to mind:

(i) *Matter*, that is, particles moving slowly compared to the speed of light, like the protons and neutrons making up ordinary matter in stars, and so forth. In this case, the energy density is, to a good approximation, just the rest energy $mc^2$. The number of these particles does not change (except possibly in the very early universe), so their density goes down inversely as the volume of the universe, that is, inversely as the *cube* of the radius $R$.

(ii) *Radiation,* including photons, gravitons and anything else (such as neutrinos), moving at or near the speed of light. In this case, the energy is about $c$ times the momentum, or $hc$ divided by the wavelength of the corresponding radiation. The expansion of the universe stretches out the wavelength proportionately with $R$. Taking this effect together with the dilution due to the increase in volume, the energy density decreases as the *fourth* power of $R$. At the present age of the universe, the energy density of radiation is negligible compared to that of matter. But in the early universe, radiation dominates.

(iii) *Vacuum energy,* the possible form of energy encountered in Section 9.10. It is due to the quantum fluctuations of the modes of vibration of the electromagnetic field and all the other fields corresponding to elementary particles. The density of vacuum energy (if it exists) is *not* diluted by the expansion of the universe, because the number of possible modes of vibration increases with the volume. Another possible source of vacuum energy is the Bose condensation of Higgs particles (Section 12.5).

Note that all the three forms of energy density listed depend upon $R$ differently from each other and from the spatial curvature term $k/R^2$ in the cosmological equation. If two of these terms have a similar size now, they must in general have quite different sizes in the past.

The first two forms of energy, matter and radiation, cause the expansion rate of the universe to slow down with time, because the expansion pulls apart the matter, and the radiation, against the attractive force of gravity. It comes as a surprise that the third item, vacuum energy, *accelerates* the expansion. The reason is that expansion does not pull vacuum energy apart. Rather it brings into being new vacuum energy, which is attracted to that already present, so that the total gravitational potential energy gets more negative, balancing the increased kinetic energy of expansion.

A frivolous analogy may help us to understand the accelerating effect of vacuum energy. Imagine a large number of highly sociable rabbits set down in a circle in the middle of Australia, initially all running away from each other (perhaps having been frightened by a man with a gun). Being sociable, they want to stay near one another and so slow down the rate of expansion of their circle. But suppose we allow for their being able to breed almost arbitrarily fast. Then they can satisfy their sociability by producing more rabbits, and their circle may expand even faster.

The vacuum energy now must be very small. It has no observable gravitational effect on the dynamics of galaxy clusters, and so it must be smaller than the rest energy of the matter in them, that is, about

$$10^{-9} \text{ joule per cubic metre.}$$

This number looks small, but we should ask, Small compared to what? In units in which $\hbar = c = 1$, this energy density is about

$$\frac{1}{(0.1 \text{ mm})^4}.$$

It is difficult to see how a length as large as 0.1 mm could be connected with the properties of the vacuum. It is tempting to guess that there must be some principle, which we do not at present understand, asserting that the vacuum energy now is exactly zero.

However, at the time of writing, there is possible observational evidence for an *acceleration* of the Hubble expansion. If this is right, there must be some form of energy in the universe other than ordinary matter and radiation, for both these inevitably slow down the expansion. Vacuum energy of a similar order of magnitude to the average matter density would suffice – another strange coincidence, that this should be true *now*. The vacuum energy might be sufficient, when added to the matter and radiation, to make up the critical density, at which the universe would be spatially flat. The new source of energy density need not be, like vacuum energy, independent of $R$: if it decreases less fast than the inverse square of $R$ it has an accelerating effect. Some people have termed this hypothetical energy density *quintessence*, borrowing the ancient term for the substance of the heavenly bodies.

How will the universe be in the far future? If the spatial curvature is negative or zero, expansion must continue forever. To see this from the cosmological equation, suppose that expansion ceased at some moment, so that $H = 0$ then. This would give a contradiction, since it would put the energy density equal to the negative or zero curvature term. If the spatial curvature is positive (closed universe) and there is no vacuum energy, on the other hand, there must come a time when the curvature term, decreasing more slowly, catches up with the energy density. Then $H = 0$ and expansion ceases. After that the universe begins to contract again and retraces its steps (at least in its spacetime structure).

Thus a spatially infinite universe will expand forever, whereas a finite one may recollapse. At present, we do not know for certain which fate awaits our descendants, but the evidence seems to favour eternal expansion.

If we follow the universe back in time, it gets smaller and smaller, so that the density of matter and radiation gets higher and higher without limit. Is this singular behaviour of the universe in the distant past just a consequence of the simplifying assumptions made, of isotropy and homogeneity, for example? In 1966, the 24-year-old Stephen Hawking realized that the early universe is like the reverse of the collapse of a star after it has formed a black hole (see Section 13.1). Just as Penrose had proved the latter must collapse to a point, Hawking was able to prove (on certain reasonable assumptions) that the universe must have started as a point singularity.

What this means physically is that, at early enough times, the effects of quantum theory on gravity cannot be negligible. Classical Einstein gravitation theory is not adequate to account for the very beginning of the history of the universe.

## 14.2 The Hot Big Bang

Nearly three-quarters of the known baryonic matter of the universe is in the form of hydrogen, about one-quarter is helium and about 2 percent consists of heavier elements, such as carbon, oxygen, nitrogen, silicon and iron. Nuclear reactions in the centres of stars (at temperatures of about $10^7$ K) fuse protons to make helium and the heavier elements, which are then dispersed in supernova explosions. But this mechanism cannot produce enough helium, which is itself mostly burnt in stars to produce the heavier elements. In 1948,

George Gamow and his co-workers suggested that nuclear fusion in the early universe, provided it was very hot, could be responsible for the helium. According to this theory, the early universe was a nuclear fusion reactor but, compared to stars, working at a higher temperature and a much lower baryon density.

When the universe was about one minute old, its size was about $10^{-9}$ what it is now, and the matter density was comparable to that of the Earth (about a thirtieth of the density in the centre of a star). Suppose the temperature of the universe was high and falling. At first neutrons and protons would be constantly changing into each other by collisions with electrons and neutrinos, but, when the temperature had fallen to about $10^9$ K, these processes would stop and there would then be about one neutron for each seven protons (because neutrons are a little heavier than protons). Collisions between neutrons and protons would produce deuterium, which would then further react to make tritium and then helium. If the baryon density was right, the nucleosynthesis would stop there, with most of the neutrons in helium, and just a small admixture of deuterium, tritium and heavier elements. The requisite baryon density is a small percentage of the critical density (defined in the last section).

Most of the thermal energy would be in radiation. *Assume* the composition of this radiation to be completely specified by saying it is in thermal equilibrium at some given temperature, just like the radiation in a furnace on Earth (though much hotter of course). Then the distribution of energy over different wavelengths is given by the Planck distribution, Figure 8.1. As the universe expands, all of these wavelengths are stretched out in proportion to the expansion factor $R$, and this is equivalent to decreasing the temperature inversely as $R$. So, when $R$ has gone from $10^{-9}$ to about $10^{-3}$ of its present value, the temperature has gone down from the order of $10^9$ K to about 3000 K. At this temperature, atoms become stable because the radiation is no longer energetic enough to knock electrons out of them. So matter becomes largely electrically neutral, and transparent to the radiation. From then on, the radiation is decoupled from all else that goes on in the universe and continues to exist with a life of its own.

At the present time, the temperature of this radiation should be a few degrees Kelvin; that is, the dominant wavelength should be a few millimetres. The subsequent history of this prediction is remarkable, and complicated. Some people realized that the *cosmic microwave radiation* might possibly be detectable. In 1964, one of these was Robert Dicke, of Princeton University, who was seeking to find the radiation. But it was two radio astronomers, Penzias and Wilson, working at the Bell Laboratories in New Jersey, who, ignorant of Gamow's prediction and Dicke's work, first, to their astonishment, detected the radiation.

By now, the cosmic microwave radiation has been very accurately measured, especially by the Cosmic Background Explorer (COBE) satellite, launched in 1989. The distribution of energy with the wavelength fits the Planck curve (Figure 8.1) very precisely. The temperature is known accurately to be

$$2.73 \, \text{K}.$$

The radiation from different directions is very nearly the same, except for a systematic variation of about 1 part in 1,000, which is attributable to the motion of the Earth. Minute variations over the sky, of the order of 1 part in $10^5$, have also been measured. As we shall see, these fluctuations have important consequences for the development of the universe.

The energy of the radiation (in a given small range of wavelengths) is at its maximum near a wavelength of one millimetre. There is about one photon of the microwave radiation per two cubic millimetres of space. The rate of flow of energy is about a tenth of a watt per square kilometre.

The discovery of the microwave radiation is of enormous importance for cosmology. We are in the midst of a furnace, made when the universe was a thousandth of its present size. Astrophysicists can measure the properties of that furnace with amazing accuracy. The concepts of thermal physics, based upon the statistical properties of the random motion of molecules (Chapter 2), seem to be applicable to the early universe (as they are to black holes; see Section 13.4).

The universe is hot. A dimensionless measure of that hotness, independent of epoch, is the radius of the observable universe divided by the typical wavelength of the microwave radiation:

$$\frac{10^{10} \text{ light years}}{1 \text{ mm}} \approx 10^{29}.$$

So, looked at in this way, the universe is very hot for its size, or very large for its temperature. This is a fundamental number characterizing the universe, which, if we were ambitious, we would like to understand.

## 14.3   The Shape of the Universe in Spacetime

Assuming that the universe is homogeneous and isotropic (on a large enough scale), and knowing the Hubble expansion rate and the temperature of the microwave radiation, we can calculate the previous history of the universe. We are still not sure whether the spatial curvature is positive, zero or negative. A positive curvature universe is easiest to visualize, so I will treat this case. Many of the features are common to all three types of curvature.

The universe is expanding now. Matter and radiation have the effect of slowing down the expansion, so the expansion rate must have been faster the further we go back into the past (assuming for the moment that there is no vacuum energy).

In trying to visualize the geometry of the universe, we meet the same difficulties as in Section 13.1: there are more than three dimensions, and the geometry of spacetime is Minkowskian not Euclidean (see Section 7.4). The first of these problems can be overcome by taking a model universe with only one space dimension (as well as the time dimension). Because of the assumed isotropy, we do not lose much by doing this. Then, since we are thinking of a spatially closed universe, in one dimension we may take it to be a circle. Simple enough, but we have to put in the spacetime part of the geometry.

To do this, we may imagine that the "universe" is embedded in a flat spacetime with two spatial dimensions in addition to time. This is not to say that the actual universe *is* embedded in higher dimensions. The geometry of the universe is something intrinsic to itself, and it can be described mathematically without reference to any embedding. But if we want to picture the geometry, embedding

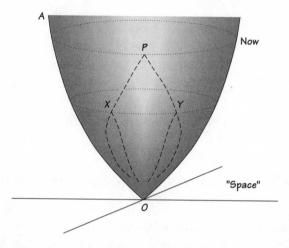

FIGURE 14.1 A model universe of one space and one time dimension embedded in one extra dimension. Spatial dimensions are plotted horizontally, and "time" vertically. The history of the universe is a surface. Cross sections at any given time are circles (shown as dotted lines). The universe starts at a point O. The history of, say, a galaxy is represented by a curve on the surface running upwards from O, for example, the curve OA. The point P represents "here and now". The histories of photons are curves on the surface at 45 degrees to the vertical "time"-axis. Four such curves are shown as dashed lines.

seems to me to be useful. The spacetime history of the universe looks like a surface embedded in this three-dimensional spacetime, as sketched in Figure 14.1.

Spatial dimensions are in horizontal planes, and "time" is plotted vertically. Cross sections at any fixed "time" are circles, standing for the spatial universe in this model. The history of a galaxy is a curve on the surface running (to a good approximation) upwards with increasing "time", and thus at right angles to the circular spatial cross sections. An example is the curve OA. The true physical time, as measured by a clock on a planet within a galaxy, is not the same as the "time" in the embedding spacetime: real time runs slower and slower the nearer we get to O.

Clearly the surface representing the history of the universe is curved: it cannot be flattened out without distorting it. (The low-dimensional model has a drawback: the spatial sections, circles, have no intrinsic curvature – they could be cut and then straightened out. In higher dimensions, they would be spheres, which do have curvature.)

The universe begins at a single event, $O$. "Before" $O$, there is no universe, no space, no time. What is more, the shape of spacetime is not even smooth at $O$: it is pointed like the vertex of a cone. One might think that this *singularity* at $O$ is a result of our assumption of homogeneity and isotropy, which is surely not exact. But Penrose and Hawking have proved, making some rather natural assumptions, that some singularity is inevitable according to Einstein's theory of gravity. However, when the universe was small enough, or hot enough, quantum theory must have been relevant to the geometry of the universe. The quantum theory of gravity is not at all well understood. So all we can say is that the region very near $O$ is not understood, and Figure 14.1 should not be believed very near $O$. What we can say with near certainty is that the universe was once very small.

If I had drawn the universe for later times, it would close up again, eventually coming to another single event like $O$, but this is a property of spatially closed universes only.

As usual in spacetime, the world-lines of photons are particularly important in understanding what goes on. The three-dimensional embedding space is just ordinary (flat) Minkowski space (as in Section 5.5), and photons move on light cones, at 45 degrees to the vertical "time"-axis. But in the physical universe, photons (like everything else) are confined to the surface in the figure, and so they move along curves on the surface that are at 45 degrees to the vertical. Through any point (except $O!$) there are just two such curves, corresponding to photons moving one way or the other round the circular space. At later times, these two curves are roughly at right angles, but for earlier times the angle between them decreases, because the surface itself gets nearer and nearer to 45 degrees to the vertical axis. This has the effect (as it must) of leading all the photon world-lines down towards $O$.

With all this in mind, we can sketch the world-lines (dashed lines in Figure 14.1) of two photons detected "here and now" at the event $P$. Suppose these to be photons in the microwave background.

They originate just before electrons and protons settled down into hydrogen atoms. Let $X$ and $Y$ be the events in which the photons were produced. Observation has shown the microwave radiation to be, to high accuracy, the same in all directions, so the temperatures at $X$ and $Y$ must be the same. Presumably, then, there must be some common cause, before $X$ and $Y$, responsible. Can this be possible?

To answer this question, we must see what parts of the universe can have any causal influence on $X$ and on $Y$. Since there are no causal influences faster than the speed of light, we need to trace back the world-lines of photons before $X$ and before $Y$. The world-lines of photons arriving at the event $X$ are shown on the diagram as two dashed lines. The bobbin-shaped region of spacetime between them is the *causal past* of $X$. The causal past of $Y$ is similar. In the figure, the causal pasts of $X$ and of $Y$ *do not overlap*. There is nothing in the previous history of the universe that influences both $X$ and $Y$. So it is very difficult to understand why the temperatures at $X$ and $Y$ should be the same.

Two warnings. First, it is not obvious that the causal pasts of $X$ and $Y$ do not overlap. This depends upon how far in the past $X$ and $Y$ are. The reader will have to take my word for it that the figure is, in this respect, not a misleading representation of the real world. Second, two dimensions are omitted in the figure. In reality, $X$ and $Y$ are replaced by a whole spherical surface in space (whence the cosmic microwave radiation has come), and, for two points on this surface separated by more than a few degrees, the causal pasts do not overlap.

Take any event in the spacetime history of the universe. It has a causal past, as defined. Only some of the galaxies in the universe have world-lines that go through this causal path. The outermost of this set of galaxies is called the *particle horizon* of the event in question. It is the boundary of that part of the matter in the universe that could have had any causal effect at that event. Thus the conclusion of the last paragraph was that the particle horizons of $X$ and $Y$ are only small fractions of the universe.

Any observer's particle horizon grows with time. Going back in time into the very early universe, it decreases indefinitely. In the future, it grows (at least until the universe, if it is bounded, begins to recollapse). Continually more and more galaxies, which previously had no influence on us, become observable (come "over the horizon").

The feature just outlined is sometimes called the *horizon problem*. What can have made the temperature of the microwave radiation the same in all directions? Why do we not fear something entirely unexpected coming "over the horizon"? Unless we know more about the very early universe, we cannot be sure how seriously to take the horizon problem. In the *very* early universe, quantum gravity must have been important. Then all notions of causality and time (even of "earliness") probably become fuzzy. Maybe, by the stage when quantum effects can be neglected, uniformity has somehow been imprinted on the universe.

## 14.4  A Simple Recipe for the Universe

We are now in a position to piece together a history of how our universe may possibly have developed. The assumptions we make are the simplest possible:

- The laws of nature are the same everywhere and at all times. In particular, the particles important in the early universe are those known from laboratory experiments (or perhaps inferred from our theories).
- The universe is homogeneous and isotropic on large enough scales.
- The universe is expanding, and the present expansion rate is roughly known (the Hubble rate $H_0$).
- The early universe was hot, and the temperature of the microwave background is a measure of the temperature of the universe.
- The number of protons and neutrons (that is, the amount of matter) in the universe now is about $10^{-9}$ of the number of photons.
- For simplicity, I will for the moment assume that all the energy in the universe is, and always has been, in the form of matter or radiation (with no vacuum energy or "quintessence").

How must the universe have "started off" to be consistent with what is known? As we trace the universe back in time, it gets smaller and hotter. Early enough, the temperature is the Planck temperature, $10^{32}$ K, the temperature that can be constructed out of the three fundamental constants $c, \hbar$ and $G$ (together with Boltzmann's constant, which converts energy to temperature). Hotter than this, the gravitational interactions of the thermal energy become strong, quantum gravity is important and we cannot in the present state of our knowledge say anything definite. So let us start at this temperature (or a little lower) and say what we must assume about the universe then.

The size of the universe then must have been at least a thousandth of a millimetre. It must be at least this size in order to give rise to the visible universe now. It may have been bigger, even infinite, but we have not yet seen any more of it. This size sounds small, but compared to the Planck length (constructed from $c, \hbar$ and $G$ and equal to $1.6 \times 10^{-35}$ metre) it is huge, roughly $10^{29}$ times as big. This enormous number is the first, and main, ingredient in our universe recipe. It is about equal to the ratio of the size (or minimum size) of the universe to the dominant wavelength of the thermal radiation. As the universe expands, the wavelengths are stretched in proportion, so the ratio $10^{29}$ does not change much (it changes by factors not too far from 1 when, for example, the quarks and antiquarks or the electrons and positrons mostly annihilate). At the present time, the size of the (visible) universe is about $10^{10}$ light years, that is, $10^{26}$ metres, and the dominant wavelength of the microwave radiation is of the order of 1 millimetre, consistent with this ratio.

(Another way to express this property of the universe is to say that, at the Planck temperature, the spatial curvature [given by $1/R^2$] is at most $10^{-58}$ of the spacetime curvature [$H^2$] due to the expansion.)

Given this hot and "large" early universe (at the Planck temperature), we can apply Einstein's cosmological equation (see Section 14.1) to it. Because of its "large" size, the spatial curvature term at first is utterly negligible, so the expansion rate is dictated by the temperature of the thermal energy. Not surprisingly, when the temperature is near the Planck temperature, the time characterizing the expansion (say the time to double in size) comes out to be of the order of the Planck time, $5 \times 10^{-44}$ second. Multiplying this by the speed of light gives an idea of the radius of the *visible* part of the universe for a hypothetical astronomer at that time. It is about $10^{-29}$ times the total size (the same numerical ratio again). Thus each "observer" knew about only a minuscule part of the whole.

This latter fact makes it hard to understand what would make the temperature the same all over the universe: the different parts seem to have no common prior cause. Nevertheless, we now stipulate the second ingredient in our recipe: that the temperature was uniform everywhere, to within 1 part in $10^5$. The pudding is very, very large and is mixed very well but not perfectly.

There is just one more ingredient: the seasoning. If the hot universe is left to itself, all particles will be present with essentially equal probabilities (bosons are actually a little more frequent than fermions). In particular, quarks and antiquarks will be exactly equal in number. If this were so, when the universe eventually cooled enough, the quarks and antiquarks (in the form of protons and neutrons and their antiparticles) would annihilate, and eventually only photons, gravitons and neutrinos would remain. There would be no matter, no stars, no planets, no life.

To avoid this false conclusion, we have to assume that the original thermal distribution is ever so slightly skewed to favour quarks over antiquarks. We need about a billion and one quarks for every billion antiquarks. The small relative excess, $10^{-9}$, that is needed here is the final ingredient in the recipe.

So three numbers are needed to specify the very early state of the universe: $10^{29}$ (the flatness of space), $10^{-5}$ (the departure from homogeneity) and $10^{-9}$ (the quark excess). These numbers cry out for explanation. It may be that some of them do not need to be fed in from the start but can be generated by the operation of physical processes shortly afterwards. In Section 14.5, I mention how this may be possible for the relative quark excess $10^{-9}$. The size number, $10^{29}$, is the subject of Section 14.7.

Let us, however, for the moment accept these three numbers as given. Then the claim is that by the operation of known or surmised physical laws, the universe will develop to what we know today. Some of the key events in this development are as follows.

(i) Soon after the time $5 \times 10^{-44}$ second (the Planck time) the universe emerges from the terra incognita of quantum gravity, with the properties just described.

(ii) There is an epoch when the universe is very simple. All particles are massless, so they all move at the speed of light. Because of asymptotic freedom (see Section 11.3), the particles at these very high energies hardly interact with one another. All types of particle (and antiparticle) are present (including gravitons probably), in numbers proportional to the number of spin states they have (fermions are three-quarters as numerous as bosons as a non-obvious consequence of the exclusion principle).

(iii) If there is a GUT (see Section 12.6), as the temperature drops to perhaps about $10^{28}$ K, there may be a phase transition, in which the $X$ particles get masses. All other particles remain massless, and the electroweak and QCD interactions display their full symmetries.

(iv) At about $10^{-11}$ second the temperature has dropped to $10^{15}$ K and the radius increased to $10^{11}$ metres. A transition takes place as the Higgs field undergoes Bose-Einstein condensation. All particles except gravitons, photons and (possibly) neutrinos get their present masses. The weak forces begin to behave quite differently from electromagnetism. Neutrinos now propagate through the universe almost freely.

(v) At $10^{-5}$ second (temperature $10^{12}$ K), yet another phase transition occurs, producing confinement of quarks and gluons: quarks begin to combine to form protons, neutrons and other baryons, and gluons no longer exist as particles in their own right.

(vi) By about 1 minute, the temperature is about $10^9$ K and the radius several light years. This temperature is equivalent to an energy less than the rest energy of an electron, so most electrons and positrons annihilate. After that, all that remains are neutrinos (now moving essentially without interaction), photons and a very few protons, neutrons and electrons (about 1 for every $10^9$ photons).

(vii) By a time of 3 minutes, about a quarter of the protons have combined with the neutrons to form helium nuclei.

(viii) When the temperature drops to about $10^4$ K, the energy density due to matter (protons and helium nuclei) has become equal to the energy density carried by the ($10^9$ as numerous) photons and neutrinos. From now on, matter not radiation controls the rate of expansion of the universe, which decelerates less fast in consequence.

(ix) Shortly afterwards, when the temperature is about 1,000 K, the electrons combine with the nuclei to form neutral atoms. This neutral gas is transparent to light, so the photons from now on proceed freely as the cosmic microwave background radiation, telling us directly about conditions at this epoch, in particular the temperature and its uniformity (except for variations of the order of 1 part in $10^5$).

(x) At last, from about $10^9$ years, gravity sets to work on the small variations in matter density. Regions of relatively higher density tend to fall inwards, and stars, galaxies, superclusters and other structures are formed.

(xi) About $1.2 \times 10^{10}$ years: Now. The temperature of the photons has dropped to 2.73 K and that of the neutrinos (at least if they have no mass) to about 2 K (the photons were boosted compared to the neutrinos when the electrons and positrons annihilated; see [vi]).

This history is plausibly reliable from stage (iv) onwards. Before then, it should be treated sceptically. The temperature is much higher than any energy yet produced, or likely to be produced, on Earth. The only experimental information on the behaviour of matter at these energies, for instance, measurements of the rate of proton decay (or upper limits on this rate), is necessarily indirect.

## 14.5 Why Is There Any Matter Now?

As the universe cools to a temperature less than that corresponding to the rest energy of a proton, protons and neutrons and their antiparticles mostly disappear from the hot universe. They annihilate with their antiparticles to produce photons, and the opposite reaction becomes increasingly unlikely. The only reason that any protons and neutrons remain now is because of the assumed small excess (1 in $10^9$) of quarks over antiquarks. Could this excess have been generated by ordinary physical processes, starting from equal numbers of quarks and antiquarks?

For this to occur, the *quark number*, defined to be

(number of quarks) − (number of antiquarks),

cannot be exactly constant. In the electroweak and QCD theories (as naively interpreted), the quark number never changes; nor is there any laboratory evidence that it changes.

But, as mentioned in Section 12.6, grand unified theories (GUTs) predict that the quark number can sometimes change, though the probability of such change is small. If a GUT is true, it is possible that the quark excess was generated in the very early universe, at or before stage (iii). The conditions for a quark excess to be generated were understood in 1966 by Andrei Sakharov, "father of the Soviet hydrogen bomb", dissident and Nobel Peace Prize winner. They are three:

- The quark number is not exactly constant.
- There is not exact symmetry under charge conjugation, $C$, or under $CP$, that is,

(charge conjugation) × (space reflection).

- The rate of change of the temperature of the universe (at the relevant time) is faster than the rate of the processes producing the quark excess.

GUTs naturally meet all these three conditions. I will now show, by an example, why these conditions are necessary, and how GUTs can satisfy them. The first condition has already been dealt with.

The second condition is also simple to understand. If $C$ were an exact symmetry, then, for any process tending to produce quarks, there would be an equally strong process tending to produce antiquarks. The same argument applies also to $CP$, because the space reflection part of $CP$ is irrelevant after adding up all directions and all spins of the particles concerned. But, in fact, both $C$ and to a lesser extent $CP$ symmetries are observed to be broken in weak interactions, even without a GUT (see Section 12.5).

The reason for the third condition is this. If the temperature is falling slowly enough, then the number of each type of particle or antiparticle is determined just by the mass (and the temperature), according to the rules of statistical physics. The $CPT$ theorem (see Section 12.2) is enough to guarantee that each antiparticle has exactly the same mass as the corresponding particle, so particles and antiparticles would be equally numerous. But if the temperature is falling faster, we cannot apply the rules of statistical physics in this simple way.

Here now is an example of how a nonzero quark number could possibly be generated. All examples of GUTs contain $X$-particles that do not have any well-defined quark number, for instance, one with charge $\frac{4}{3}$, which can undergo either of the two decays

$$X \to u + u, \quad X \to \bar{d} + e^+.$$

Here $u$ is an up quark (charge $\frac{2}{3}$), $\bar{d}$ is an antidown quark (charge $\frac{1}{3}$) and $e^+$ is a positron. The decay products in these two decay modes have different total quark numbers, 2 and $-1$ respectively.

The $X$ particles are very heavy, perhaps 1 percent or so of the Planck mass. Their mass defines a temperature, say $T_X$, equivalent to the rest energy of the $X$. When the temperature of the universe is higher than $T_X$, the $X$ particles are in thermal equilibrium along with all other particles, and their decay rate is slow compared to the universe's expansion rate. As the universe cools to about $T_X$ (at about stage [iii] in Section 14.4), the $X$ particles begin to decay through the two decay modes just mentioned (among others), but by then the $X$s are no longer in thermal equilibrium, so they are not being regenerated from collisions of lighter particles. Clearly, the $X$ decays generate (in general) a non-vanishing quark number.

But what about the antiparticles, $\bar{X}$, and their decays? When the temperature is above $T_X$, the $\bar{X}$ are present in thermal equilibrium in equal numbers to the $X$. If $CP$ symmetry were obeyed, the quark number generated by the decays of the $\bar{X}$ would exactly cancel that from the $X$. But because $CP$ is violated, this cancellation does not happen. (Although the *total* decay rates of $X$ and $\bar{X}$ are equal, the distributions among the various modes are not.) Thus a net nonzero quark number is expected to result. Unfortunately, the amount of quark number depends upon the details of the GUT theory used, and at present there is no way to make a quantitative comparison with the known value.

## 14.6   How Do We Tell the Future from the Past?

Chapter 2 left us with a mystery. The laws of nature do not distinguish the past from the future (I ignore the very small amount of time reversal non-invariance mentioned in Section 12.2, assuming, perhaps wrongly, that this is not important), yet the behaviour of complex systems seems very clearly to make such a distinction. Play a video of a drop of dye diffusing through water and it will look just like that. Play it backwards and the dye will appear to come together and form a drop: something we would never see in nature. Play a video of a very close-up view of the motions of just a few molecules of water and dye (if that were possible), and we would not know whether it were played backwards or forwards.

Shuffle a new (ordered) pack of cards and it will almost certainly become disordered. Shuffle a disordered pack, and it is most unlikely to become perfectly ordered. There is no mystery about this. If the all packs of cards in the world were just randomly ordered, we would practically never come across an ordered pack, so we would not notice the apparent dissymmetry between past and future. The reason we notice anything is that we do have access to ordered packs of cards, when we buy a new pack or when someone deliberately goes through a pack and orders it.

So the question seems to be, Why, in the universe, do we encounter statistically unlikely situations, such as a drop of dye in a glass of water? Why are there mountains that have not yet eroded? Why is the Sun hotter than the Earth, so that plants on Earth can absorb sunlight? Why does the Sun contain hydrogen that has not yet "burnt" to form heavy elements like iron? Why, in general, is the universe not in a state of statistical equilibrium?

Maybe if the laws of nature provide no answers to these questions, the "initial conditions" of the universe do. Given a cosmological model, we can try to see whether it gives the answers we seek. The hot big bang cosmology provides, if not exactly *initial* conditions, at least a description of the universe at very early times (say just after [i] in Section 14.4). At first sight, that very early universe seems to be just what we do *not* want – something in exact thermal equilibrium, a completely disordered and featureless universe, with particles and their motions distributed in a purely random, statistical manner, determined just by the temperature. How, out of this, do structures and local departures from equilibrium (for example, our Solar System) develop?

There is a simple example that may be somewhat analogous. Take a glass containing, at some given temperature, a supersaturated solution of, say, sugar. *Supersaturated* means that the solution is unstable: left alone, with care, the solution can stay unchanged for some time, but a small disturbance, like the addition of a tiny crystal of sugar, would cause a lot of sugar to crystallize out of the solution – an apparent increase in order, since crystals are more ordered than liquids. Actually, the total disorder (the total entropy) does increase, because heat is released during the crystallization, so that the molecules of the remaining solution get more random thermal motion, which more than offsets the loss of disorder due to the crystal formation.

But why does the supersaturated solution not necessarily crystallize immediately? This is because, for crystallization to begin, a number of sugar molecules have to get near enough to each other for the attractive forces between them to take effect – an unlikely occurrence, unlikely enough to delay crystallization. There is a sort of probability bottle-neck to be got through. Once a very small crystal has formed, other sugar molecules can attach themselves to it one by one.

A video of the crystallization of a supersaturated solution, if played backwards, would look something like a video of sugar dissolving in water. A thermometer in the solution would distinguish the two processes.

So could the development of the universe from the fireball state be in any way analogous to the spontaneous crystallization of a supersaturated solution? For this to be so, the early cosmic fireball must not really be the totally random state that it appears.

There are several factors that seem to be relevant to this question; here are four of them:

- The hot big bang universe is *not* in thermal equilibrium, for this would require the temperature to be unchanging. But as the universe expands, the temperature falls. Whenever this rate of cooling is faster than some relevant physical process, equilibrium can be lost. We have already met examples. The universe passes too fast through the nuclear reaction epoch ([vii] of Section 14.4) for all the hydrogen to burn to form heavy nuclei like iron: some helium forms and much unburnt hydrogen remains. Similarly, at an earlier stage (see Section 14.5), the speed of cooling may allow the development of a very small excess of quarks over antiquarks, even though equilibrium would demand equal numbers.

- The temperature of the fireball cannot have been exactly the same everywhere: there were variations of the order 1 part in $10^5$ in the temperature of the cosmic microwave background. These may be the seeds out of which grow galaxies and other structures.

- Given the existence of gravity, it is quite difficult to see what a state of true equilibrium would be like. Matter subject to gravitational attraction has a peculiar *instability*: it may condense into clumps, converting the gravitational potential energy into kinetic energy in the process. Eventually, black holes may form.

- Finally, a point connected to the last one: the hot big bang universe has the matter and radiation of the universe in approximate thermal equilibrium. But what about spacetime itself? In the model, it acts mainly as background in which the particles move. And a very simple and special background at that: homogeneous and isotropic to a good approximation. But in Einstein's theory of gravity, spacetime is just as much a dynamical thing as the motion of particles. Should not spacetime share in the randomness? Partly this is allowed for by the presence, in thermal equilibrium, of gravitons, on the same footing as photons, and so on. Gravitons are just (quantized) ripples in spacetime. But is this enough? Are there, for example, small black holes present in the very early universe? The behaviour of particles at very high temperatures is supposed to be fairly simple, because the forces between them (electroweak and strong, or maybe GUT forces) decrease at high energies. Gravitational forces, by contrast, are still increasing with energy, as we approach the Planck energy.

It may well be that the initial expanding fireball universe, for all its apparent randomness, is really a very unlikely state (like an ordered pack of cards), and the "arrow of time" is marked by the increase in randomness from this starting point.

But if this is so, there is a tricky paradox to be faced. Suppose that there is enough energy in the universe so that it is has positive spatial curvature and is closed. Then Einstein's laws demand (assuming there is no vacuum energy or "quintessence") that the universe will expand to a maximum radius and thereafter contract again. At the maximum size, the cosmic background radiation will attain its minimum temperature. During the contraction, the microwave wavelengths will be diminished, just as they were stretched during expansion. The energies of the photons, and so the temperature, will rise again. The universe will contract, and its temperature rise, without limit for so long as quantum gravity can be neglected. It looks as if the universe must again reach a fireball state, exactly like the one from which it began, but contracting instead of expanding.

If this is correct, there seems to be symmetry in the cosmology between the initial and final fireball states. How then could the cosmology account for the "arrow of time"? There are three possibilities:

- The arrow of time flips over at the time when the universe reaches its maximum size. I find this very difficult to believe. A star at this time would have to stop radiating light and instead have light pouring into it from space. Stephen Hawking tells us, in *A Brief History of Time*, that he rejects this possibility.
- The universe is *not* in fact spatially closed, so we need not worry about what would happen if it were. To be convinced

by this argument, I would need to understand exactly why the closed universe is self-contradictory.

- The final fireball, although superficially like the initial one, really has important differences. These differences would have to be such that the final state was more probable than the initial one. The fluctuations at the end might be less than at the beginning, or there might be black holes existing at the end but not at the beginning.

## 14.7 Inflation

The hot big bang theory of the early development of the universe (Section 14.2) left some baffling questions unanswered. Why was the universe initially so *flat*; that is, why, at the Planck temperature was the size so much larger (by a factor of $10^{29}$) than the Planck length? Why is the universe so *uniform*, although different parts of the early universe could have had no causal connection with each other (Section 14.3)? Maybe these questions would be answered if we understood the quantum gravity era before the Planck time, but there has been another interesting attack on the problems, called the theory (or theories) of *inflation*.

The modern inflationary theories of cosmology originated in 1980 from the work of the American physicist Alan Guth. The idea is to make use of the expansive power of *vacuum energy*. We saw in Section 14.1 that vacuum energy, unlike rest-mass energy or radiation energy, has the unexpected property of making the universe expand faster, in spite of the attractive nature of gravity. We also saw that the vacuum energy in the universe *now* must be extremely small; otherwise it would have noticeable effects. But what if there were a period in the early universe when there was a substantial vacuum energy? Let us first see what the effect would be, then discuss how such a vacuum energy might come about.

Vacuum energy causes the universe to expand exponentially; that is, it doubles in size after every repetition of a fixed period of time (say $10^{-34}$ second). Such an exponential expansion very quickly produces an enormous increase in size. I do not know whether the name *inflation* was intended to refer to the universe's blowing up like a balloon or to refer to exponential monetary hyperinflation.

(The German currency inflated by a factor of $10^{12}$ between 1913 and 1923.) Both similes are apt.

Suppose the universe started neither flat nor uniform, then inflated to an enormous size. Only a small part of this giant universe would be within our horizon now (light from elsewhere would not have had time to reach us). So we would be seeing only a tiny patch (much magnified) of the original universe, small enough to be very nearly flat and small enough to be causally connected and so uniform. The pre-inflationary universe would be largely hidden from us, and what we observe now insensitive to its details.

How could such a temporary vacuum energy come about? In GUT theories (Section 12.4), we expect phase transitions as the universe cools, in which some Higgs particles undergo Bose-Einstein condensation (analogous, for example, to the superfluid phase transition as liquid helium is cooled). But phase transitions are sometimes temporarily delayed if the cooling happens fast enough. For example, water vapour can be temporarily cooled below 100 degrees Celsius, without turning to water, existing in a *supercooled* state. Such a state is unstable, like that of a boulder precariously balanced on the top of a hill. Any disturbance may produce condensation into an expanding water drop (a raindrop, for example).

(This effect was put to use in the Wilson cloud chamber, one of the first devices to detect the tracks of charged particles, invented in 1911 by C.T.R. Wilson. The charged particles pass through supercooled water vapour, knocking electrons out of some of the atoms in their paths. The resulting electric charges trigger the formation of water drops in the wake of the particle. Wilson was led to his invention from his interest in cloud formation in the mountains of his native Scotland.)

Below 100 C, the lowest energy state of water is the liquid state. The unstable, supercooled gas has more energy, which is released if it liquefies.

Returning now to the expanding, cooling early universe, suppose that a Bose-Einstein condensation phase transition is delayed, leaving the universe temporarily in a supercooled state. Then the vacuum contains extra energy, and this vacuum energy is postulated as the cause of inflation. So long as the vacuum remains supercooled, inflation carries on. Eventually, the delayed transition does take place, and inflation stops. Then the vacuum energy is released as heat, that is to say, in the form of ordinary radiation and matter. Practically everything we see now comes from this "reheating": anything in the universe prior to inflation is so diluted as to be negligible.

(As mentioned in Section 14.1, there is evidence that the Hubble expansion is accelerating *now*, so there may be some vacuum energy – or "quintessence" – now. But this, if it exists, must be far less than during the inflationary era.)

This is the rough idea of inflation. It very naturally leads to a flat universe, flatter perhaps than the universe actually is. But there are variations of the inflationary scenario that produce a less flat universe. I don't think there is, at the time of writing, a consensus about the best version of inflationary theory.

Inflation is believed to help explain the universe's known departures from homogeneity and isotropy. At some stage in the early universe, there must have been small perturbations (departures from homogeneity), which were the seeds from which, by the action of gravity, galaxies and other large-scale structures grew. Such perturbations should have had their effect on the cosmic microwave background radiation, and indeed in the 1990s very small differences (of size about 1 part in $10^5$) were measured in the radiation coming from different directions.

Any microscopic perturbations that were present before inflation would be stretched to astronomical scales by the inflation. It is even suggested that *quantum fluctuations* in the very early universe could be transformed by inflation into *classical* density perturbations. In a flat Minkowski spacetime background, quantum fluctuations do not give density perturbations. The quantum vacuum energy (see Section 9.10) of, say, electromagnetism is exactly uniform: the quantum fields fluctuate, not their energy distribution. Nevertheless, when quantum theory is combined with more exotic spacetime geometries, it is harder to see what happens. We know this from the Hawking radiation from black holes (Section 13.4). During inflation, there is a horizon of constant size, and, as the wavelengths of the quantum fluctuations are stretched by the inflation, they may become comparable to that horizon size. Current research along these lines seems to be successful in explaining the observed deviations from isotropy in the microwave background. But it also forces us to think hard about the thorny question of the interpretation of quantum mechanics (see Section 8.11).

## 14.8   Conclusion

Microwave astronomers can "see" the universe as it was when it was some million years old – about a ten thousandth of its present age. From the present abundance of helium, we can be fairly confident about the minute-old universe. For the first second, we must rely on our theories of elementary particles. If current ideas are right, the very early universe, as it cooled, may have seen a number of "phase transitions", in each of which the nature of the vacuum changed. In the very early, very hot universe, the particles and their interactions may have been much simpler and more symmetric than they are today. But earlier still we encounter the barely understood realm where quantum gravity rules.

We are not certain about the nature or amount of much of the matter in the universe, and whether it is sufficient eventually to halt the expansion. We do not know when, if ever, vacuum energy plays an important role. We are not certain why the universe appears to be so flat and so uniform, why there is any matter remaining and why that matter is now organized in a structured way. Research on all these questions is going on at an amazing rate. But profound questions, about quantum gravity, for instance, remain to be answered.

# 15

# QUERIES

Newton's *Principia* is written (in Latin) in a lofty, austere way, as if to allow the reader no opportunity to disagree. The *Opticks* published (in English) 17 years later is more human. It ends with several *Queries*, in which Newton speculates, without claiming certitude.

I have ventured to borrow Newton's word as the title of this chapter. It is meant to be a warning to the reader that I am now venturing off the fairly well-beaten track followed in the preceding 14 chapters. The speculations that follow are not mine, of course. I have chosen ideas that seem to me to have attracted the attention of the greatest numbers of physicists. I don't suppose any of them is exactly right as it stands. Some may be completely wrong. But I hope that some of them have some truth in them. Only time, and experiment, will tell.

## 15.1  Hidden Dimensions: Charge as Geometry

I begin with a speculation that is almost certainly wrong. My excuses are that it is very pretty and that string theory (Section 15.3) makes use of some of the same ideas.

Around 1915, there seemed to be two beautiful theories: Maxwell's electromagnetism and Einstein's gravity. An obvious dream would be to try to unify them and, in the process, perhaps to find a geometrical basis for electromagnetism similar to Einstein's geometrical theory of gravity. With hindsight, we now know this vision to have been a mirage, because electromagnetism is only a part of the electroweak forces, but even so let us look at one intriguing idea that came up in those days.

There is one big obstacle to be overcome in any attempt to include other forces with gravity. The great beauty of Einstein's theory is that there is no gravitational force: all that happens is that *all* masses follow geodesics in spacetime. But, in an electromagnetic field, charged and neutral particles do *not* move the same way. If the shape of spacetime is to account for the motions of charged as well as neutral particles, charged ones must somehow know something about the spacetime that the neutral ones don't. I will now describe one way in which this might be done.

The idea came first from the German Theodor Kaluza, who communicated it to Einstein in 1919. Einstein wrote, "At first glance I like your idea enormously". The idea was worked out more fully in 1926 by Oskar Klein, a Swede.

Kaluza-Klein theory was as follows. Einstein's spacetime is enlarged to five dimensions, four of them space-like and one time-like. The reason we are not aware of the extra space dimension is that it is "curled up" very tightly, as if into a tiny tube. In fact, as we shall see, the circumference of this tube has to be about $10^{-32}$ metre.

It is easy to visualize a two-dimensional space in which one dimension is curled up: it is just a very thin cylinder, like a drinking straw. It is a bit harder in more than two dimensions. One way to think of it is illustrated in Figure 15.1, for a three-dimensional spacetime, with one space dimension curled up.

Using five-dimensional spacetime, all the features of electromagnetism get geometrical interpretations. The electromagnetic fields are connected with curvatures that involve the fifth dimension along with the other dimensions.

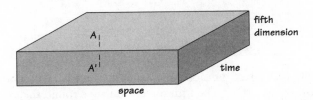

FIGURE 15.1  A three-dimensional spacetime, with one space dimension curled up. It is decreed that points like *A* and *A'* are to be regarded as being the same. So, if you follow the path from *A* to *A'*, you get back where you started from (like going round a cylinder).

Charged particles are ones that go round and round in the fifth, curled up dimension (as well as moving in ordinary space, of course). Neutral particles do not move in the fifth dimension. Then the geodesics along which charged particles move are different, because they explore the extra curvature connected with the fifth dimension.

The charge is proportional to the angular momentum of the motion round the curled up fifth dimension. In quantum theory, angular momentum is quantized, so *charge is quantized*. This explained why all electric charges were multiples of the charge on the electron (now we would say the charge on a quark). It comes out that $\frac{1}{137}$, the dimensionless measure of the strength of electromagnetism, is (apart from a numerical factor)

$$\left[\frac{\text{Planck length}}{\text{circumference of curled up dimension}}\right]^2$$

(it must be something like this for dimensional reasons), and this is why the size of the fifth dimension must be so small.

Gauge transformations (local changes of the phase of a charged particle's wave function – see Section 8.12) correspond to the freedom to change, at different points of spacetime, the origin from which the distance round the fifth, curled up, dimension is measured. This is illustrated in Figure 15.2.

Nowadays, there is an alternative explanation of the quantization of charge in gauge theories like GUTs (Section 12.6), so the Kaluza-Klein idea loses some of its point. Pretty as the idea is, it

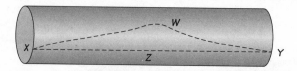

FIGURE 15.2 Redefinition of the fifth coordinate in a two-dimensional example. The fifth dimension runs round the cylinder. One dimension of ordinary spacetime runs along the cylinder. The distance round the cylinder could be measured from the line $XZY$, but it could be just as well measured from any other curve, $XWY$, for instance. This freedom is the interpretation of gauge-invariance in Kaluza-Klein theory.

seems not to lead anywhere. It makes no new predictions. There is no explanation why the five-dimensional spacetime should have the structure assumed, with just one dimension curled up very tightly. One reason for mentioning Kaluza-Klein here is to prepare the reader for the use of higher dimensions in string theory (Section 15.3).

## 15.2 Supersymmetry: Marrying Fermions with Bosons

Symmetries of various sorts play an important role in physics. There are two main kinds of symmetry: spacetime symmetries and others, often called *internal* symmetries, that have nothing obviously related to spacetime.

The spacetime symmetries include symmetry under translations, which expresses the fact that the laws of nature are the same everywhere, and rotation invariance, which says that the laws make no reference to any particular direction. In Einstein's special theory of relativity, time is included along with space in these symmetries.

An example of an internal symmetry occurs in QCD (Section 11.3). The three colours of quarks have identical properties, and the laws of nature are unchanged under transformations that mix one colour with another. What is more, this symmetry is a local *gauge* symmetry (meaning that the transformations can be done independently at each point of spacetime.)

It would be nice if internal symmetries had a geometrical meaning, as spacetime ones do. It would also be nice somehow to combine spacetime and internal symmetries. The preceding section gave a historical example of an attempt at such a combination.

In the early 1970s a possibility was discovered for combining the two sorts of symmetry in a most surprising way. Like so many discoveries in physics, this one was made independently by more than one group (three in this case) at about the same time. There were two pairs of collaborators in the Soviet Union and a third pair in the United States. This symmetry is called *supersymmetry*.

Usually, symmetry transformations mix similar things together, for example, the three colours of quark, which have otherwise identical properties (like mass, spin and electric charge). Supersymmetry transformations mix things that are in some ways as different as chalk and cheese: fermions and bosons. Remember that bosons are particles such that any number can be put at the same point (or, more accurately, in the same quantum state), but fermions are particles such that only one can be put in any given quantum state. Bosons have spin 0, 1, 2 and so on (in units of $\hbar$); fermions have spin $\frac{1}{2}$, $\frac{3}{2}$, and so on. If electrons and protons and neutrons were not fermions, stars would not exist. If photons were not bosons, light (in the ordinary sense) would not exist.

It seems at first sight absurd to think of a transformation that mixes a boson with a fermion. Zumino, one of the discoverers of supersymmetry, relates how, on his first meeting with Feynman, that great physicist said to him: "So you're the guy who thinks that fermions are the same as bosons". How, then, is this remarkable trick pulled off?

I will take the simplest possible example. Take a neutrino (a fermion) with no mass and two possible spin directions, along the direction of motion or opposite to it. Take, in addition, two hypothetical particles with zero spin (bosons), and also with no mass. Then it is possible to find a field theory for these particles, with interactions between them, which possesses supersymmetry, that is, with equations that are unchanged under transformations of the particles into each other, including bosons into fermions.

Supersymmetry becomes a little less mysterious if we realize the type of transformation involved. Ordinary symmetry transformations can be applied again and again. A circle is unchanged on rotation through 1 degree. Applying this again, we get 2 degrees, and so on. This process never comes to a stop (when we get to 360 degrees, we go round again; we do not stop). Supersymmetry is not like this. Any transformation can be applied only once.

The possible transformations in the simple model just mentioned can be represented as in Figure 15.3. In this model, the fermions and bosons both have zero masses. In more complicated examples, the particles can have mass, but there must always be fermions and bosons with the *same* mass transforming into each other under the supersymmetry. Sadly, in nature, there are no known examples of fermions and bosons with identical masses (except possibly photons and neutrinos with zero masses). So, does that not sink supersymmetry without trace?

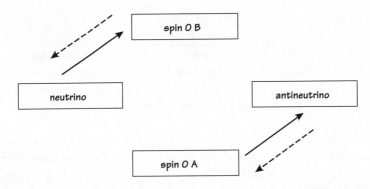

FIGURE 15.3    The possible transformations in a simple supersymmetry model. The four particle states are indicated by the rectangular boxes. The two sorts of arrow represent two supersymmetry transformations. Neither can be used more than once.

We know from electroweak theory (Section 12.5) that there can be symmetries that are broken by the Bose-Einstein condensation of Higgs particles in the vacuum, but that are still useful. If supersymmetry is to be relevant to nature, it must be assumed that it is broken in the same sort of way. In particular, particles that transform into each other may still acquire different masses. The supersymmetry would become most obvious at very high energies, at which the masses are comparatively unimportant. But the supersymmetry would also have indirect effects at low energies.

What particles should transform into each other supersymmetrically? The most popular scheme, uneconomical as it may seem, is to postulate the existence of undiscovered particles as supersymmetric partners for each of the known ones. For example, a spin 0 "selectron" as partner for the electron (spin $\frac{1}{2}$), "squarks" as partners for quarks, a "photino" as partner for the photon and so on. These hypothetical particles are assumed to be heavy enough to have escaped detection up to now. The hope is that some of them will soon be found in experiments at somewhat higher energies.

Why should people believe in schemes like these? There are two main reasons. The first is that supersymmetry deals with the problem of the apparently infinite energies coming from the quantum fluctuations of the fields corresponding to all the particles. As we saw in Section 9.10, this energy is positive for bosons but negative for fermions. As supersymmetry relates bosons to fermions, it is not surprising that it can cause cancellation between the two types of vacuum energy. Cosmology (see Section 14.1) puts a very tight bound on the amount of vacuum energy in the universe now, so it is very important to understand why it should be at most very small.

In general, supersymmetry tends to cause cancellations between quantities that otherwise might be large, and this helps to make the predictions of quantum field theory insensitive to what happens at very high energies.

The second main success of supersymmetry is more particular; it relates to GUTs. As explained in Section 12.6, these theories depend upon the prediction that the weak, electromagnetic and strong coupling strengths should approach one common value at very high energies. Assuming the existence of only the known particles, calculations show these coupling strengths not quite to meet. If supersymmetry, with its extra particles, is assumed, the three strengths meet up almost perfectly.

There is a way of looking at supersymmetry that is mathematically pretty but physically rather mysterious. It introduces the idea of *superspace*. In Section 9.5, I explained the use of mathematical quantities called *Grassmann variables*, which are not ordinary numbers but obey strange looking rules like

$$a^2 = 0.$$

In ordinary three-dimensional space, a point is determined by three numbers, its coordinates, say, $(x, y, z)$. In superspace, a "point" is specified by three ordinary numbers like these, *but also by* some Grassmann variables as well. We cannot attach any direct physical interpretation to the dependence on the Grassmann variables, but nevertheless superspace provides a compact and elegant way of calculating with supersymmetry.

Superspace provides a sort of "geometrical" interpretation of internal symmetries. The transformations that mix the Grassmann co-ordinates are connected with internal symmetries. Supersymmetry connects these with rotations in ordinary space.

## 15.3   String Theory: Beyond Points

In this book, I have written about *elementary particles*, electrons, photons, quarks and so on. The adjective *elementary* here distinguishes these objects from others that are clearly composite: atoms made of nuclei and electrons, protons made of quarks and so on. But "elementary" particles are not without structure. For example, the electron's magnetic strength has corrections of over a tenth of a percent as a consequence of a cloud of virtual electrons and positrons that it carries about with it.

Nevertheless, quantum field theory starts from fields for idealized *point* particles, then calculates the more complicated structures of physical particles. To speak of a "point" particle means that its state is entirely determined by specifying its position together with its spin direction. In contrast, to specify the state of an atom, one needs to describe its internal condition as well as the position of its centre.

People have often wondered whether the idea of a true point was realistic, but it proved rather difficult to think of anything else that was consistent with all the laws of physics. Around 1970, an alternative to the point was discovered: the *string*. Whereas a point has no dimensions, a string has one dimension. To specify the state of a string, one must say where it lies in space, that is to say, give the position of a curve in space. These strings are supposed to be infinitely thin: they have no structure at right angles to their length.

Why a string, not a membrane, or a splodge or anything else? The step from a point to a string turns out to be manageable by rather elegant mathematics. Objects of higher dimension seem to be less tractable.

A string may be closed, a little loop, or open, with two ends. Its length is not fixed. Rather it is elastic and at any moment adopts a length depending on its state of motion. The simplest example to picture is a spinning, pulsating loop, in which the centrifugal force balances the tension to prevent the string from shrinking. In the case of an open string, the very ends of the string move with the speed of light; otherwise the tension acting at the end point could not be balanced.

A point particle is characterized by one main property, its mass. Related to a string, there is a similar quantity, the mass per unit length. Multiplying by $c^2$, one gets the energy per unit length, which is the same thing as the tension in the string. Using the fundamental constants $c$ and $\hbar$, we can construct a length

$$\sqrt{\frac{c\hbar}{\text{tension}}},$$

which fixes the minimum length of the string. In popular string theories, this length is taken to be of order of magnitude of the Planck length ($10^{-35}$ metre) or a little longer. The reason for this will be given in a moment. Clearly strings are much too tiny to be observed as strings. The mass per unit length is of the order of the Planck mass divided by the Planck length. Expressed in ordinary units, this is a force of $10^{44}$ newtons!

String theories first arose out of attempts to understand the excited states of mesons. But in the early 1970s the strategy shifted completely. It was realized that a closed string has a state of motion (a rotating ellipse) with spin $2\hbar$ and zero total mass. This is naturally identified with the graviton, so the aim became more ambitious: to construct a unified theory of gravity with all the other forces of nature. In general, strings moving in their lowest quantum states are to be identified with ordinary particles. Higher states of excitation have much higher masses, comparable perhaps with the Planck mass.

In spite of this, there are, at the time of writing, new ideas about applying string theory to the particular subject of QCD (the theory of forces between quarks in hadrons), in a partial return to the original motivation. The theory seems to be protean enough to have more than one application.

Although a string sounds like a fairly simple physical object, it turns out to be hard to find a complete theory consistent with all the requirements of relativity and quantum theory. In fact, when consistent theories were first found, in the early 1980s, the discovery caused enormous excitement among particle theorists. The consistent theories had two important features: supersymmetry had to be built in, and higher dimensions were needed.

A rough way to see how string theory can be made supersymmetric is to think of the string as moving in superspace, as defined at the end of the last section. As usual, supersymmetry is helpful because it makes possible cancellations between bosons and fermions.

Even when supersymmetry is incorporated in this way, to be consistent string theories need more than four dimensions of spacetime. For example, a typical consistent superstring theory employs a superspace with 32 Grassmann dimensions and 10 "ordinary" spacetime dimensions. At this point, one might well think that the whole thing is a fairy tale, and nothing to do with the physical world.

But there is a way forward, This is to use the Kaluza-Klein trick. In Kaluza-Klein theory (Section 15.1) we start with five dimensions of spacetime but compactify one of them by rolling it into a little cylinder. In string theory, we may compactify the six unphysical space dimensions, in some generalization of the cylinder. This may be even turned to advantage, because the choice of compactification can build in more, interesting physical structure. As with Kaluza-Klein theory, we ought ideally to be able to predict the manner of the compactification, not just guess at it. No one has yet achieved this.

If an unphysical dimension is compactified into a cylinder, a string can relate to that dimension in two possible ways. It can move as a whole round the cylinder, just as a particle can. Such a motion has kinetic energy. But also, a closed string can be wound round the cylinder, as an elastic band is wound round a rolled up newspaper. Then the string is kept stretched and cannot shrink shorter than the circumference of the cylinder. It consequently has stored potential energy. These two sorts of configuration, with their two sorts of energy (kinetic and potential), are related to one another in an interesting way. They are called *dual* to each other.

The history of a moving particle is represented in spacetime by its *world-line* (see Section 5.4). Similarly, the history of a moving string is represented by its *world sheet*. A slice through this sheet at any given time gives the position of the string at that time. In the case of a closed string, the world sheet is a sort of tube.

A string may break into two pieces, or two strings may fuse into one. Figure 15.4 shows how processes like this look in spacetime diagrams. Two strings (closed ones, in this example), *A* and *B*,

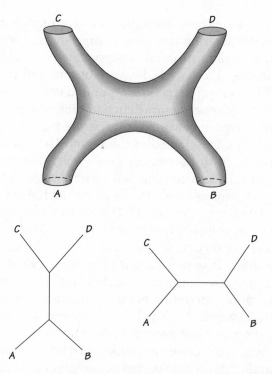

FIGURE 15.4    A world sheet string diagram and two related Feynman diagrams. Time is plotted vertically. *A* and *B* are approaching at early times, and *C* and *D* are receding at late times.

approach each other and merge. Later, the single string breaks up again into two, *C* and *D*, which move apart. The complete world sheet looks a bit like a balloon with four nozzles attached to it. In quantum theory (in Feynman's formulation, Section 8.8), one must sum over all such histories. This sum will include cases in which the balloon is long and thin in one direction or another.

Suppose the incoming and outgoing strings *A, B, C, D* are in states corresponding to physical particles. Then the process going on is the collision of two particles, *A, B*, to produce two others, *C, D*. In ordinary quantum field theory, such processes can be represented by Feynman diagrams (Section 9.8). Two such possible Feynman diagrams are drawn in the lower part of the figure. They can be thought of as extreme cases of the string diagram, in which

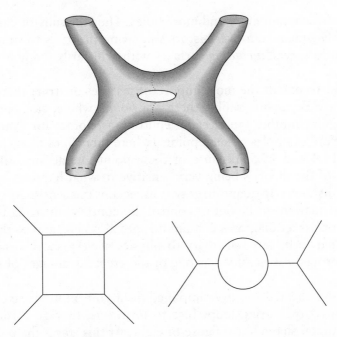

FIGURE 15.5    Corrections to the diagrams in Figure 15.4. At the top is a string world sheet with a hole in it. At the bottom are two examples of Feynman diagrams obtained by stretching out the world sheet in appropriate ways.

the world sheet is stretched out very long and thin in different ways. Note that a *single* string diagram corresponds to more than one (two in this case) Feynman diagram.

In the case of ordinary quantum field theory, the Feynman graphs in Figure 15.4 give only the first approximation in calculations about the collision process. To improve the approximation, graphs like those at the bottom of Figure 15.5 must be included. In string theory, there is a corresponding type of world sheet: that at the top of Figure 15.5. It has a hole in it, so the surface is like an inner tube with four nozzles attached to it. By stretching it so that parts become long and thin, it can be made to resemble either of the Feynman diagrams below it (amongst others).

There is a series of other Feynman diagrams giving yet smaller corrections to the collision amplitude. The corresponding string diagrams contain more and more holes. Thus the rule for carrying out a Feynman sum-over-histories in string theory is to sum over all world sheets, *including* surfaces with arbitrarily many holes in them.

Now to one of the most important features of string theory. In ordinary Feynman graphs, there are *points* where, say, one particle emits another (an electron emitting a photon, for example). There are examples of such points (where three lines meet) in Figures 15.4 and 15.5. Because of these point interactions, there is a risk of the theory's being very sensitive to what happens at very small distances. In quantum gravity especially, this sensitivity causes the calculations to be out of control. By contrast, in string theory the worldsheet diagrams contain no special points, and so there is expected to be less sensitivity to physics at very short distances. This property is heralded as one of the great advantages of string theory.

(I have cheated and oversimplified the argument in the last paragraph. As two string loops fuse to form one, they go through a transitional shape like a figure of eight. At this stage, the point at the centre of this 8 is a definite event, that is, a point in spacetime. There are no such special points on the world sheets in Figures 15.4 and 15.5, because I have drawn them without regard to the orientation of the sheet everywhere with respect to the light cone. The claim is that the correct way to calculate the Feynman sum-over-histories is by using smooth surfaces like the ones in the figures. We came across a similar claim in Section 13.7.)

String theory has not up to now made any clear contact with experimental results. Nonetheless, many of the brightest theoretical physicists are convinced that it has such beauty and promise that it must represent the way forward. Certainly it has yielded an extraordinarily rich and fascinating mathematical structure, which seems always to be offering glimpses of more hidden treasures.

One of the successes of string theory is to create a complete and consistent model of a black hole. I say "complete and consistent" because quantum theory is included and Hawking radiation (Section 13.4) accounted for. The model black hole has a large number of possible internal states – the right number to account for the hidden information demanded by black hole entropy

(Section 13.5). I say "model" because (at the time of writing) there is no theory of any black hole that might exist in the real world. More complications have to built in before that is achieved. The ideas going into these models include all those mentioned in the preceding sections of this chapter and in the next section.

## 15.4 Lumps and Hedgehogs

A dichotomy that has run through physics from the time of the ancient Greeks is that between point particles ("atoms") and the continuum. One strand of thought is that there is nothing except "atoms and the void". Another attaches importance to some continuous "aether".

One may ask the question whether particles perhaps are not really points but are constructed as some small, stable configuration of a continuum. In fluid flow, vortices may form and retain their identities for some time. Could particles be something like vortices in the aether? In the nineteenth century, a vortex theory of atoms was seriously proposed. (This is not to be confused with Descartes's vortex theory of planetary motion, mentioned in Section 1.7).

The discovery of quantum theory has subtly modified this dichotomy, because, as we have seen in Chapter 9, the quantum theory of a field (a continuous thing) is *equivalent to* the quantum theory of a number of particles. But one may still ask whether there exist small, stable configurations of *classical fields*, and whether these might not masquerade as "particles".

(There is yet another possible source of confusion. In the last section, I described string theory, whereby strings are not points but are extended in one direction. But these strings are assumed to exist as fundamental objects, and not to be constructed out of any continuum, so they are not the subject of this section.)

In the physics of solids and liquids, small, stable constructions certainly do exist. For example, in superfluid liquid helium, there can be vortices, which are thin and persistent. These vortices are characterized by two attributes (see Section 10.4). Down the core of a vortex the fluid reverts to its non-superfluid phase. And, on encircling the vortex, the phase angle of the order parameter increases by 360 degrees (or a multiple thereof). The flux tubes in superconductors (see Section 10.6) are similar and in addition have a magnetic field trapped within them.

These vortices or flux tubes are not particle-like, because they are long in one direction. But very thin films of superfluid helium have been studied, and in them a vortex running the (very short) distance across the film is particle-like. As the temperature of such a superfluid film is raised, more and more of these particle-like little vortices form and are responsible for the breakdown of superfluidity. (Of course, superfluid helium is not really a continuum but is made of helium atoms. The vortices are large compared to atoms but much smaller than the macroscopic area of the helium film.)

The analogy with superconductivity helped to stimulate the invention of the electroweak theory in particle physics (Section 12.5). Are there any constructs similar to vortices and flux tubes in electroweak theory? There is one such object that arises rather naturally, and that is a *magnetic pole*. In Section 3.5, I pointed out that in Maxwell's theory there was a *duality* between electricity and magnetism, but yet, as a matter of observation, electric charges exist but isolated magnetic poles (say, a north pole without a south pole) have not been discovered.

In certain field theories with Bose-Einstein condensation of a Higgs field, monopoles can arise as follows. The vortices in superfluids and flux tubes in superconductors are related to what happens to the phase angle of the order parameter as you circle round the vortex or tube. If you circle round once (going through 360 degrees), the phase angle increases by 360 degrees (or a multiple thereof). So there is a relation between two angles, an angle in ordinary physical space and the phase angle of the order parameter. It is like an elastic band round a cylinder. The band might just lie on top of the cylinder (so it can be lifted off), or it could circle round it once, or twice or more times. The angle going round the band is linked to the angle going round the cylinder.

By relating angles like this, one can create only long thin objects, not point-like ones (in three-dimensional space). But if one wraps a sphere round a sphere one can create a point-like configuration – something with a point centre. For example, a balloon can lie on top of a football (American soccer ball), and then it can be lifted off and shrunk. But if the football is forced inside the balloon, and the balloon then sealed up, it is trapped: the balloon cannot be removed or shrunk. It is harder to visualize, but a balloon may be wrapped twice or more times round a sphere.

To realize this possibility, we need something in a two-dimensional space (like a spherical surface) rather than a phase angle that runs in one dimension (round a circle). In a field theory with three Higgs fields (undergoing Bose-Einstein condensation), there is just this possibility. The three fields live in a three-dimensional space (not physical space, but a mathematical construct), and their values may be restricted to run over a sphere in that space. The object thus constructed has at its centre a point at which Bose-Einstein condensation breaks down, and on any sphere about that point the values of the three fields "wrap" themselves around.

Now suppose these Higgs fields are part of some electroweak or GUT theory, which includes the electromagnetic field. Then the object just described turns out to be a magnetic pole: magnetic lines of force emerge from it radially in all directions. Where do these lines of force emerge from? Not from the very centre: there is no point pole there. Rather, the source is smeared out over the region in which Bose-Einstein condensation dies away on approaching the centre.

Such an isolated magnetic pole is called a *monopole*, to emphasize the difference from the two poles of a magnet (a *dipole*).

This way of constructing monopoles was discovered independently in 1974 by 't Hooft in Holland and Polyakov in Russia. The objects have sometimes been called *hedgehogs* because the directions of the spines vary at different points on a hedgehog in a vaguely similar way to the "directions" of the Higgs fields at different points round a pole.

The strength of this magnetic monopole is very simple. It is given by

$$\text{(magnetic pole strength)} \times \text{(electric charge)} = 4\pi c\hbar,$$

where by *electric charge* I mean the charge associated with the Higgs field, let us say the charge on the electron.

The simplicity of this formula is no accident. To see this, let us check that it is consistent with what we know about the quantum theory of electrically charged particles moving in magnetic fields (Section 8.12). If, say, an electron moves around in a closed path (getting back where it started from), its wave function's phase changes by an angle that depends on the magnetic flux through that closed path (times the charge on the electron). For this rule to make sense, the "flux through the closed path" must be unambiguously defined. To measure the flux, one chooses a closed surface with the path as its boundary and counts the number of lines of force through the surface. But if a magnetic monopole is present, the result depends upon the choice of surface, as seen in Figure 15.6, where the surface might be drawn to the right or to the left of the pole.

The difference in flux between the two choices is proportional to the pole strength. As a result of this ambiguity, there is an ambiguity in the phase of the wave function of the electron of amount

$$\frac{1}{ch} \times \text{(magnetic pole strength)} \times \text{(electron charge)}.$$

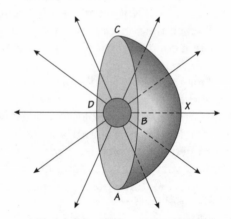

**FIGURE 15.6**  An electron moves in a closed path *ABCD* near a magnetic pole. The magnetic lines of force are shown. To calculate the magnetic flux through *ABCD*, one may choose the hemisphere going through *X*. Alternatively one may choose a similar hemisphere on the other side of the pole. The two answers are different: in the first case the lines of force are going from left to right through the hemisphere; in the second case, from right to left. The difference in flux is the total number of lines of force emerging from the pole, which is a measure of the pole strength.

But the phase of a wave function is an *angle*, and a change of 360 degrees, that is, $2\pi$ radians (or a multiple thereof), makes no difference. Thus the ambiguity is unimportant provided that

$$(\text{magnetic pole strength}) \times (\text{electron charge})$$
$$= 2\pi c\hbar \times (\text{whole number}).$$

The monopole constructed here certainly satisfies this condition (with the "whole number" equal to 2).

The consistency condition, relating magnetic monopole strength to electric charge strength, was first derived by Dirac as long ago as 1931. It is a wonderful example of the interaction of electricity and magnetism with quantum theory.

There is no evidence that magnetic monopoles exist. There are reasons to believe that they should have a large mass, so that might explain why they have not been made on Earth. But as the early universe cooled, and Bose-Einstein condensation of Higgs fields took place (see Section 14.4), monopoles should be formed, as "defects" in the condensate (a little like imperfections in a crystal). It is an important fact about our universe that such monopoles have escaped detection on Earth.

*If* monopoles exist anywhere, the symmetry between electricity and magnetism is complete. It is an interesting sort of symmetry, which has been given the name *duality*. The rule connecting electric charge and magnetic monopole strengths involves a whole number, so it is called a *quantization of charge*. If any monopole exists, any possible electric charge must be quantized to be a whole number multiple of

$$\frac{2\pi c\hbar}{(\text{magnetic pole strength})}.$$

This would be a possible explanation of the observed quantization of charge. Other possible explanations have been mentioned (Sections 12.6 and 15.1).

We know that the electric charge on the electron is smallish, in the sense that

$$\frac{(\text{charge})^2}{4\pi c\hbar} \approx \frac{1}{137}.$$

The rule connecting monopole strength to electric charge implies that the corresponding quantity

$$\frac{(\text{magnetic pole strength})^2}{4\pi c \hbar}$$

is at least $\frac{137}{4}$, that is, quite large. One consequence of this is that calculations about monopoles are difficult.

In string theory using (compactified) 10-dimensional spacetime, there are many possibilities for monopoles and generalizations to more dimensions. At the time of writing, duality in string theory is much discussed. It might, for example, afford a way of translating a difficult, strong coupling problem into an easier, weak coupling one.

## 15.5 Gravity Modified – a Radical Proposal

An idea put forward during the 1990s is so startling that I cannot resist mentioning it, however slender may be the chance of its being true.

Einstein's geometric theory of gravity is so compelling that physicists are naturally reluctant to tamper with it. Yet it has not been tested at distances of about a millimetre or less. To see the difficulty of making such a test, note that the gravitational acceleration due to a mass of one gram and one centimetre away from it is some $10^{-10}$ of the acceleration due to the Earth's gravity. The idea is to modify gravity theory at distances of perhaps about one millimetre, yet retain the principle that gravity is a manifestation of curvature.

Suppose, to take a particular possibility, there are five dimensions of space and one of time, but that two of the spatial dimensions are "curled up" with a size of the order of a millimetre. Suppose that gravity extends to all this six-dimensional spacetime, but that all the other forces and particles of nature confine themselves to a very thin slice in the extra two dimensions. (It is possible that confinement to such thin slices may be a consequence of string theory.) Figure 15.7 gives a very rough indication of what happens.

My discussion is framed in terms of Newtonian gravity, though it ought to be possible to understand it in terms of spacetime curvature. Gravitational lines of force emerge from a mass at a point $P$.

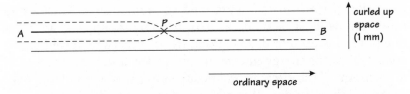

FIGURE 15.7   An indication of how gravity might be distributed in a spacetime with extra curled up dimensions. One dimension of ordinary space is plotted horizontally, and an extra dimension vertically. Everything except gravity is supposed to be confined to the very thin slice *AB*. The dashed lines represent gravitational lines of force emerging from a mass at *P*.

They extend into the whole of the enlarged space. Far from *P*, that is, at distances much more than 1 mm, they distribute themselves very nearly as they would in ordinary three-dimensional space, so there is the usual inverse-square law of force. However, most of the lines of force lie outside the thin slice *AB*, so they are not able to produce a force on other masses (which all lie in *AB* by hypothesis) and are "wasted". Gravitation is effectively much diluted. So, to account for the observed extreme weakness of gravitation, the fundamental strength of gravitation need *not* be very small. It may, for example, be characterized by a mass equal to about 1,000 proton masses, not the $10^{19}$ proton masses of the Planck mass. The weakness observed, for distances much larger than 1 mm, is due to the dilution of the lines of force in the extra dimensions.

But at distances comparable to 1 mm, the lines of force are distorted, as in the figure, and the observed gravitational force gets stronger. This is one of the striking effects to be expected according to theories of the kind described.

Consider a very high-energy collision between elementary particles, with energies of the order of 1,000 times the rest energy of a proton (such as may be achieved in a few years). At these energies, the fundamental strength of gravity (in six-dimensional spacetime) is not weak. Therefore a graviton may be emitted as well as ordinary particles. An example of a process like this is

electron + positron → photon + graviton.

The graviton has a wavelength much less than 1 mm and is emitted from the slice $AB$ in the figure, and so is unobservable (since all observations must be made in $AB$). It therefore appears as if just one photon is produced, something that is otherwise forbidden by conservation of energy and momentum (for instance, if the total momentum of the electron and positron together is small, the photon would have to have the same small momentum and therefore a small energy – $c$ times the momentum – too, whereas the energies of the electron and positron are large). Of course, energy is really conserved: it just escapes with the invisible graviton. The occurrence of crazy-looking processes like this is another striking prediction of this type of theory.

At the least, ideas like the one just described encourage physicists to question basic laws of physics and to undertake experiments to check them.

# APPENDIX A

# THE INVERSE-SQUARE
# LAW

Newton's law of gravitation (Section 1.7) has an inverse-square dependence on the distance. Double the distance means a quarter the force, and so on. This particular dependence on distance (as opposed to, say, an inverse cube or some other rate of decrease) has several special features, which I will explain in this appendix.

First of all, there is an interesting mathematical *analogy* with another physical situation. I stress that it is only an analogy: it is not meant to be a physical explanation of gravity, or anything like that.

The analogous situation is this. Imagine a large tank of still water. In the middle, put a small device near the center that sucks out water at a steady rate. (See Figure A.1.)

Where the water goes does not concern us: we can imagine it being drawn out through a narrow pipe that does not significantly interfere with the rest of the water in the tank. As a consequence of sucking out of the water, there must be a flow of water from distant parts of the tank towards the device, which I shall henceforth call (for want of a better word) an *extractor*. We also want to arrange that the sucking action of the extractor is the same in all directions. Then the flow in the tank will be the same in all directions. We want to work out the speed at which the water in the tank is flowing towards the extractor, and in particular how this speed depends upon the distance from it. Imagine any sperical surface in the fluid with its centre at the extractor. The total flow of water in through that surface is given by multiplying the area of the surface by the speed of flow at the surface. But also, this total flow is the same

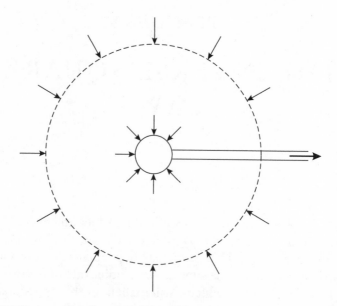

FIGURE A.1    An "extractor" sucking out water in a large tank. The flow in the tank is radially inwards. The total flow in through the surface of the larger sphere is the same as that into the extractor and is therefore independent of the size of the sphere.

fixed amount for any such spherical surface, independent of its size, because it is just equal to the rate at which the extractor is sucking out water. (We have actually made an assumption here: that the water cannot be compressed or expanded. This is a good approximation for water under normal circumstances.)

Thus we get the equation

$$(\text{area} \times \text{speed}) = (\text{total flow}),$$

or

$$\text{speed} = \frac{\text{total flow}}{\text{area}}.$$

The next thing we need to remember is that the area of a surface goes up with the square of its size. In fact, the area of a sphere is $4\pi$ times the square of its radius, though this number $4\pi$ is not actually important to us. In the case of the spherical surfaces we have imagined in the water tank, the radius is just the distance from the extractor. So the last equation becomes

$$\text{speed} = \frac{\text{total flow}}{4\pi \times (\text{distance})^2}.$$

This is tells us how the speed depends upon the distance and is nothing but the inverse-square law.

This little deduction about the flow of a liquid uses two things. The first is the *spherical symmetry* that we have assumed to hold: that is, the fact that the flow is the same in all directions from the extractor. The other is the *conservation* (and incompressibility) of water: water does not appear or disappear anywhere in the tank except at the small extractor, so that the same total amount flows through any closed surface surrounding that device.

In our analogy, the extractor corresponds to a small mass and the speed of the flow at a point corresponds to the gravitational force that would be exerted on another small mass there.

We can use this analogy to give a simple proof of the theorem that, when he proved it, enabled Newton to complete his work on gravitation. This is the theorem that gives the gravitational field produced by a *large* body (like the Sun), provided it has spherical symmetry. We assume that this large body is made up of a lot of small "particles", each of which produces a gravitational force going down according to the inverse-square law. In order to use the analogy, imagine a lot of small extractors arranged in a spherically symmetric way (say, uniformly throughout the inside of a sphere).

We want to find the consequent flow of water. Inside the sphere, amongst the extractors, the flow will be rather complicated. But suppose we go to any point outside. We can then use exactly the same argument as before, depending on spherical symmetry and conservation of fluid. The result will be the same: a flow directed towards the centre of the distribution of extractors, and falling off as the square of the distance from that centre. So the spherically symmetric distribution of extractors behaves, as seen from outside that distribution, as if there were just one stronger extractor at the centre.

Because of the analogy with gravitation, we can say the same about a spherically symmetric distribution of small masses: at points outside it gives the same gravitational force as if all the mass were concentrated at the centre.

This tells us the gravitational force exerted by a large spherical body on a small one. Using Newton's second law (action and reaction are equal and opposite), we can find the force exerted by the small body on the large one. It too will be just as if all the mass of the large one were concentrated at its centre.

Finally, combining our two results, we conclude that the force between the two spherical bodies (say, the Earth and the Moon) is the same as if each of them had all its mass concentrated at its centre. Thus an apparently rather complicated situation is replaced by a simpler one.

The inverse-square law is essential for this demonstration. The theorem is not true in general for any other law of force.

The analogy we have used here shows us that the mathematics of the flow of an (incompressible) fluid is the same as the mathematics of Newton's gravitational force. This is true generally, not just in the simple case of spherical symmetry.

The attraction or repulsion between two electrically charged particles also, like gravity, obeys the inverse-square law. Therefore, the same mathematics of incompressible fluid flow provides an analogy for electrical forces too. We can even make an analogy for the two opposite kinds of electricity, positive and negative. We do this by imagining two sorts of devices in the fluid, one sucking in fluid (as we have assumed up to now) and one exuding fluid out. We may call these *extractors* and *sources*.

The electrical field lines introduced in Section 3.1 have an interpretation in the fluid analogy: they represent the direction of flow, and the number of them going through any surface gives the quantity of fluid flowing through that surface.

I emphasize again that all this is merely an analogy. I have not proved that gravity must obey an inverse-square law or explained why it has an inverse-square law. But we can perhaps say that the fluid flow analogy suggests that the inverse-square law is a natural sort of thing. Also, to prevent any possible misunderstanding, Descartes's theory of gravitation as due to whirlpools in the "material of the heavens", although it involves the flow of some sort of fluid, has nothing whatsoever to do with the analogy above.

There is another special property of the inverse-square law, most conveniently stated about electric charge. Take an electrically conducting closed shell (of any shape), with electrical charge on it and with a cavity inside. Assume that it has settled down so that nothing is changing with time. Then the electric field is zero everywhere in the cavity.

To prove this, we assume that there is a field somewhere inside and show that this leads to a contradiction. There can be no electric field within the conducting shell, for if there were, it would cause charge to flow, contrary to the assumption that nothing is changing with time. Therefore any field line within the cavity must form a closed loop. This would contradict the fourth rule in Section 3.1, because the electric field would do work on a charge taken around such a loop.

This property was used, by Priestley, for example, to verify the inverse-square law indirectly.

Finally, there is another respect in which the inverse-square law has very special consequences. This is related to the shape of the

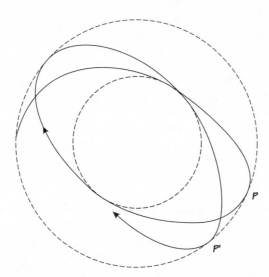

FIGURE A.2   A possible orbit in an attractive force that does not obey the inverse-square law. The orbit does not "close up". The points like $P$ and $P'$ are called *perihelions*. The fact that $P'$ does not coincide with $P$ is "the precession of the perihelion".

orbits of bodies moving under the inverse-square attractive force. These orbits have the form of ellipses (or hyperbolas or parabolas). It was Newton's great triumph to prove this. But what would happen if the distance dependence of gravity were not the inverse square one, say, falling off a little less fast or a little faster? What would be the shape of the orbits then? They would not be ellipses, of course. But in general they would look quite different. They would not even "join up", never getting back to having the same direction at the same place. An example of a possible orbit is shown in the Figure A.2, to be contrasted with the elliptic case. Thus the property of "joining up", which one could easily take for granted, is in fact a very special one. It is one of the fortunate "accidents" that went towards making the Solar System intelligible.

We saw in Chapter 7 that, in Einstein's theory of gravity, the force law is *not* exactly inverse-square (though it is very close to that), so the orbits do not exactly "join up".

# APPENDIX B

# VECTORS AND COMPLEX NUMBERS

This appendix gives a brief explanation of two important mathematical ideas: vectors and complex numbers.

First, *vectors*. There are many physical quantities that naturally have a direction associated with them, in addition to their magnitude. Examples are velocities, forces, electric fields. These are vectorial quantities, and mathematicians say that they can be "represented by a vector". But the simplest example of a vector is the geometrical displacement (in a straight line) from one point to another. (The Latin word *vector* means "carrier", as in the usage *insect vector*). We may use this example to illustrate the mathematical properties of vectors.

Mathematicians use symbols to denote vectors, and it is conventional to use boldface type, **v**, **E**, and so on, to emphasize that they are not ordinary numbers. Are there mathematical operations that can be performed on vectors, addition, multiplication, and so forth? There is a very simple and natural definition of addition. It is illustrated in Figure A.3.

Any vector (in ordinary three-dimensional space) can be made up as the sum of three vectors, one in each of three specified independent directions. Thus a vector needs three ordinary numbers in order to specify it – the lengths of the three "component" vectors.

Multiplication of vectors is a more complicated question. There is no way to define an operation of "multiplication" acting on a pair of vectors, which has *all* the properties of multiplication of ordinary numbers (like $x \times y = y \times x$).

Now *complex numbers*.

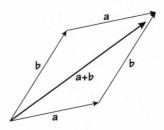

FIGURE A.3 The symbols **a** and **b** represent two displacement vectors. The sum **a** + **b** is defined to represent the single displacement, which has the same effect as the two original ones done in succession. The order does not matter.

Phrases like *irrational numbers, surds, transcendental numbers* and *imaginary numbers* show that mathematicians have struggled to understand various new sorts of "numbers" that they wanted to use but had difficulty in defining.

Most people accept the whole numbers 1, 2, 3, and so on, as intuitively obvious. For many practical applications, what we need are what mathematicians call the *real numbers*. These are decimals that do not necessarily stop. For example,

$$\sqrt{2} = 1.4142136\ldots$$

is a decimal that "goes on forever". This just means that, if we need it to any specified degree of accuracy, we can in principle calculate it to the required number of decimal places. But it never stops or repeats itself.

The simplest representation of real numbers is as distances along a straight line, measured from some origin O. If we call distances to the right of O *positive*, then we can represent *negative* numbers as distances to the left.

Complex numbers can be thought of as things that are represented by points in a *plane*, thus generalizing the one-dimensional line to the two-dimensional plane. Another way to think of complex numbers is as two-dimensional vectors, that is, vectors confined to some given plane. One way to specify a complex number is illustrated in Figure A.4. We give the distance from the origin O, which I will call the *length*, and the angle with the horizontal axis

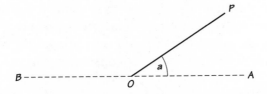

FIGURE A.4   A complex number as representing a point in a plane with respect to an origin O. The point P represents a typical complex number. The length of the line OP is called the *length* of the complex number. The angle marked as a is called the *phase angle* of the complex number. Real numbers are a special case of complex numbers; they are those represented by points on the line BOA.

(on the page of the book), which I will call the *phase angle*. The length is by definition a positive real number.

But, to get a useful definition of something worthy to be called a "number", we should be able to define *addition* and *multiplication*, and these operations should obey the same rules as they do for ordinary numbers.

The addition of complex numbers is the same as for vectors, illustrated in Figure A.3. We have to shift one of the numbers, say, **b** in the figure, so that it is measured from the tip of the other one, **a**.

The definition of *multiplication* is not quite so simple. It may be expressed as the following rule:

- To multiply two complex numbers, multiply their lengths and add their phase angles.

Thus if one complex number has length 2 and angle 20 degrees and the other has length 3 and angle 30 degrees, then multiplying them together gives the complex number with length 6 and angle 50 degrees.

It turns out that with these two definitions the ordinary rules of addition and multiplication, like those exemplified, are obeyed. The most difficult of these rules to verify is

$$c \times (a + b) = (c \times a) + (c \times b).$$

(Complex numbers are not usually denoted by boldface type.) One way to see why the rule works is the following. Take the triangle representing the addition of *a* and *b*, as in Figure A.3. Then scale this figure up by the length of *c* and rotate the figure through the angle that is the phase angle of *c*. Clearly the same scaling and rotation are done to all three sides of the triangles in the figure. This proves the rule.

Complex numbers are useful and important because they are the biggest generalization of numbers possible without sacrificing some of the laws of arithmetic.

Since complex numbers generalize real numbers, one can do things with complex numbers that are impossible with real ones. For example, square roots of negative numbers exist among complex numbers (but not among real ones). In fact all roots (square, cube and so on) of all numbers exist among complex numbers. Complex numbers may be said to be "complete" in this sense.

Figure A.5 is an example. Here the points represented by *U* and *V* represent the two square roots of −4, which is represented by the point *B*.

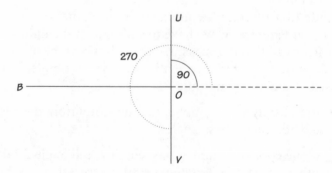

FIGURE A.5   The complex square roots of −4. The line *OB* represents −4. It is a complex number of length 4 and angle 180 degrees. The complex number represented by *U* has length 2 and angle 90 degrees. If you square it (multiply it by itself), you get the complex number with length $2 \times 2 = 4$ and angle $90 + 90 = 180$ degrees, that is, −4, as required. Thus *U* represents a square root of −4. But *V* represents another such square root. *OV* also has length 2, but angle 270 degrees. On squaring, one gets an angle $270 + 270 = 540$. But $540 = 180 + 360$ and 360 is a complete rotation; so 540 degrees is the same thing as 180 degrees.

It was in order to give a meaning to the square roots of negative numbers that complex numbers were invented. The complex number with length 1 and angle 90 is usually denoted by the letter $i$, and it is a square root of $-1$. The other square root has length 1 but angle 270 degrees. It may also be written $-i$. Any complex number can be written in terms of real numbers and $i$, but we will make no use of this notation.

# APPENDIX C

# BROWNIAN MOTION

The object of this appendix is to explain the property of Brownian motion mentioned in Section 2.5. We will start with a simple model situation (whose relevance may not be immediately apparent). Two people toss a coin for money. If a head comes up, he pays her one pound. If a tail comes up, she pays him one pound. On average, if the game is played many times, each player's winnings and losses will tend to cancel out.

Now suppose that each day the players play a sequence of a certain fixed number of tosses. Call this number $n$. At the end of the day, they note how many pounds he has paid her. Call this amount the winnings, $W$. Note that if he has won, $W$ is a negative number. Now square this: $W^2$. (Never mind what a "square pound" means.) Note that the square of a number is always positive, whether the number itself is positive or negative. Now average over lots of days. The average value of $W$ will be zero (if it were not, we would suspect that the coin was biased). But the average value of $W^2$ is certainly not zero, because each value of $W^2$ is positive (or perhaps zero). In fact, it can be shown that the average value of $W^2$ is just the number $n$ (the number of games each day).

I will illustrate this with two examples. First suppose $n = 2$. The possible outcomes of the games are HH, HT, TH, TT, where $H$ stands for head and $T$ for tail. For each of these outcomes, the values of $W$ (the winnings) are 2, 0, 0, −2. So the values of $W^2$ are 4, 0, 0, 4. Since the four outcomes are equally likely, to find the average add up the four possible values and then divide by 4. So the average value of $W^2$ is

$$\frac{4+0+0+4}{4} = 2.$$

This is just the value of $n$, as claimed.

Let us try $n = 3$. The possible outcomes are HHH, HHT, HTH, THH, HTT, THT, TTH, TTT. The winnings are $3, 1, 1, 1, -1, -1, -1, -3$, so the values of $W^2$ are $9, 1, 1, 1, 1, 1, 1, 9$. The average of these is

$$\frac{9 + 1 + 1 + 1 + 1 + 1 + 1 + 9}{8} = 3,$$

and again this is the value of $n$.

What has all this to do with Brownian motion? Consider someone (drunk perhaps) walking on a long narrow paved footpath. At each step she goes from one paving stone to one of the two neighbouring stones and is equally likely to step forward as backward. Suppose, after $n$ steps, she ends up $W$ stones forward of where she started (allowing negative values of $W$ to represent backward motion). The chances are exactly like the coin tossing, and again the average value of $W^2$ is equal to $n$.

This argument extends to the case of someone on a large chessboard, who at each step moves to one of the four adjacent squares with equal likelihood. Now one can show that the average value of the square of the *distance* is equal to the number of steps (counting the length of the side of a square as the unit of distance). The reason for this is Pythagoras's theorem, which says that the square of the distance is the sum of the squares of the number moved north and the number moved east. The idea generalizes to three dimensions too.

If the walker takes steps at a constant rate,

$$n = \text{number of steps per minute} \times \text{time (in minutes)},$$

and

$$[\text{distance}]^2 = [W \times \text{length of each step}]^2.$$

Using these two relations, together with

$$\text{average } (W^2) = n,$$

we arrive at

$$\text{average } [(\text{distance})^2]$$

$$= (\text{time}) \times (\text{number per minute}) \times (\text{length of step})^2.$$

Thus the average of $W^2$ is proportional to the time, the result stated in Section 2.5.

# APPENDIX D

# UNITS

When we learn arithmetic we learn rules with numbers like 6 × 20 = 120, but in applications we are always concerned with numbers *of* something or other. For example, if one had six packets of biscuits with 20 biscuits in each packet, the equation might be written out in the form

(6 packets) × (20 biscuits per packet) = 120 biscuits.

This example refers to whole numbers, but the same sort of thing holds for decimal numbers, for example:

(81 cubic metres) × (118.610 pence per cubic metre)

= 96.07 pounds

(this is, of course, an approximation, correct to the nearest penny). Here, cubic metres and pence and pounds are examples of *units*. In the first example, I suppose one might say that biscuits and packets were units.

In both examples, the units on both sides of the equations match up consistently. In the second equation, the cubic metres on the left "cancel out" (the *per* means "divided by"), and the money units are common to both sides of the equation.

These remarks are rather trivial and self-evident. But in science, dealing with units, although sometimes annoyingly tricky, can also be useful and important.

A list of units that might be used at some time or other in science would be long. It would include

miles, millimetres, light years, hours, years, nanoseconds, acres, litres, gallons, fluid ounces, ounces, grams, joules, calories, watts, volts, amps, coulombs, pascals, rads, degrees celsius, kelvin, bits, bytes

Fortunately one can reduce the number of units one needs to bother with. Firstly, this can be done in an obvious way, by agreeing, for example, to express all lengths in metres and never to use yards. (It is a trivial matter to switch from metres to millimetres, and so on, if that is convenient.)

The number of units can also be reduced by going back to a possible process of measurement. For example, quantity of electric charge can be deduced by measuring the force between two (equal) electric charges that are some known distance apart and using Coulomb's law (Section 3.1). We can, if we wish, *choose* to measure charge in terms of a unit that makes Coulomb's law simple. If someone else chooses to use another unit of charge, like the traditional coulomb, then to convert from one to another is a tedious matter, but nothing more. In this way, all the units used in electricity and magnetism can be reduced to units of *time, length* and *mass*.

The progress of science reveals new connections between things and therefore tends to reduce the minimum number of independent units that are necessary. For example, heat originally seemed to be something on its own, not reducible to mechanical terms, and so needing its own unit (say, the calorie). But then it was realized that heat was just a form of energy (see Section 2.4), so it could be measured in joules – a joule is just shorthand for

$$\text{kilogram (metre)}^2 \text{ (second)}^{-2}.$$

What, then, is the significance of the equation

$$1 \text{ calorie} = 4.252 \text{ joules}?$$

A calorie was defined to be the heat required to raise the temperature of 1 gram of water by 1 degree Celsius. The degree Celsius itself is defined so that 0 and 100 degrees correspond to melting ice and boiling water. The number 4.252, therefore, is something that depends in a complicated way upon detailed properties of water and mercury. It has no fundamental or universal significance at all. If we want to understand units in relation to the basic laws of physics, it is clearly much better to use joules than calories.

In a similar way, temperature was understood in terms of the average energy per molecule in random heat motion, and so temperature too could be measured in joules. The conversion factor is Boltzmann's constant, $1.38 \times 10^{-23}$ joule per kelvin. This is perhaps an inconveniently small number, but from a fundamental point of view this is a small price to pay.

In this way, by about the year 1900, there seemed to be three fundamental types of unit, those of time, length and mass.

Sometimes deductions about physical laws can be made by considerations about how units must consistently match up. This type of argument is called *dimensional*. I will give the usual textbook example.

What is the law that determines the period of oscillation of a pendulum? Suppose it is an idealized pendulum: a small weight suspended by a thin, light rod, making small swings, hanging from a frictionless support, and the whole enclosed in a vacuum. Then what could the period depend upon? Here we have to make some physical judgements about what is relevant. The *length* of the pendulum certainly is relevant. The mass and the weight of the pendulum bob might also be relevant, but we know that according to Newton the mass cancels out in motion under gravity. What is left is the acceleration due to gravity at the surface of the Earth, a quantity that is known to be about 10 metres per second per second. The period might depend upon the length of the swing, but for small swings it is found (as Galileo is supposed to have observed) that there is no such dependence (to a good approximation).

With these assumptions, the only possible equation such that the units "match up" is

$$\text{period} = \text{a constant} \times \sqrt{\frac{\text{length}}{\text{acceleration due to gravity}}}.$$

Here "a constant" is a number that cannot be determined from this dimensional argument alone. (It is in fact equal to $2\pi$.) To see why the equation must be like this, suppose as an example that the square root were omitted. Suppose you change from measuring time in minutes to measuring it in seconds. Then the number giving the acceleration due to gravity is decreased by the factor $60^2$, and the equation would increase the number for the period by $60^2$, whereas it should increase by 60 – clearly nonsense.

This type of argument is rather surprising. It seems to get something for nothing. As I have emphasized, some physical judgement, or, if you like, assumptions go into it. There are actually several "constants of nature" that the period might *conceivably* have depended on: the speed of light, the charge on an electron, the mass of a proton and so on. Take the speed of light, for example. This is something that turned up all over the place in Chapters 3, 4, 5, 6 and 7. Why should it not turn up here? Again, this comes down to physical judgement. The laws of nature seem to be such that, for many processes involving only speeds small compared to the speed of light, there is a good approximation that makes no reference to the speed of light. There are small corrections to this approximation that involve the ratio of the speeds concerned to the speed of light (by assumption, a small ratio). The speed of a pendulum bob *is* certainly very small compared to the speed of light.

For this reason, these "dimensional" arguments are not as miraculous as they might appear: a lot of more or less hidden assumptions go in.

Now about "universal constants of nature". These are numbers, measured in physics, which are expected to be relevant to a very wide class of physical phenomena. Thus for example the mass of the Earth is *not* such a universal constant, because it would not be relevant to physical processes outside the Solar System. Even the mass of the hydrogen atom (the simplest atom) would not now be regarded by most people as a universal constant, because there are many other forms of matter apart from the electrons and protons in hydrogen.

The simplest example of a universal constant is the speed of light, about

$$c = 3 \times 10^8 \text{ metres per second.}$$

Not only is this the speed of all electromagnetic radiation, it is the maximum speed of anything and helps define the geometry of Minkowski spacetime (see Chapter 5).

The second example of a universal constant is Planck's constant

$$h = 6.626 \times 10^{-34} \text{ kilogram (metre)}^2 \text{ (second)}^{-1}.$$

This number is relevant to anything that involves quantum theory, that is, to anything of molecular size and less, but also to many macroscopic systems.

Both these constants of nature are dimensional: that is, they have units, as set out earlier. They depend upon arbitrary things like the definition of the metre. We can, if we want, choose to use *natural units* in which, by definition, $c = 1$ and $h = 1$. If we do so, the three basic units of time, length and mass are reduced to one. That is to say, there remains only one unit to be defined in an arbitrary, "man-made" way, say, the kilogram.

Actually, the officially defined units go some way in this direction. The second is defined in terms of the period of of electromagnetic radiation emitted by atoms of the heavy metal caesium. But the metre is *defined in terms of the second* to be the distance travelled by light in

$$\frac{1}{299,792,458} \text{ second.}$$

This number in the bottom of this fraction (an accurate version of the speed of light) is, of course, chosen so that this metre agrees as well as possible with the old definition in terms of a standard bar of metal. It is no longer possible to "measure the speed of light". Rather, what one can do is, by means of the definition, to relate some distance in which one is interested to the second.

Anyone with a laboratory and some caesium can set up her own standards of time and length according to the definitions. In contrast, at present the kilogram is still defined to be the mass of a standard piece of platinum-iridium alloy stored in a vault in Paris, which can be maintained and copied with an accuracy of 1 part in $10^8$. Before long this definition will probably be replaced by one involving Planck's constant $h$, so that $h$ *by definition* has a certain numerical value when expressed in terms of seconds, metres, and kilograms. The new definition will become convenient when certain quantum electronic measurements can be carried out as accurately as 1 part in $10^8$. When this happens, laboratories anywhere will be able to reconstruct the standard of mass without consulting Paris.

Finally, there is the gravitational constant, which appears in both Newton's law of gravitation (Section 1.7) and in Einstein's

(Section 7.5). It is (giving it its conventional letter $G$)

$$G = 6.6726 \times 10^{-11} \, (\text{metre})^3 \, (\text{kilogram})^{-1} \, (\text{second})^{-2}.$$

This number is not relevant to as wide a class of phenomena as are $c$ and $h$. But if we do count $G$ as a fundamental constant we can get rid of *all* the artificial units, and any reference to a particular atom like caesium, by choosing units so that $G = 1$ as well $c = 1$, $h = 1$.

Another way of expressing this is to say that nature provides the *natural units*: of time

$$\sqrt{\frac{hG}{c^5}} = 1.35 \times 10^{-43} \text{ second}$$

of length

$$\sqrt{\frac{hG}{c^3}} = 4.04 \times 10^{-35} \text{ metre}$$

and of mass

$$\sqrt{\frac{hc}{G}} = 8.61 \times 10^{-9} \text{ kilogram}.$$

The first two of these natural units are very small, even compared to the lengths and times relevant to nuclear physics. The third is very large compared to, say, the mass of a hydrogen atom.

Whenever all these three fundamental constants are relevant, we cannot make any "dimensional arguments" of the kind illustrated by the pendulum formula. In fact that particular example would have been invalidated if either $c$ or $h$ were relevant to the pendulum.

The idea of "natural units" occurred to Planck in 1899, actually the year before he introduced his constant $h$. He wrote:

It is possible to give units for length, mass, time and temperature which, independently of special bodies and substances, retain their meaning for all times and for all cultures, including extraterrestrial and extrahuman ones.

The three natural units are often called *Planck units*.

# GLOSSARY

*The numbers in parentheses refer to relevant parts of the book, for example, (2.3) is Section 2.3 and (5) is Chapter 5.*

ACCELERATION: The rate of change of velocity with time. Since velocity has a direction, so does acceleration (that is, they are both vectors).

ACTION (6, 8.8): A quantity defined in terms of time-averages of energies. The principle of least action is one of the most fundamental ways of formulating laws of classical physics. The action also plays a key role in quantum theory.

ACTION-AT-A-DISTANCE: (A different use of the word *action* from the technical one in the last item.) Describes a theory that just gives the force between two distant particles, without any mention of anything going on in the intervening space. Chief examples are Newton's law of gravity (1.7) and Coulomb's law of electrostatics (3.1). For moving particles, an action-at-a-distance theory is harder to make. The theory of electromagnetism due to Faraday and Maxwell (3) is not of action-at-a-distance type.

ADAMS, JOHN COUCH (7.1): 1819–1892. While an undergraduate at Cambridge, he predicted, from the perturbations of the orbit of Uranus, the existence of an eighth undiscovered planet. This was ignored by the astronomer royal, Sir George Airy, until Leverrier announced similar results. Neptune was discovered by Johann Galle in Berlin in 1846. Adams subsequently refused a knighthood and the position of astronomer royal.

AETHER: (Also written *ether*.) Some "substance", with extremely different properties from ordinary matter, that has from time to time been supposed to be the seat of physical effects like electricity and magnetism and light. The Michelson-Morley experiments (5.2) in the 1880s failed to detect any speed of the Earth relative to the aether, and by the time of Einstein's 1905 theory of relativity the aether was no longer a viable concept, at least in any of its older forms.

ALPHA-PARTICLE: The nucleus of a helium atom, made up of two protons and two neutrons.

AMPÈRE, ANDRÉ MARIE (3.3): 1775–1836. French physicist and mathematician. His father was guillotined in Lyon in 1793. Appointed an inspector-general of universities in 1808 by Napoleon. Most noted for his work between 1820 and 1827 on the magnetic forces between electric currents. (His name has been given to the SI unit of electric current. The ampere is defined so that two parallel currents flowing 1 metre apart exert a force of $2 \times 10^{-7}$ newton on each other. Most household electrical devices use a current of a few amps.)

ANTISHIELDING (11.3): The effect whereby the strength of QCD increases with distance, oppositely to QED and contrary to the way electric charge is shielded in an ionic crystal.

ATOM: Historically, a very small particle as a constituent of matter (sometimes defined to be "indivisible"). Now, a very small (size in the region of $10^{-10}$ metre) body that is the smallest part of a chemical element. Atoms consist of a number of electrons orbiting about a nucleus. The chemical properties are determined by the number of electrons.

BARDEEN, JOHN (10): 1908– . American physicist working for many years at the Bell Telephone Laboratory. Coinventor of the transistor and contributor to the BCS theory of superconductivity. Twice Nobel Prize winner.

BARYON (11, 12): A strongly interacting fermion. The lightest baryons are protons and neutrons.

BELL, JOHN (8.10): 1928–1990. Physicist, born in Belfast and spending much of his career at the European laboratory CERN in Geneva. Best known for his critique of quantum theory and his inequality, which distinguishes quantum theory from classical substitutes for it.

BLACK HOLE (13.1): A region of spacetime containing so much mass that no matter or electromagnetic radiation can escape from it. A black hole is bounded by a *horizon*. A black hole is uniquely determined by its mass, angular momentum and electric charge.

BOHM, DAVID (8.10, 8.12): 1917–1992. Theoretical physicist who, after encounters with the Un-American Activities Committee, left the United States for Brazil, Israel and then London. Noted for his unorthodox thoughts about quantum theory and for his realization that in quantum theory charged particles can be influenced by passing near a magnetic field, without going through it.

BOHR, NIELS (8.3): 1885–1962. Great Danish theoretical physicist. The father of quantum mechanics, he oversaw its development and interpretation. Also made key advances in nuclear physics. Went to Manchester, where Rutherford worked, in 1912. Appointed professor in Copenhagen in 1916 and became head of a new Institute of Theoretical Physics there in 1921. Heisenberg visited him in occupied Denmark in September 1941 – a meeting that left Bohr unhappy. He fled Denmark dramatically in September 1943 for Sweden and thence London and on to the United States. After the war, he campaigned with little success for openness about nuclear weapons.

BOLTZMANN, LUDWIG EDUARD (2.7): 1844–1906. Austrian physicist, working at Graz, Vienna, Munich and Leipzig. He showed how random motions of molecules explained all the effects connected with heat and interpreted entropy as a measure of randomness. His work was controversial to his contemporaries. He died by suicide.

BORN, MAX (8.6): 1882–1970. At Göttingen from 1921 to 1933, he led one of the great centres of quantum theory. He then left Germany and took up a professorship in Edinburgh in 1936; he returned to Göttingen in 1953.

BOSE, SATYENDRA NATH (9.3): (Pronounced to rhyme with *hose*.) 1894–1974. Indian physicist. Realized that photons must be indistinguishable.

BOSE-EINSTEIN STATISTICS (9.3): The way to count distinct quantum states for an assembly of photons or other bosons.

BOSE-EINSTEIN CONDENSATION (10.3): A state with a macroscopic number of free bosons in a single quantum state, and a generalization of such a state for interacting bosons.

BOSONS (9.3): Particles like photons and helium atoms. Particles of any one such kind are indistinguishable from each other.

BOYLE, ROBERT (2.1): 1627–91. Irish. Did his important work in Oxford. Worked with air pump on the pressure of air. Believed in the atomic theory. Author of *Sceptical Chymist*. Founder member of the Royal Society.

BRAHE, TYCHO (1.6): (1546–1601) Danish nobleman astronomer. Observed the supernova of 1572 and the comet of 1577. Showed that the supernova was outside and the comet within the Solar System. Given island of Hven by the king, Frederick II, for his great observatories of Uraniborg (Urania was the Greek muse of astronomy) and Stjernborg ("Castle of the Stars"). Spent the last years of his life near Prague, with the emperor Rudolph II as patron. Kepler became his assistant.

BRITISH ASSOCIATION FOR THE ADVANCEMENT OF SCIENCE (2.4): Founded 1831. Meets annually in different places. Famous Victorian discussions took place at its meetings (most notably the debate on evolution between T. H. Huxley and Bishop Wilberforce at Oxford in 1860).

BROWNIAN MOTION (2.5): The "wandering" motion seen through a microscope of very fine particles suspended in a liquid. It is caused by the buffeting by the molecules of the liquid moving with their random thermal motion. The theory of Brownian motion was given by Einstein (Appendix C).

$c$: The speed of light in a vacuum, very nearly $3 \times 10^8$ metres per second.

CALORIC (2.1): In an obsolete theory of heat, the name given to the "fluid" that was supposed to be heat.

CALORIE (2.4): A unit sometimes used to measure heat. Equal to 4.252 joules.

CASIMIR EFFECT (9.10): The small attraction between two metal plates, which can be attributed to a difference in the quantum ground state energies of the modes of oscillation of the electromagnetic field between the plates and outside the plates.

CASSINI (1.7): Four generations of directors of the Paris Observatory: Giovanni Domenico (1625–1712) went to Paris from Italy and became director in 1669; Jacques (1677–1756), César-François (1714–1784) and Jean Dominique (1748–1845) succeeded him. Jacques's measurements seemed, incorrectly, to show the Earth to be prolate, not oblate.

CELSIUS, ANDERS (1.7): 1701–1744. Swedish astronomer. Deviser of thermometric scale. Took part in expedition to Lapland in 1736 to determine shape of the Earth.

CENTRE-OF-MASS: (Also called *centroid*). The mass-averaged centre of a distribution of matter. For example, the centre-of-mass of two particles 5 metres apart, one of mass 4 grams and the other of 1 gram, is on the line joining them and 1 metre from the heavier one (and so 4 metres from the lighter).

CHARGE: As a noun, a quantity of electricity. (The word is now also used for other similar quantities, usually ones that are conserved.) As a verb, the process of putting a charge on an object. (Compare "charge your glasses".)

CHATELET, GABRIELLE-EMILIE, Marquise du: 1706–1749. French intellectual. Mistress of Voltaire. Studied Newton's work with Maupertuis and Clairaut. Translated the *Principia* into French, finishing it just before dying in childbirth.

CLAIRAUT, ALEXIS-CLAUDE (1.7): 1713–1765. French mathematician and astronomer. He developed Newton's theory of the motion of the Moon.

CLASSICAL: This adjective in physics conventionally means "non-quantum", that is, something in which the effects of quantum theory are neglected. Of course, all of physics up to about 1900 was "classical", because quantum theory was not discovered then. There are many situations, usually on scales a lot larger than atomic dimensions, in which the classical approximation is a very good one.

CLIFFORD, WILLIAM (7.4): 1845–1879. English mathematician.

COLOUR (11): Quarks of each "flavour" (that is, of each mass and charge) occur in three distinct kinds. What distinguishes them is called (facetiously) "colour".

CONDENSATE (10): In Bose-Einstein condensation, the name given to the set of bosons that have condensed.

CONFINEMENT (11.3): The property of the forces of QCD that prevents free quarks from being isolated.

CONSERVATION LAW: A law that states that some quantity does not change with time, that is, "is conserved". Examples are conservation of (total) electric charge (3.1), momentum and energy (2.3, 5.11).

CONSTANT: This word is sometimes used as a noun by mathematicians, meaning a quantity that does not vary when some other quantities do. An example: "Although the price of a loaf went up and the daily wage increased, the ratio of the daily wage to the price of a loaf remained *a constant*, and that constant was equal to 45". (In this example it would have been more natural to end "remained constant at 45".) Sometimes "a constant" may actually be a constant only when certain conditions vary, but not others. In the example, the "constant" might be 45 in France but 66 in Germany.

CONSTANT OF NATURE: An extreme case of "a constant", but one that is thought not to vary under *any* circumstances. An example is the gravitational constant (*G*), which appears in Newton's law of gravitation (1.7). As far as anyone knows, this constant has the same value everywhere in the universe and at all times. With some "constants of nature" there is a question about how far they tell us about nature and how far they merely reflect our choice of measurement units.

CONTINUUM: Any continuous "substance" that has no atomic structure, so that there is no substructure even at very small scales. The continuum view of matter (or anything else) is opposed to the atomic view. Modern quantum theory has shown that the two views are not really mutually exclusive.

COOPER PAIR (10.6): In a superconductor, a pair of oppositely moving conduction electrons that are loosely bound. The existence of such pairs was proposed by the American physicist Leon Cooper (1930–).

COPERNICUS, NICOLAUS (1.4): 1473–1543. Polish. Studied in Italy. Became a (lay) canon at Frauenberg Cathedral. Argued that a Sun-centred theory of planetary motion was superior to the Earth-centred one, provided that the stars are far enough away that their apparent motion is inappreciable.

COULOMB, CHARLES (3.1): 1736–1806. French military engineer and physicist. Established by careful experiments the inverse-square law of electrical attraction and repulsion. (The SI unit of electric charge, the coulomb, is named after him. The coulomb is defined as the charge carried when a current of 1 ampere flows for 1 second.)

CRITICAL TEMPERATURE (10): Used in various senses, for example, the temperature above which a gas cannot be liquefied by pressure, the temperatures below which helium becomes superfluid and below which some materials become superconducting.

CURVATURE (7.2): This has a technical mathematical definition applied to surfaces and spaces that are not flat. In particular, they are said to have *intrinsic curvature* if they cannot be flattened out without mutilating them.

DESCARTES, RENÉ (1.7): 1596–1650. Influential French mathematician and philosopher. He did not accept Newton's action-at-a-distance theory of gravity and propounded an alternative, but unsuccessful one. Explained the rainbow as caused by sunlight refracted on entering and leaving raindrops and reflected off the surface from inside. Tutor to Queen Christina of Sweden.

DIFFRACTION (4.6): Effects resulting from the wave nature of light. Light does not travel exactly in straight lines. For example, light passing through a small hole, whose size is not large compared to the wavelength of the light, spreads out and produces complicated patterns.

DIMENSION (Appendix D): In science, one of the usages of this word is connected with the units in which quantities are measured. Thus the sentence "Speed has the *dimensions* of length divided by time" means that speed is measured in units of metres per second, or miles per hour (or any unit of length per any unit of time). Any equation should be dimensionally correct; that is to say, the dimensions of the left-hand side should be the same as those of the right-hand side.

DUFAY, CHARLES (3.1): 1698–1739. French chemist and experimenter on electricity.

EDDINGTON, SIR ARTHUR (7.7, 13.1): 1882–1944. Director of Observatory at Cambridge. One of the first to develop the physical theory of the structure of stars. He led an expedition in 1919 to test Einstein's theory of gravitation by observing the deflection of light from a star during an eclipse of the Sun.

ELECTRON (5.1): Light negatively charged particles that are among the constituents of atoms. Their arrangement and movement account for most of the electrical, optical and chemical properties of matter. The electron was discovered by J. J. Thomson in 1897. The word *electron* had been suggested for a smallest unit of electric charge in 1891 by the Irish physicist George Johnstone Stoney (1826–1911).

ENERGY (2.3, 5.11, 6.8): A quantity connected with moving bodies or with changing fields whose importance is that (under suitable circumstances) it does not change with time. It is made by adding kinetic energy to potential energy. For a particle, kinetic energy depends on the speed, and potential energy depends upon the position. The unit of energy is the joule, which is kilogram $\times$ (metre)$^2$ $\times$ (second)$^{-2}$.

ENTROPY (2.6, 2.7, 13.5): A measure of the "randomness" of the motions in a physical system. More precisely, entropy may be defined as the logarithm of the number of "small-scale configurations" of the system that are allowed, given a knowledge of its "large-scale configuration".

EQUILIBRIUM: When a system settles down so that it no longer changes (or at least its macroscopic properties no longer change) with time, it is said to be in equilibrium. It may be difficult to know whether the equilibrium is exact, or whether very slow changes may eventually take place.

EULER, LEONHARD (6.2): 1707–1783. Enormously prolific Swiss mathematician. Worked in Basel, St. Petersburg and Berlin. Developed mechanics and was among the first to formulate the principle of least action.

EVENT (5.3): This word is used technically in relativity theory to mean something at a definite time and a definite position. An event is represented by a point in a spacetime diagram.

EXCITATION (10): A state in which a system has energy above the minimum possible. An example of an excitation in a metal is an electron with an energy above the Fermi energy.

FERMAT, PIERRE DE (4.4): 1601–1665. French mathematician and government lawyer. Appears in this book in connection with the "principle of least time" in optics, the forerunner of similar principles in the mechanics of particles (see Chapter 6). Famous in mathematics for his "last theorem", which was proved only in 1995.

FERMI, ENRICO (8.13): 1901–1954. Great Italian physicist. Worked in Rome from 1927 to 1938, when he left for the United States. He led the group who made the first self-sustaining neutron chain reaction in 1942.

FERMI ENERGY (10.5): The energy up to which quantum states must be filled in order to accommodate a given number of fermions (for example, conduction electrons in a metal) consistently with the exclusion principle.

FERMION (8.13): A particle (or composite system) subject to Pauli's exclusion principle. The wave function for a number of identical fermions must change sign when the positions (and spins) of two are interchanged.

FEYNMAN, RICHARD (8.8, 9.8, 10.3): 1918–1988. Born in New York, he spent much of his career at the California Institute of Technology in Pasadena. His influence pervades most of theoretical physics from the 1940s onwards. Charismatic and unconventional.

FFOC (6.5): My acronym for *flux for an open curve*. A generalization of the magnetic flux (number of field lines) through a closed curve, and the basis of the idea of gauge-invariance.

FIELD (3.1): For example, the *electric field* is the term used to describe the "electrical state" of some region of space. The information necessary to provide this description is, for every point in the region, the force that would be exerted on an electric "test charge" at that point. Since force has direction associated with it (that is, is a *vector*), so does the field at any point. Magnetic field is defined similarly, but by using a (hypothetical) "test magnetic pole". Although this is the operational definition of *field*, many physicists (most notably Faraday) have thought of the field as something happening in space; so far there has been no successful definition of this "something" in terms of anything else.

FIELD LINES (3.1): (Also called *lines of force.*) A representation of an electric or magnetic field (or other field) by means of curves, whose direction shows the direction of the field and whose number (crossing unit area) corresponds to the strength of the field.

FLAVOUR (11, 12): The different kinds of quark ($u$, $d$, $c$, $s$, $t$, $b$), distinguished by their masses and electric charges, are said (facetiously) to be different "flavours".

FLUX (3.4): In electricity and magnetism, the flux through a closed loop is defined to be the number of field lines going through that loop. (If there are electric charges present, so that field lines may terminate, one must choose a surface across the loop and talk about the flux through this surface. The flux so defined depends upon the position of the surface relative to the charges.)

FORCE (1.7): Something like a push or a pull that tends to make an object move. More precisely, Newton's second law of motion states that force is equal to mass times acceleration. This may be taken as a definition of force. To get useful information, one then needs some law or equation that determines the force, for example, Newton's law of gravitation or Coulomb's law of electric force (3.1). The unit of force is a newton, which is kilogram × metre × $(\text{second})^{-2}$.

FRANKLIN, BENJAMIN (3.1): 1706–1790. American publisher, statesman and scientist. Experimented on static electricity and lightning.

FRESNEL, AUGUSTIN JEAN (4.6): 1788–1827. French physicist. Consolidated the wave theory of light by showing how to explain diffraction and by realizing that transverse waves are necessary to explain polarization effects.

FREQUENCY (4.1): The number of oscillations per unit time in, for example, wave motion.

FRIEDMANN, ALEXANDER (14.3): 1888–1925. Russian scientist. Contributed to fluid dynamics and cosmology. First to show that Einstein's theory of gravity allows expanding universe cosmologies (published in 1922).

**G**: The gravitational constant.

GALILEO GALILEI (1.1, 1.5): 1564–1642. Professor of mathematics at Pisa 1589, moved to Padua 1591. Became mathematician to Grand Duke Cosimo de Medici in 1610. Improved and used telescope. Championed the Copernican (Sun-centred) theory of Solar System. Showed that the pendulum was suitable for controlling a clock. Realized that a body would remain in motion without any force. Found the law of fall under the Earth's gravity. His *Dialogue on the Great World Systems* and *Discourses Concerning Two New Sciences* are each in the form of a dialogue, a popular Renaissance form. His confrontation with the Inquisition in his later years is an irresistible subject for dramatists.

GALVANI, LUIGI (3.3): 1739–1798. Italian anatomist. Detected chemically produced electricity by its action on a nerve in a frog's leg. His name is remembered in the word *galvanized*, used both technically (*galvanized iron*) and metaphorically (*galvanized into action*).

GAP (10): A gap between the energy of the ground state of a system and its first excitation; also called *gap energy*. Superconductors have such a gap.

GAUSS, KARL FRIEDRICH (7.2): 1777–1855. One of the greatest of mathematicians, at least the equal of Archimedes and Newton. Supported by the duke of Brunswick until he became director of the Observatory at Göttingen. He developed vast areas of new mathematics; a lot of what he had done was only discovered among his papers after his death. He was a virtuoso of numerical calculations. He made his mark on planetary astronomy, geodesy and magnetic measurement. He was a coinventor of the electromagnetic telegraph.

GELL-MANN, MURRAY (11, 12): 1929– . Like Feynman, spent most of his career at the California Institute of Technology. A dominant contributor to the modern theory of elementary particles. Frighteningly clever, he includes among his "hobbies" the studies of languages and of birds.

GEODESIC (7): The nearest thing to a straight line on a curved surface or in a curved space. It is a curve of least distance (or stationary distance) between two given points. On a spherical surface, geodesics are great circles.

GILBERT, WILLIAM (3.2): 1544–1603. Physician to Elizabeth I and James I. Systematically studied magnetism. Wrote *De Magnete*, 1600.

GLOBAL: Sometimes used of physical laws that directly relate things at different parts of space, not just in a very small region of space. Opposed to *local*.

GLUON (11.3): A quantum of the QCD fields (analogous to the photon in QED). Gluons do not normally occur as separate particles, but they have indirect effects.

GRAVITATIONAL CONSTANT (1.7): The universal number that appears in Newton's law of gravitation (and also in Einstein's) and dictates the strength of gravitation, conventionally called G.

GRAVITATIONAL MASS (1.7): The name sometimes given to the masses that appear in Newton's law of gravitational force. This is to bring out the logical distinction between this "mass" and the "mass" (inertial mass) that appears in Newton's third law of motion. Experimentally, the gravitational mass turns out to be equal to the inertial mass to a very high degree of accuracy, so the distinction is only made to emphasize that (on Newton's theory) the two might not have been equal. In Einstein's theory of gravity (7), it is obvious that the two must be the same.

GRAVITY: The attractive force between any two masses. In particular, the force (the weight) that the Earth exerts on objects near it. Gravity obeys Newton's law (1.7) to a good approximation (when bodies are not moving too fast and when the gravitational force is not too big). Einstein's theory (Chapter 7) is a much better account of gravity.

GROUND STATE: The state of a system that has least energy.

$h$ (8.1): Planck's constant. More often used is $\hbar = \frac{h}{2\pi}$.

HADRON: A baryon or a meson, that is, any particle made of quarks.

HAMILTON, SIR WILLIAM ROWAN (6.4): 1805–1856. Irish mathematician. Professor of astronomy at Trinity College, Dublin, and astronomer royal of Ireland. Reformulated

Newtonian mechanics and found general principle of least action. Discovered a beautiful new algebra called *quaternions* (which might or might not have applications to physics), but it is now known that Gauss had preceded him.

HEAT (2): The energy of random motion of molecules, or more generally random time variations of physical quantities. Heat used to be measured in special units, such as the calorie, but is now measured in joules (like any other form of energy).

HEAVISIDE, OLIVER (3.5): 1850–1925. With no university education, he developed the theory of cable telegraphy. He suggested that a reflecting layer in the upper atmosphere would account for the successful transmission of radio waves despite the curvature of the Earth.

HEISENBERG, WERNER (8.4): 1901–1976. One of the heroes of the quantum revolution. Invented quantum theory in 1925 and the uncertainty principle in 1927. Professor at Leipzig 1928. Leading theoretician with the (unsuccessful) German uranium fission programme during the 1939–45 war. Snatched by U.S. Army on 3 May 1945 (four days before German surrender) and held with other German fission scientists near Godmanchester for six months, before being settled in Göttingen.

HELIUM II (10): Liquid helium in its superfluid state.

HELMHOLTZ, HERMANN VON (2.3): 1821–1894. Professor of physiology at Königsberg, Bonn and Heidelberg, and of physics at Berlin. Worked on the mechanisms of hearing and vision. Formulated principle of conservation of energy. Had a theory of electromagnetism that differed from Maxwell's. Encouraged his pupil Hertz to experiment on electromagnetic waves. He had great prestige, somewhat like Kelvin's in Britain.

HIDDEN VARIABLES (8.10): Hidden variable theories are attempts to remove the uncertainties of quantum theory by postulating the existence of quantities (the "hidden variables") that do have definite values but are hidden from us. Local hidden variable theories, in which the variables are associated with local regions of space, seem to be ruled out by experiment.

HILBERT, DAVID (7.5): 1862–1943. Very important German mathematician. He appears in this book as publishing the final mathematical form of Einstein's theory of gravitation in 1915 just before Einstein. He also made rigorous the mathematics required in quantum theory.

HISTORY (8.8): This word is sometimes used to mean the sequence of changes in a physical system over some given period. Thus the history of a particle is just its motion. In quantum theory, the word is used in a slightly generalized sense.

HORIZON (13.1, 14.3): This word is used in two related senses in Einstein's theory of gravity (both senses suggesting a limit of visibility). (a) *Event horizon*. The surface round a black hole through which no matter or radiation can escape. (b) *Particle horizon*. In cosmology, for a given observer at a given time, the world-lines of the farthest galaxies from which light can have reached that observer during the history of the universe.

HUYGENS, CHRISTAAN (4.6): Pronounced "hoykhenz". 1629–1695. Dutch. Discovered Saturn's rings. Put wave theory of light on a strong foundation. Invented the pendulum clock.

IDEAL GAS (2.5, 10.3): A mathematical idealization: a gas of molecules with arbitrarily small forces between them. A real gas at sufficiently low density behaves approximately like an ideal gas.

INELASTIC: A collision between two bodies (such as billiard balls) is called inelastic if the kinetic energy of their motion is less after the collision than before. The lost energy is converted into some other form of energy, for example, heat. In practice, a collision between billiard balls is almost elastic: the kinetic energy of their motion is almost as much after the collision as before.

INERTIAL MASS (1.7): See *mass*.

INTERFERENCE (4.6): If a wave is split into two parts that then recombine, the two waves may combine constructively at some points and cancel at others, producing alternations in intensity. This is characteristic of a wave motion.

INVARIANCE: This word is used in physics especially of properties that are unaltered when certain changes are made. Invariance principles express the idea that some things are irrelevant to the laws of nature. For example, "rotational invariance" says that results of experiments in a closed laboratory are independent of the orientation of that laboratory, that is, unchanged after it is rotated. (In so far as the Earth's gravitation is felt in the laboratory, it is not "closed".)

ION: An atom or group of atoms that has an electric charge as a result of gaining or losing some electrons. In solution some compounds partly break up into ions, for example, sodium chloride (table salt) into positively charged sodium ions (which have lost one electron) and negatively charged chlorine ions. The name *ion*, from the Greek for "going", was one of several words coined at the request of Faraday by the formidable Cambridge polymath William Whewell (1794–1866).

JOSEPHSON, BRIAN (10.6): 1940– . Predicted, in 1962, striking effects in the flow of current between two narrowly separated superconductors.

JOULE, JAMES (1.4): 1818–1889. From Manchester. Established the equivalence of heat and other forms of energy. (The unit of energy, a *joule*, is named after him.)

K (2.6): Denotes the kelvin (or absolute) temperature scale. Zero Celsius is 273.15 K.

KALUZA, THEDOR F.E. (15.1): 1885–1954. German mathematical physicist, at Göttingen from 1935. Studied at least 15 languages. Successfully swam for the first time after reading a book about swimming.

KEPLER, JOHANNES (1.6): (1571–1630) German Protestant. Studied theology at Tübingen. Became mathematician in Graz. Joined Brahe in Prague and succeeded him there on Brahe's death in 1602. Observed supernova of 1604 (near a rare, astrologically noteworthy conjunction of Saturn, Jupiter and Mars). Inheriting Brahe's astronomical observations, he found his three laws of planetary motion (1.6), including the law that planets move in ellipses not circles.

KINETIC ENERGY (2.3): See *energy*. Kinetic energy is the energy of motion.

KLEIN, OSKAR B. (15.1): 1894–1977. Versatile Swedish theoretical physicist.

LAPLACE, PIERRE SIMON: 1749–1827. French mathematician and scientist who greatly advanced Newtonian mechanics. He held public office under Napoleon and Louis XVIII. There is a well-known story that when Napoleon told him he had not mentioned the author of the universe in his *Mécanique Céleste*, Laplace replied, "Sire, I had no need of that hypothesis".

LASER (9.3): Acronym for *light amplification by stimulated emission of radiation.*

LATTICE: A regularly repeating array, as of the atoms in a crystal.

LEE, TSUNG DAO (12.2): (1926–) Theoretical physicist. Left China for Chicago in 1946. Analyzed mirror symmetry of weak interactions.

LEIBNIZ, GOTTFRIED WILHELM (2.3): 1646–1716. German mathematician, philosopher, lawyer and diplomat. Proposed kinetic energy (*vis viva*) as an important quantity connected with motion. He invented the calculus independently of Newton and was involved in controversy with him. He invented a calculating machine.

LEMAÎTRE, GEORGES (ABBÉ) (14.1): 1894–1966. Belgian Catholic priest, cosmologist. Professor of astronomy at Louvain. In 1927, he discovered that Einstein's theory of gravity permits an expanding universe cosmology.

LEPTON: Fermionic, spin $\frac{1}{2}$ elementary particles other than quarks, including electrons, muons, taus and neutrinos. (From the Greek for "small".)

LEVERRIER, URBAIN JEAN JOSEPH (7.1): 1811–77. French astronomer. Predicted (just after Adams) the existence of Jupiter. Also appreciated that the orbit of Mercury suffered perturbations, which he attributed to another undiscovered planet but that was eventually explained by Einstein's 1915 theory.

LIGHT CONE (5.4): The "surface" in spacetime giving the history of a flash of light. A slice through the light cone corresponding to a given time is a sphere, which grows in size as the time increases. If only two space dimensions are considered, the light cone is a cone in the ordinary sense. If only one space dimension is considered, the light cone becomes just two straight lines.

LOCAL: A word sometimes used to characterize laws or equations that refer to very small (as small as we please) regions of space. For example, it is sufficient (for most purposes) to apply Maxwell's equations of electromagnetism to very small regions. The fields over large regions can be deduced from these local equations.

LOGARITHM (2.6): In mathematics, the operation of finding the logarithm of some number is the inverse of the operation of raising a number to the power of another number. Thus $3 = \log_{10} 1,000$ means that $10^3 = 1,000$. Logarithms are used in this book only in the definition of *entropy*.

LONGITUDINAL (4.2): Types of wave motion, like sound waves in fluids, in which the material motion is to and fro in the direction of the wave. Opposed to *transverse* waves.

LORENTZ, HENDRIK ANTON (5.1, 5.6): 1853–1928. Dutch. Appointed as the first professor of theoretical physics at Leyden at age 24 and remained there until 1912, when he became director of an institute at Haarlem. He worked out much of the electromagnetic theory of electrons. In 1904, he found the Lorentz transformations of space and time, which were to be part of Einstein's theory of special relativity in 1905. Lorentz did not attend scientific conferences outside Holland until 1897, but after 1900 he was so highly esteemed that he became the natural choice for chairman at many international physics conferences.

MACH, ERNST: 1838–1916. Austrian. Professor of, in succession, mathematics, physics and philosophy (and member of the upper chamber of the Austrian parliament). His positivist philosophy of science (and critique of absolute space) influenced Einstein (temporarily at least) and also the founders of quantum theory. He studied shock waves produced by high-speed projectiles, and because of this his name is used to denote the speed of an object as a fraction of the speed of sound.

MAP (7.4): In this book, I use the word *map* to signify any representation of a curved surface (more generally, space) on a flat surface (more generally, space). Such a representation inevitably introduces distortions.

MARCONI, GUGLIELMO (4.8): 1874–1937. Developed the technology of radio and founded a company to commercialize it. Went to England in 1896.

MASER (9.3): Acronym for *microwave amplification by stimulated emission of radiation*.

MASS (1.7): The property of a body that appears in Newton's third law of motion ("Force equals mass times acceleration"). What saves this from being a tautological definition is the *additive* property of mass: that is, if we take a lump of metal defining the kilogram, manufacture an identical lump and then put the two together, the composite lump will have a mass of 2 kilograms. Mass as defined here is sometimes called *inertial mass*, to distinguish it from *gravitational mass*.

MAUPERTUIS, PIERRE-LOUIS (1.7, 6.2): 1698–1759. French scientist. Championed Newtonianism in France, particularly Newton's calculations of the oblateness of the Earth. Instigated, and took part in, expedition to Lapland to measure the shape. The first to try to use a principle of least action in mechanics. Later he became the subject of Voltaire's mockery as a philosopher who tried to deduce the existence of God from a line of mathematics. Frederick II of Prussia made him president of the Berlin Academy in 1746.

MEISSNER EFFECT (10.6): The expulsion of magnetic fields from the bulk of a superconductor. Discovered by Meissner and Ochsenfeld in Berlin in 1933.

MESON (11): A strongly interacting boson. Examples are pions and kaons.

METRIC (7.4): For a *map*, the information that gives the true distance (or proper-time) between two nearby points on the map.

MICHELL, JOHN (13.1): 1724–1793. Briefly professor of geology at Cambridge; later Yorkshire clergyman. Estimated distance of stars and found evidence for double stars. Argued for possible existence of black holes.

MOLECULE (2.5): A small body that is the least part of a chemical compound. Molecules are made up of atoms. For example, a molecule of carbon monoxide consists of one oxygen atom and one carbon atom. In a gas, the spacing between the molecules is large compared with the size of a molecule (except at high pressure or low temperature).

MOMENTUM (1, 5.11, 6.8): The momentum of a body is its velocity times its mass. Because velocity has a direction associated with it, so does momentum (that is, it is a vector). Newton's third law of motion may be expressed as "Force equals rate of change of momentum". Because of Newton's second law ("Action and reaction are equal and opposite"), the total momentum of a closed system of particles does not change with time. When there are electric and magnetic fields varying with time, the momentum in these fields has to be included too.

NEWTON, SIR ISAAC (1.7) (4.5): 1642–1727. At Trinity College, Cambridge from 1661 to 1701, except for a period at Woolsthorpe during the plague years of 1664–6 – a period in which he had many of his great ideas. He propounded the laws of motion and his law of gravity and deduced the elliptical form of planetary orbits – a theory that stood for nearly two and a half centuries. He founded the science of fluid mechanics. He showed that white light was made up of coloured light and did much else in optics. He is often judged to be one of the three greatest mathematicians of all time (the others are Archimedes and Gauss). In 1696 he became warden of the Mint, and later master of the Mint. He was president of the Royal Society from 1703 until his death. He devoted much effort to theology and to alchemy. Touchy and obsessive in his quarrels. (The SI unit of force is the newton.)

OERSTED, HANS CHRISTIAN (3.3): 1777–1851. Professor of physics at Copenhagen. In 1820 he discovered that electric currents produce magnetic effects and thus began the unification of electricity with magnetism.

ORDER PARAMETER (10): When solids and liquids are cooled, they may pass into states with long-range order. The quantity that defines this order is the order parameter. For example, in magnetized iron, the local direction of the magnetization defines the order parameter.

PARALLAX (1): The difference in apparent direction of an object viewed from different places, particularly stars, planets and so on, viewed from different points as the Earth rotates and as it circles the Sun. The determination of parallax, and the difficulty thereof, was very important in the history of astronomy.

PARTON (11.2): Any constituent of baryons and mesons (that is, a quark or a gluon).

PASTEUR, LOUIS (12.2): 1822–1895. Founder of microbiology. Appears here as the discoverer of the handedness of organic molecules.

PAULI, WOLFGANG ERNST FRIEDRICH (8.13): 1900–1958. Born in Vienna. Professor in Zurich, but in Princeton during the war. Friend of Carl Jung. Formulated the exclusion principle in 1924, postulated the existence of the neutrino in 1933 and proved in 1940 that particles must be bosons or fermions according as their spin is an integral or half-integral multiple of $\hbar$. Enormously respected, he has been called "the living conscience of theoretical physics".

PERIOD (4.1): The time for a complete oscillation, for instance, in a wave motion.

PERPETUAL MOTION (2.3, 2.6): A hypothetical machine that would go on working forever, without access to any external source of power, and perhaps giving out some useful power – in other words, a free lunch. Towards the end of the eighteenth century, the impossibility of such a machine was taken as axiomatic, and this axiom was used to found the science of thermodynamics.

PHASE: Has two distinct technical meanings in physics, each related to its ordinary sense of a "stage" of development. First, different states of the same chemical substance, for example, ice and water, are sometimes called *phases*. Changes from one phase to another (like melting of ice) are called *phase transitions*. Second, the stage of a wave, whether it is at crest or trough or somewhere definite in-between, is called the *phase*. This is related to the next entry.

PHASE ANGLE: In mathematics, a complex number is represented by a line in a plane. This line has a *length* and a direction, defined by its angle to some standard direction. This angle is usually called the *phase* of the complex number. I use the expression *phase angle* in order to emphasize that the phase is an angle. In quantum theory, wave functions are complex numbers, and so have phase angles.

PHASE SPACE (2.8): A rather unrelated use of the word *phase*. A mathematical space in which velocities (or momenta) as well as positions are represented.

PHASOR (4.6): I use this to mean a way of representing the local state of a wave by a little arrow, whose direction signifies whether one is at a crest or trough or somewhere in between.

POINCARÉ, JULES HENRI (2.8, 5): 1854–1912. Amazingly creative French mathematician and physicist. (Cousin of the French president Raymond Poincaré.) Elected to Académie Française in 1908. He appears in this book in two places. He introduced a new, geometrical way of thinking about dynamical motions, which was the first step towards the theory of chaos. He came very close to discovering the theory of relativity just before Einstein in 1905, but he seems never to have understood it, or accepted it, with Einstein's clarity. He gave the name *Lorentz transformations* to the rules for comparing different observers' measurements of lengths and times (see Section 5.8).

POLARIZATION (4.6): This word is used in more than one sense in physics. Applied to a transverse wave motion, it means the direction of the transverse vibration. In the case of electromagnetic waves, the direction of polarization is conventionally defined to be the direction of the electric field.

POTENTIAL ENERGY (2.3): See *energy*.

POYNTING, JOHN HENRY (3.5): 1852–1914. Professor of physics at Birmingham. Showed that radiation had momentum and could exert pressure.

PREECE, SIR WILLIAM: Engineer-in-chief of the General Post Office in the latter part of the nineteenth century and so influenced both the electric telegraph and the beginnings of wireless. He refused to accept Heaviside's calculation that self-inductance would reduce distortion in telegraph signals along cables. He backed Marconi's wireless systems. He disagreed with Lodge's ideas about lightning conductors.

PRESSURE (2.5): The force on a unit area of a surface. In a fluid, the pressure is the same in all directions. In a gas, the pressure is due to bombardment of a surface by randomly moving molecules. The SI unit of pressure is the pascal (after Blaise Pascal), which is one newton per square metre. Atmospheric pressure (at sea level) is about 100,000 pascals and this is called a *bar*. Atmospheric pressure on weather maps is given in millibars: units of 100 pascals. Pressure used to be expressed in terms of millimetres of mercury; for example, atmospheric pressure is about 760 mm of mercury, meaning that it can support a column of mercury 760 mm high. (Note that the cross-sectional area of the mercury column does not matter, because the pressure is the force per unit area.)

PRIESTLEY, JOSEPH (3.1): 1733–1804. Unitarian minister and librarian. Chemist, who discovered oxygen and other gases. Supported the French Revolution and took refuge in the United States in 1794.

QCD (11.3): Quantum chromodynamics: the theory of the strong forces between quarks, constructed by generalizing the gauge-invariance of QED to the more complicated "colour" gauge-invariance.

QED (9.7): Quantum electrodynamics: the quantum theory of electromagnetism.

QUANTUM (8): As a noun, a discrete quantity of something, for example, energy or angular momentum. As an adjective, pertaining to quantum theory.

QUANTIZE (8): To force a quantity to have a discrete set of values, for example, the energy of a hydrogen atom. More generally, to quantize a classical theory is to produce the corresponding quantum theory.

QUARK: A fractionally charged, fermion constituent of baryons and mesons.

RAY (4.3): A very thin beam of light. To some approximation, light rays remain thin and travel in straight lines.

REFRACTION (4.3): The bending of light (or other wave such as sound) on passing from one transparent medium to another.

RIEMANN, GEORGE FRIEDRICH BERNHARD (7.2): 1826–1866. Hugely innovative Göttingen mathematician and theoretical physicist. For our purposes his main importance is that, following Gauss, he developed the mathematics of curved space and speculated that physical space might be curved. He also predicted the existence of shock waves.

RÖMER, OLE CHRISTENSEN: 1644–1710. Danish astronomer. Discovered the finite speed of light from observations of the satellites of Jupiter.

ROYAL INSTITUTION: Planned and partly created by Benjamin Thompson (Count Rumford). Initially financed by a group of "proprietors" paying 50 guineas each. George III was patron.

One aim was to improve agriculture. Had a lecture room, a laboratory and a "repository of useful devices". Appointed Davy and then Faraday, two men who were masterly lecturers as well as being great scientists.

ROYAL SOCIETY: Founded in 1660, it was one of the first "scientific academies" (the Paris Academy of Sciences was founded in 1666). Partly inspired by the ideas of Francis Bacon (1561–1626), which favoured empiricism over tradition. Newton was president from 1703 until his death in 1727.

RUMFORD, COUNT: See *Thompson*.

SAKHAROV, ANDREI (14.5): 1921–1989. Russian nuclear physicist and cosmologist. Played important role in Soviet hydrogen bomb project. Dissident. Nobel Peace Prize 1975. Exiled to Gorky 1980–6.

SALAM, ABDUS (12.3): 1926–1996. Pakistani theoretical physicist, who worked mainly in London and Trieste. One of the discoverers of the unified theory of weak and electromagnetic forces.

SCALAR: A scalar quantity, in contrast to a *vector*, is one that has no direction associated with it. A scalar is completely defined by a single number. For example, mass is a scalar, whereas velocity is a vector.

SCHRÖDINGER, ERWIN (8.5): 1887–1961. He and Heisenberg were the first to construct, in 1925, complete quantum theories. Educated in Vienna, he became a professor in Zürich in 1922 and in Berlin in 1926. He spent 1934–6 in Oxford but returned to Austria (Graz) until 1939. Then he went to Dublin, to the Institute of Advanced Studies created there with him in mind by the prime minister Eamon de Valera (were it had two schools: Celtic Studies and Mathematical Physics). In 1956, he returned to Vienna. Like Einstein, he was never satisfied with quantum theory. His book *What Is Life?* (1944) may have affected the development of molecular biology. Both Crick and Watson have attested to its influence on them.

SCHWARZSCHILD, KARL (13.1): 1873–1916. German observational astronomer and theoretical physicist. Director of Potsdam observatory. In the last year of his life, he calculated exactly, from Einstein's theory of gravitation, the metric of spacetime outside a spherical star. This was the beginning of the idea of a black hole.

SNEL, WILLEBRORD (4.3): 1580–1626. Dutch physicist. The law governing refraction of light is called *Snel's law*.

SOLITON (10.4, 15.4): A form of fluid flow or a configuration of fields that has a restricted size and that persists in time.

SPACETIME (5): Space and time considered together, as in relativity theory. Spacetime has four dimensions (three of space and one of time). (Diagrams normally omit one or two of the space dimensions.)

STATE: A word sometimes used in its everyday sense (as a condition of something), but also in a mildly technical sense in quantum theory, whereby a "state" of a system is a condition defined as precisely as the uncertainty principle allows.

STATIONARY (6.3): Suppose a quantity depends upon some variables (for instance, the height above sea level as depending upon latitude and longitude). Suppose there is some value of these variables such that the quantity is not changing there. Then this is called a *stationary point*. Particular examples are maxima (tops of mountains) and minima (bottoms of lakes). But there are other possibilities, like the top of a pass, which are stationary but are neither maxima or minima.

STRONG INTERACTIONS (11): The forces that hold atomic nuclei together. The forces among protons, neutrons and pions, and more generally between all baryons and mesons. These forces are called *strong* because they are not characterized by a small number, such as the $\frac{1}{137}$ of QED. They act over a short range of about $10^{-15}$ metre.

SUPERCONDUCTIVITY (10.6): The flow, without resistance, of electric currents in metals and other materials when cooled below their critical temperatures.

SUPERFLUIDITY (10.3): The flow, without friction, of liquid helium when cooled below its critical temperature.

SUPERNOVA (13.2): A word coined in 1934 meaning a star that increases in brightness very rapidly and then, for some weeks, is very bright indeed, perhaps several hundred million times as bright as the Sun. Chinese astronomers recorded such events in 1006, 1054, 1181, 1572 and 1604, and the latter two were observed in Europe. Supernova explosions are thought to occur when a star becomes unstable and collapses under gravity, either because it has used up its nuclear fuel or because it accretes mass from elsewhere.

SUPERPOSITION (4.1): Sometimes the laws governing a physical system are simple enough that two solutions can be added together to give another. For example, given two possible electromagnetic waves, if one adds the two electric fields and the two magnetic fields, one gets another possible wave (were it not so, one beam of light could not pass through another). When superposition is possible, it is a very powerful tool. Sometimes superposition is not exactly true but is a very good approximation.

TEMPERATURE (2.2): Intuitively, "the intensity of heat". For a low-density gas, the temperature is related to the average kinetic energy of a molecule. More generally, temperature is defined in terms of heat and entropy (2.7). Temperature is conventionally denoted by the letter $T$. The modern unit of temperature is the kelvin, but it might also be measured in joules, taking account of a conversion factor $k$ (Boltzmann's constant). Zero on the Celsius scale is about 273 kelvin. At 0 kelvin (*absolute zero*), there is no random motion.

THOMPSON, BENJAMIN; LATER COUNT RUMFORD (2.4): 1753–1814. Born in America, went to England and spent 14 years in Bavaria, where he was made a count by the elector (he was also knighted by George III). He championed the motion theory of heat. He founded the Royal Institution but left it 3 years later (amidst some controversy). He was an enthusiast for the application of science to social improvement. Founded the English Garden in Munich. A colourful life. (See the biography by Sparrow.)

THOMSON, JOSEPH JOHN (5.1): 1856–1940. Appointed third Cavendish professor (after Maxwell and Lord Rayleigh) in Cambridge at the age of 28. Discovered the electron in 1887. Discovered that the neon has two isotopes (that is, atoms of different mass but identical chemical properties).

TIDAL FORCES: The forces on the different parts of a body due to the variation of the gravitational field across that body. The tidal forces due to the gravitation of the Moon and the Sun cause the tides in the Earth's oceans (see Section 1.7). In Einstein's theory of gravity, tidal forces are directly related to the curvature of spacetime.

TOWNES, CHARLES HARD (9.3): 1915– . American physicist who invented the theory of the maser and laser and constructed the first maser.

TRANSVERSE (4): Describes wave motions in which the material motion is at right angles to the direction of propagation of the wave. Opposed to *longitudinal*. Sound waves in solids can be longitudinal or transverse. Light waves are purely transverse.

VACUUM: This word is used in two slightly different senses. First, people talk about *the* vacuum, meaning an idealized state of some region when there is no matter and no radiation present. This is never exactly realized, but *a* vacuum would be some approximation to it, made by pumping out as much air (and anything else) as possible. From the seventeenth century onwards, people have been able to produce higher and higher vacua, enabling more and more new effects to be observed.

VACUUM ENERGY (9.10, 14): Energy that might reside in the vacuum even in the absence of observable particles. Related to the Casimir effect. If present, it would tend to speed up the expansion of the universe.

VECTOR (Appendix B): Any quantity that needs for its specification a direction as well as a magnitude. Examples are velocity, acceleration, force, momentum, and electric and magnetic fields. There is a natural way to define "addition" of vectors.

VELOCITY: The rate of change of position with time. Since the change in position has a direction associated with it, so does velocity (that is, velocity is a vector). The word *speed* is sometimes used in the sense of "magnitude of velocity", that is, the velocity with its direction disregarded.

VOLTA, ALESSANDRO (3.3): 1745–1827. Professor of natural philosophy at Parvia. Made the first electric battery ("voltaic pile") around 1799, thus allowing the study of current electricity. (The SI unit of electric potential, the volt, is named after him; 1 volt is 1 joule per coulomb.)

WAVELENGTH (3.1): In a wave motion, the distance (at any given time) from one crest to the next.

WEINBERG, STEVEN (12.3): 1933– . Prolific U.S. theoretical physicist. One of the discoverers of the unified theory of weak and electromagnetic forces. Author of excellent textbooks and books for the general reader.

WEYL, HERMANN: 1885–1955. German mathematician and mathematical physicist. Developed the theory of symmetries (for example, rotational symmetry) in quantum theory. With Einstein in Zürich and Princeton.

WORK (2.3): The work done by a force on a moving particle is the force times the distance moved in the direction of the force. The work is equal to the increase in kinetic energy of the particle. The unit of work is a joule.

WORLD-LINE (5.3): In relativity theory, this means the history of the positions occupied by a particle. In a spacetime diagram, the world-line is represented by a curve.

WU, CHIEN-SCHIUNG (12.2): (1912–) Moved from China to California in 1936. Discoverer of mirror symmetry breaking in radioactive decay.

YANG, CHEN NING (11.3, 12.2): (1922–) Born in China, he began his research in Chicago. With Lee, analyzed mirror symmetry in weak interactions. Invented non-Abelian gauge theories (with Mills), the basis of much recent progress in particle physics.

YOUNG, THOMAS (4.6): (1773–1829) British physician, physiologist, physicist and Egyptologist. Established the wave theory of light by observing *interference*.

# BIBLIOGRAPHY

Badash, L. *Kapitza, Rutherford and the Kremlin*. New Haven, CT: Yale University Press, 1985.

Barrow, J. *The Origin of the Universe*. London: Weidenfeld and Nicholson, 1994.

Baumgardt, C. *Johann Kepler, Life and Letters*. New York: Philosophical Library, 1952.

Beeson, D. *Maupertuis: An Intellectual Biography*. Oxford: The Voltaire Foundation at the Taylor Institute, 1992.

Begelman, M., and Rees, M. *Gravity's Fatal Attraction*. New York: Scientific American Library, 1995.

Bondi, H. *Relativity and Common Sense*. London: Heinemann, 1965.

Cantor, G. N., and Hodge, M.J.S. *Conceptions of Ether*. Cambridge: Cambridge University Press, 1981.

Carnot, Sadi. *Reflexions on the Motive Power of Fire*, translated by Robert Fox. Manchester, Manchester University Press, 1986.

Cassidy, D. C. *Uncertainty*. New York: W. H. Freeman, 1991.

Chandrasekhar, B. S. *Why Things Are the Way They Are*. Cambridge: Cambridge University Press, 1998.

Chandrasekhar, S. Verifying the Theory of Relativity. *Notes and Records of the Royal Society*, 1975–1976. 30:249,

Chandrasekhar, S. *Newton's Principia for the Common Reader*. Oxford: Clarendon Press, 1995.

Chandrasekhar, S. *Truth and Beauty*. Chicago: University of Chicago, 1987.

Cronin, J. *Pasteur, Light and Life*. *Physics World* 11: 23–4, 1998.

Cross, J. A. *Electrostatics, Principles Problems and Applications*. Bristol: Adam Hilger, 1987.

Descartes, René. *Principia Philosphiae*, translated by E. S. Haldane and G. R. T. Ross. Cambridge: Cambridge University Press, 1931.

Dubos, R. *Pasteur and Modern Science*. Madison, WI: Science Tech Publishers, 1988.

Elkana, Y. *The Discovery of the Conservation of Energy*. London: Hutchisons Educational Press, 1974.

'Espinasse, M. *Robert Hooke*. London: Heinemann, 1956.

Feynman, R. *The Character of Physical Laws*. Harmondsworth: Penguin, 1992.

Feynman, R. *QED: The Strange Theory of Light and Matter*. Harmondsworth: Penguin, 1990.

Feynman, R. *Six Not So Easy Pieces*. London: Harmondsworth: Penguin, 1999.

Frayn, M. *Copenhagen*. London: Methuen, 1998.

Galilei, Galileo. *Dialogue on the Great World Systems*, translated by Thomas Salusbury, 1661, revised Girgio de Santillano. Chicago: University of Chicago Press, 1953.

Hall, T. *Carl Friedrich Gauss*, translated by A. Froderberg. Cambridge, MA: MIT Press, 1970.

Heilbron, J. L. *The Dilemmas of an Upright Man*. Berkeley: University of California Press, 1986.

Hofstadter, D. R. *Gödel, Escher, Bach*. New York: Vintage Books, 1980.

Hoskin, M. *Cambridge Illustrated History of Astronomy*. Cambridge: Cambridge University Press, 1997.

Hoyle, F. *Nicolaus Copernicus*. London: Heinemann, 1973.

Hunt, B. J. *The Maxwellians*. Ithaca, NY: Cornell University Press 1991.

Lasoto, J.-P. Unmasking Black Holes. *Scientific American* 280:40, May 1999.

Lloyd, G.F.R. *Greek Science after Aristotle*. New York: Norton, 1973.

Lorenz, E. *The Essence of Chaos*. Seattle: University of Washington Press, 1993.

Maxwell, J. C. *A Treatise on Electricity and Magnetism*, 3rd ed. Mineola, NY: Dover Publications, 1954.

Mendelssohn, K. *The Quest for Absolute Zero*. New York: Taylor and Francis, 1977.

Milburn, G. J. *Schrödinger's Machines*. New York: W. H. Freeman, 1997.

Millar, D.I.J., and Millar, M. *The Cambridge Dictionary of Scientists*. Cambridge University Press, 1996.

Misner, C. W., Thorne, K. S., and Wheeler, J. A. *Gravitation*. New York: W. H. Freeman, 1973.

Moore, W. *A Life of Erwin Schrödinger*. Cambridge: Cambridge University Press, 1994.

Newton, I. *Mathematical Principles of Natural Philosophy*, translated by A. Motte revised by F. Cajorie. Berkeley: University of California Press, 1934.

Newton, I. *Opticks* (based on the 1730 edition). Mineola, NY: Dover Publications, 1952.

Nicholson, M. *Science and Imagination*. Ithaca, NY: Cornell University Press, 1956.

Pais, A. *Niels Bohr's Times*. Oxford: Clarendon Press, 1991.

Pais, A. *Subtle Is the Lord*. Oxford: Clarendon Press, 1982.

Peterson, I. *Newton's Clock: Chaos in the Solar System*. New York: W. H. Freeman, 1993.

Plutarch. *De Facie in Orbe Lunae*, translated by H. Chernis. London: Loeb Classical Library, Heinemann, 1957.

Rees, M. *Before the Beginning*. New York: Simon & Schuster, 1997.

Ronchi, Vasco *The Nature of Light*. London: Heinemann, 1970.

Sakharov, A. *Memoirs*. London: Hutchinson, 1990.

Silk, J. *A Short History of the Universe*. New York: Scientific American Library, 1994.

Sparrow, W. J. *Knight of the White Eagle: A Biography of Benjamin Thompson, Count Rumford*. London: Hutchinson, 1964.

Squires, E. *The Mystery of the Quantum World*. 2nd ed. Bristol: Institute of Physics Publishing, 1994.

Stephenson, F. R., and Clark, D. H. Historical Supernovas. *Scientific American* 234:100, 1976.

Stewart, I. *Does God Play Dice?* 2nd ed. Harmondsworth: Penguin, 1997.

't Hooft, G. *In Search of the Ultimate Building Blocks*. Cambridge: Cambridge University Press, 1996.

Thoren, V. E. *The Lord of Uraniborg: A Biography of Tycho Brahe*. Cambridge: Cambridge University Press, 1990.

Wali, K. C. *Chandra*. Chicago: University of Chicago Press, 1991.

Weinberg, S. *The First Three Minutes*. New York: Basic Books, 1988.

Whitehead, A. N. *Science and the Modern World*. New York: Macmillan, 1926.

Whittaker, E. T. *History of the Theories of Aether and Electricity*. London Nelson, 1951.

Wilson, C. *How Did Kepler Discover His First Two Laws?* *Scientific American* 226:93, 1968.

Wood, A. *Thomas Young, Natural Philosopher*. Cambridge: Cambridge University Press, 1954.

# INDEX